Heteroligand Molecular Systems

Heteroligand Molecular Systems

Bonding, Shapes and Isomer Stabilities

Alexander A. Levin and Pavel N. D'yachkov
Kurnakov Institute of General and Inorganic Chemistry
Russian Academy of Sciences, Moscow, Russia

Translated from the Russian by Victor A. Sipachev

London and New York

First published 2002 by Taylor & Francis
11 New Fetter Lane, London EC4P 4EE

Simultaneously published in the USA and Canada
by Taylor & Francis Inc,
29 West 35th Street, New York, NY 10001

Taylor & Francis is an imprint of the Taylor & Francis Group

This book has been produced from camera-ready copy
supplied by the authors
Printed and bound in Great Britain by
MPG Books Ltd, Bodmin

Originally published in Russian in 1990 as
ELECTRONNOE STROENIE, STRUCTURA I PREVRASHCHENIYA GETEROLIGANDNYKH MOLEKUL by Nauka, Moscow

British Library Cataloguing in Publication Data
A catalogue record for this book is available from the British Library

Library of Congress Cataloging in Publication Data
A catalog record for this book has been requested

ISBN 0-415-27235-1

CONTENTS

PREFACE

Heteroligand molecules and complexes built of a central atom surrounded by peripheral atoms and atomic groups, or ligands, of different kinds make up one of the most numerous and important classes of molecular systems. Most frequently, we come across such systems in inorganic, coordination, and organometallic chemistry. These are rapidly developing areas of science having a wide range of practical applications, which by itself makes studying heteroligand systems, whose number far exceeds that of homoligand molecules and properties are extremely varied, a topical problem.

There is, however, a more fundamental reason for our interest in heteroligand molecular systems. When we turn from homoligand to heteroligand compounds, there appear new problems that do not arise with simpler homoligand molecules. Such is the problem of mutual influence of ligands exemplified by the *trans*-influence well known in coordination chemistry (this problem is touched upon in almost every textbook on inorganic chemistry, however, in such a way that the impression is created that we deal with some enigmatic phenomenon of an unknown nature). One of the key problems of the structure of heteroligand molecular systems is that of the relative stability of isomers, because the very existence of one or another form of a heteroligand system is determined by the relative stability of competing molecular structures. Studies of the relative stabilities of isomers in the ground and electronically excited states of molecules is of importance for elucidating the mechanisms of many thermochemical and photochemical processes.

Interest of researchers in these and related problems has led to the accumulation of a wealth of diffraction, thermochemical, and spectral data on the structure, isomerism, and isomerization of heteroligand systems. These data need to be summarized and described from a general standpoint. This book aims at fulfilling this task.

Theoretical treatment of the problems specified above is performed from a general and at the same time simple and natural point of view, through constructing adiabatic potentials (potential energy surfaces) of heteroligand molecular systems, because the shape of the potential surface determines the geometry and energy characteristics of any chemical system. For this purpose, we use perturbation theory methods, in which the adiabatic potential of the initial homoligand system is assumed to be known, and passage from the homoligand system to its heteroligand derivative is treated as perturbation. It is pertinent to mention here that the method of perturbation has long and successfully been applied in diverse fields of theoretical chemistry, from organic quantum chemistry (perturbation molecular orbital theory) to coordination chemistry (theory of vibronic interactions, which will be used in constructing adiabatic potentials of heteroligand molecules and complexes). An important advantage of this approach in comparison with computations is the explicit analytic form of perturbative treatment results. This makes theory comprehensible and verifiable and, most importantly, ensures understanding of the nature of phenomena, whereas computations are mere numerical experiments, which are capable of revealing property patterns but incapable of explaining them in any general terms.

From this standpoint, we give a detailed description of a wide range of problems related to the mutual influences of ligands and relative isomer stabilities, which arise with typical heteroligand molecules and complexes of nontransition elements and transition metals. Actinide metal compounds are also included. First, we consider quasi-octahedral σ-bonded molecules and complexes, which are the simplest for theoretical treatment (Chapter 3), and quasi-square-planar and quasi-tetrahedral molecules and complexes (Chapter 4). In these chapters, we discuss heteroligand molecules and complexes with filled electronic shells derived from closed-shell homoligand systems. Chapter 5 treats σ-bonded derivatives of homoligand systems with open electronic shells. In Chapter 6, heteroligand complexes containing multiply bonded ligands alongside σ-bonded ones are considered. Special consideration is given to heteroligand derivatives of homoligand molecules that contain ligands nonequivalent by symmetry. These problems are the object of a separate discussion in Chapter 7, because they can be treated in terms of first-order perturbation theory, whereas the derivatives of homoligand systems with ligands equivalent by symmetry, which are considered in the preceding sections, require the use of second-order perturbation theory. Note that, in these chapters, all conclusions on the behavior of various molecular systems refer to isolated molecules and complexes. Extending these conclusions to molecules and complexes in solution or to polymeric formations in the solid phase is a mere extrapolation, although frequently used in structural chemistry. Intermolecular interactions are usually ignored. Cooperative couplings between molecular structural units in crystals are only considered in the concluding eighth chapter dealing with somewhat unexpected applications of the theory of heteroligand systems to order-disorder ferroelectrics and related materials, which form molecular crystals with structural units linked by H-bonds.

The two first chapters contain the necessary minimum of information about quantum chemistry including perturbation theory applications to chemical problems. This theoretical background is meant to be used in the chapters that follow. In writing this introduction, our purpose is to make the book most widely available, even to those whose knowledge of the theory of chemical bonding does not go beyond a higher-school course in physical chemistry. More details about the scope of this book can be inferred from the list of contents.

The book was written for two categories of readers. It is intended for a broad audience of scientists and teachers dealing with coordination, inorganic, and organometallic chemistry and also with the theory of the structure of molecules. We also address higher-school senior and post-graduate students who specialize in these fields of chemistry. We hope that every reader will find this book interesting in some respect; synthetic chemists and specialists in X-ray structure analysis will find a theoretical explanation of phenomena familiar to them, and theoretical chemists will find a description of one more domain of quantum chemistry applications.

The English version is based on the monograph (A. A. Levin and P. N. D'yachkov, *Electronnoe stroenie, structura i prevrashcheniya geteroligandnykh molekul* (The Electronic and Geometric Structure and Transformations of Heteroligand Molecules), Moscow: Nauka, 1990) published in Russian. For the English edition, the book was considerably revised and corrected, a number of sections were written anew, and some sections were omitted. The illustrative material was enlarged to make the presentation clearer. The list of references has been augmented, largely by including works published after 1990.

Many people contributed to this and Russian editions. We are thankful to Drs. A. P. Klyagina and S. P. Dolin for long-standing cooperation and Profs. I. B. Bersuker, B. G. Vekhter, V. Z. Polinger, and B. S. Tsukerblat for fruitful discussions of the vibronic aspects of the theory of the structure of molecular systems. Professor M. A. Porai-Koshits called our attention to a number of problems of stereochemistry of complexes and made especially helpful comments. We are grateful to the participants of the workshops headed by Profs. Yu. A. Buslaev and A. A. Ovchinnikov for stimulating discussions. We also gratefully record our obligation to Prof. M. E. Vol'pin and N. M. Shagina (Editor, Moscow Branch, Gordon and Breach Publ.) for the role they played in promoting the English edition. The authors thank the Russian Foundation for Basic Research, which financially supported studies of the problems described in Chapter 8 (project no. 96-03-32348).

A. A. Levin
P. N. D'yachkov

NOTATION

A_i, A_{ig}, A_{iu}	one-dimensional A-type irreducible representations
a_i, a_{ig}, a_{iu}	molecular orbitals (MOs) corresponding to one-dimensional irreducible representations
$a_{\Gamma,\gamma}$	coefficients of separation of substitution operator into symmetrized components
A, A'	linear and quadratic vibronic constants in the E–e problem
A_{ij}^{ν}, A_j^{ν}	linear vibronic constants
A_{nm}^{ν}	orbital linear vibronic constants
B_{ig}, B_{iu}	one-dimensional B-type irreducible representations
b_{ig}, b_{iu}	MOs corresponding to one-dimensional B-type irreducible representations
$c_{\mu j}$	coefficients of atomic orbitals (AOs) in MOs
c_i	coefficients of central atom AOs in MOs
D_{4h}	symmetry group of a square and a tetragonally distorted octahedron
Det	determinant
E, E_g, E_u	two-dimensional irreducible representations
$E_2^{(1)}(Q)$	component linear in Q in the expression for adiabatic potential
$\mathcal{E}(Q)$	molecular adiabatic potential
$\mathcal{E}_i$	energies of many-electron states
$\mathcal{E}_{\text{tot}}$	total molecular energy (including kinetic energy of nuclei)
$\mathcal{E}_{\text{J-T}}$	Jahn–Teller stabilization energy
e, e_g, e_u	MOs corresponding to two-dimensional irreducible representations
F_{trans}, F_{cis}, ΔF	forces acting on *trans* and *cis* ligands and their difference
G	point symmetry group of a molecule

$G(ij)$	operator of interactions between electrons i and j
H	effective one-electron Hamiltonian for an arbitrary molecular configuration
H_0	effective one-electron Hamiltonian for molecular configuration Q_0
H_S	one-electron substitution operator
H_Γ^γ	symmetrized components of one-electron substitution operator
$\mathcal{H}$	electronic Hamiltonian for an arbitrary molecular configuration
$\mathcal{H}_0$	electronic Hamiltonian for the initial (high-symmetry) configuration of a molecule
$\mathcal{H}_0^{(1)}$; $\mathcal{H}_0^{(2)}$	sums of one- and two-electron operators in $\mathcal{H}_0$
$\mathcal{H}_S$	many-electron substitution operator
$\mathcal{H}_S^l$	many-electron substitution operator for replacement of ligand l
$\mathcal{H}_\Gamma^\gamma$	symmetrized components of substitution operator
$\mathcal{H}_{\text{tot}}$	total molecular Hamiltonian including kinetic energy of nuclei
J_{ij}	interaction parameters of the Ising model
K_B, K_E, K_T	force constants for vibrations of B, E, and T symmetry types
K_i	isomerization constants
K_ν	force constants
l_i	coefficients of ligand group (symmetrized) AOs in MOs
M_i	masses of atomic nuclei
N	number of electrons in a molecule
$N(M–L)$	overlap population for the central atom–ligand bond
N_n	number of atomic nuclei in a molecule
n_i	MO populations
O_h	octahedral symmetry group
P_μ, P_i	electronic populations of AOs
Q	set of nuclear coordinates for an arbitrary molecular configuration
Q_0	set of nuclear coordinates for the initial (high-symmetry) molecular configuration
Q_1	totally symmetric normal coordinate

Q^f	set of normal or symmetrized nuclear coordinates corresponding to a minimum of the adiabatic potential of a heteroligand system
Q^f_ν	nuclear coordinates from set Q^f
Q_v	normal or symmetrized coordinates of atomic nuclei
q_i	coordinates of electrons
R	central atom–ligand interatomic distance
S_0	mean value of many-electron substitution operator for the ground-state wave function
S_j	matrix elements of many-electron substitution operator between ground and excited states
S^c_j, S^c_{mn}, S^t_j, S^t_{mn}	matrix elements of substitution operator for *cis* and *trans* isomers
S_{nm}	matrix elements of one-electron substitution operator in the basis of MOs
$S_{\mu\nu}$	overlap integrals between AOs
T_d	tetrahedral symmetry group
T_e, T_n	kinetic energy operators for electrons and nuclei
T_i, T_{ig}, T_{iu}	three-dimensional irreducible representations
t_i, t_{ig}, t_{iu}	MOs corresponding to three-dimensional irreducible representations
U	one-electron potential energy operator in Hamiltonian H
U_0	one-electron potential energy operator in Hamiltonian H_0
V_{ee}, V_{en}, V_{nn}	electron–electron, electron–nucleus, and nucleus–nucleus interaction operators
W	many-electron vibronic interaction operator
w	one-electron vibronic interaction operator
X_i, Y_i, Z_i	Cartesian coordinates of atomic nuclei
x_i, y_i, z_i	Cartesian coordinates of electrons
α, β	spin functions
α_μ	Coulomb integrals
$\beta_{\mu\nu}$	resonance integrals
β'	derivative of a resonance integral
Γ, Γ_i, Γ_n	irreducible representations
$\Gamma_i \times \Gamma_j$	direct product of irreducible representations
$[\Gamma_i^2]$	symmetrical part of direct product
δ_{ij}	Kronecker symbol
Δ	Laplacian operator

$\Delta\alpha$, $\Delta\alpha_\mu$	changes in Coulomb integrals caused by ligand substitution
$\Delta\beta$, $\Delta\beta_{\mu\nu}$	changes in resonance integrals
$\Delta\mathcal{E}_{t\to c}$, $\Delta\mathcal{E}_{c\to t}$	*trans* → *cis* and *cis* → *trans* isomerization energies
$\Delta\mathcal{E}_E$	extrastabilization energy for E term
ΔR_{trans}, ΔR_{cis}, ΔR	changes in *trans* and *cis* bond lengths and their difference
ε_i	one-electron levels
ζ_i	nuclear charges
η_i	overlap populations between central atom AOs and ligand group AOs
π_i	ligand π orbitals
ρ, φ	polar coordinates
σ_i	ligand σ AOs
$\sigma_i^{(z)}$	pseudospins
Φ, Φ_μ	Slater determinants
φ_μ	ligand group orbitals
χ_μ	atomic orbitals, central atom orbitals
Ψ_0	ground-state many-electron wave function of a molecular system
Ψ_i	many-electron wave functions
${}^{(1)}\Psi_i$, ${}^{(3)}\Psi_i$	singlet and triplet state wave functions
Ψ_{tot}	molecular wave function depending on electronic and nuclear coordinates
ψ_i	molecular orbitals
ω_j	energy gaps between ground and excited many-electron states
ω_{nm}	energy gaps between MOs

CHAPTER 1

THE ELECTRONIC STRUCTURE OF MOLECULES

The fundamentals of the theory of heteroligand molecular systems will be discussed starting with Section 2.4. Chapter 1 and the first three sections of Chapter 2 are introductory and, for reader's convenience, summarize some of the principal concepts of quantum chemistry including chemical applications of perturbation theory. This is done as the further presentation requires. For more complete information, the reader is referred to books on quantum mechanics [1–3] and quantum chemistry [4–12] and also to the monographs and reviews cited in the list of references.

1.1. Total and Electronic Hamiltonians

1. The quantum theory of molecules is usually divided into the theory of structure and dynamics, and this is exactly how the textbook [13] is titled. The problems that we discuss in this book do not concern dynamical matters, such as internal motions of atoms, collisions of molecules, their dissociation, interaction with radiation or particles, etc. Our only concern will be the structures of molecules. This is also true of those book sections that treat of chemical reactions, because we are not interested in the development of reactions in time, but confine ourselves to estimating their heat effects. Accordingly, we will only consider stationary states of molecules, that is, states characterized by definite total energy values. It has, on occasion, been stated that in quantum mechanics, stationary states of an arbitrary molecule as a system of many particles, electrons and nuclei, are described by wave functions Ψ_{tot}, which are time-independent and only depend on coordinates of all electrons $q = \{q_1, q_2, \ldots\}$ and all nuclei $Q = \{Q_1, Q_2, \ldots\}$ of the system. However more precisely, the wave function of a stationary state, $\Psi_{\text{tot}}(t)$, depends on time, but this function only differs from Ψ_{tot} by a simple standard multiplier, that is, $\Psi_{\text{tot}}(t) = \Psi_{\text{tot}} \exp(-i\mathcal{E}_{\text{tot}}t/\hbar)$, where $\mathcal{E}_{\text{tot}}$ is the total energy of a molecule in the given state, $i = \sqrt{-1}$ is the imaginary unit, and $\hbar$ is the Planck constant. It follows that studying stationary states reduces to finding time-independent wave functions Ψ_{tot} and the corresponding energies $\mathcal{E}_{\text{tot}}$.

Some points should be mentioned concerning coordinates q_i and Q_j, which are the arguments of the function Ψ_{tot}. Coordinate q_i is a set of three spatial (usually, Cartesian) electron coordinates (x_i, y_i, z_i) and a spin variable determining the projection of the electron spin onto axis z. As nuclear coordinates Q_i, we can use Cartesian atomic coordinates (X_i, Y_i, Z_i). We may also use normal coordinates, as is traditional for vibrational spectroscopy, or symmetry coordinates (Section 2.2); spins of nuclei are usually not considered in the theory of the structure of molecules.

2. Wave functions Ψ_{tot} and energies $\mathcal{E}_{\text{tot}}$ of stationary states are found by solving the time-independent Schrödinger equation

$$\mathcal{H}_{\text{tot}}\Psi_{\text{tot}} = \mathcal{E}_{\text{tot}}\Psi_{\text{tot}}, \tag{1.1}$$

where $\mathcal{H}_{\text{tot}}$ is the operator acting on Ψ_{tot} and called the total Hamiltonian of a molecular system. This Hamiltonian is usually represented as the sum of five operators,

$$\mathcal{H}_{\text{tot}} = T_e + T_n + V_{ee}(q) + V_{en}(q, Q) + V_{nn}(Q). \tag{1.2}$$

Here, the individual terms have the form

$$T_e = \sum_{i=1}^{N} -\frac{1}{2}\left(\frac{\partial^2}{\partial x_i^2} + \frac{\partial^2}{\partial y_i^2} + \frac{\partial^2}{\partial z_i^2}\right), \tag{1.3}$$

$$T_n = \sum_{i=1}^{N_n} -\frac{1}{2M_i}\left(\frac{\partial^2}{\partial X_i^2} + \frac{\partial^2}{\partial Y_i^2} + \frac{\partial^2}{\partial Z_i^2}\right), \tag{1.4}$$

$$V_{ee} = \frac{1}{2}\sum_{i\neq j}^{N} \frac{1}{\sqrt{(x_i - x_j)^2 + (y_i - y_j)^2 + (z_i - z_j)^2}}, \tag{1.5}$$

$$V_{en} = -\sum_{i=1}^{N}\sum_{j=1}^{N_n} \frac{\zeta_j}{\sqrt{(x_i - X_j)^2 + (y_i - Y_j)^2 + (z_i - Z_j)^2}}, \tag{1.6}$$

$$V_{nn} = \frac{1}{2}\sum_{i\neq j}^{N_n} \frac{\zeta_i\zeta_j}{\sqrt{(X_i - X_j)^2 + (Y_i - Y_j)^2 + (Z_i - Z_j)^2}}. \tag{1.7}$$

According to (1.3) and (1.4), the action of operators T_e and T_n on wave functions reduces to the summation of wave function second partial derivatives with respect to the Cartesian coordinates of electrons (T_e) and nuclei (T_n) taken with the coefficients $-1/2$ and $-1/2M_i$. Operators V_{ee}, V_{en}, and V_{nn} multiply Ψ_{tot} by the functions of the coordinates of electrons and nuclei specified in the right-hand sides of equations (1.5)–(1.7). Operators (1.3)–(1.7) are written in atomic units. In this system of units, the mass of the electron, the absolute charge of the electron, and the Planck constant ($\hbar$) equal unity. Operators T_e and T_n correspond to the kinetic energy of electrons (T_e) and nuclei (T_n). The other operators describe attraction between electrons and nuclei (V_{en}), mutual repulsion of electrons (V_{ee}), and mutual repulsion of nuclei (V_{nn}); ζ_i are the charges of nuclei; N and N_n are the number of electrons and the number of nuclei in the system; and M_i are the masses of nuclei. Of course, (1.2) is not the most general Hamiltonian form. However, this Hamiltonian will be our point of departure. We will use various approximations to (1.2), such as are sufficient for handling the problems considered below.

3. In the theory of the structure of molecules, we can considerably simplify calculations by observing that coefficients $1/M_i$ in operator (1.4) are exceedingly small, of the order of 10^{-3}–10^{-4}. This offers a possibility of solving the Schrödinger equation (1.1) with the use of the adiabatic approximation or its generalizations to degenerate or pseudodegenerate electronic terms. (More exactly, we will deal with the "simple adiabatic approximation" or its generalization usually applied to analyze Jahn–Teller effects [14, 15].)

Without going into the theory underlying these approximations (see, e. g., [14]), note that they allow us to study the behavior of electrons and nuclei in three steps. First, the T_n operator in (1.2) is ignored, the positions of the nuclei are fixed, and motions of the electrons are studied. Next, while motions of the nuclei, that is, the T_n operator in (1.2), are still ignored, the spatial arrangement of the nuclei is not considered fixed. At this step, the sum of the energy of the electronic subsystem of the molecule and the energy of repulsion of the nuclei is studied as a function of nuclear coordinates. The obtained dependence is called the adiabatic potential of the molecule. Adiabatic potentials are often calculated numerically with the use of computational techniques of quantum chemistry [16–21]. In the model approach applied to Jahn–Teller effects, this problem is solved by perturbation theory methods with the use of the data obtained in the first step. Lastly, at the third step (but we shall stop short of reaching it), the T_n operator is included, and the problem of motions of the nuclei in the obtained adiabatic potential is solved.

4. Apart from the adiabatic approximation, we will use the valence approximation frequently employed in theoretical chemistry. The terms electrons and nuclei are taken to mean valence electrons and atomic cores rather than all electrons and "naked" nuclei. However,

for brevity, we will use these terms without specially stating that most often, valence electrons and atomic cores are implied.

1.2. Electronic Hamiltonian and Electronic Terms

1. Let us ignore operator T_n in Hamiltonian (1.2). Equation (1.2) for $\mathcal{H}_{\text{tot}}$ then becomes

$$\mathcal{H} = T_e + V_{ee} + V_{en} + V_{nn}. \tag{1.8}$$

This operator is called the electronic Hamiltonian, because it only includes the coordinates of electrons as dynamic variables. Nuclear coordinates are contained in (1.8) as parameters. The time-independent Schrödinger equation $\mathcal{H}\Psi = \mathcal{E}\Psi$ corresponding to Hamiltonian (1.8) describes stationary states of the electronic subsystem of a molecule, or electronic terms, as functions of nuclear coordinates. At the first step, we must fix these coordinates in $\mathcal{H}$, that is, put $Q = Q_0$, $Q_0 = \{Q_1^0, Q_2^0, \ldots\}$. (As is usual, we will count the coordinates of nuclei from point Q_0 in the corresponding multidimensional space of configurations and frequently write 0 instead of Q_0 and Q instead of $Q - Q_0$; $Q - Q_0 = \{Q_1 - Q_1^0, Q_2 - Q_2^0, \ldots\}$). Configuration Q_0 is usually selected to be the characteristic configuration of a molecular system.

For high-symmetry rigid molecules such as CH_4, SF_6, $PtCl_4^{2-}$, etc., where the atoms are involved in small-amplitude vibrations about definite equilibrium positions, the characteristic configuration is assumed to coincide with the equilibrium nuclear configuration, for which energy $\mathcal{E}$ is minimum. The equilibrium configurations of the specified molecules are characterized by the T_d, O_h, and D_{4h} point groups, respectively:

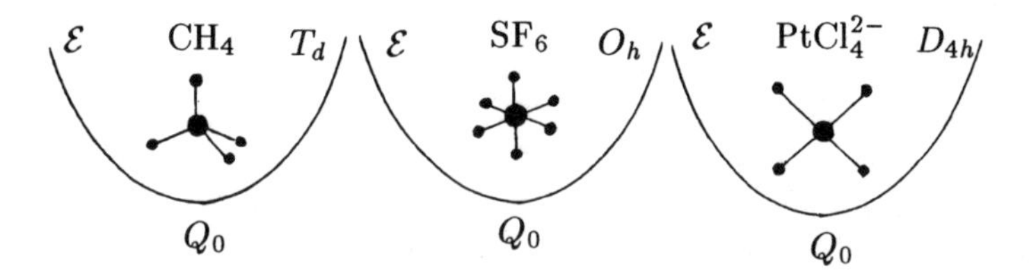

Another, also numerous, class of molecules includes low-symmetry species such as H_2O, NH_3, or distorted octahedral transition metal complexes. The Q_0 configuration can then be selected differently depending on the purposes of study. Often, the low-symmetry equilibrium nuclear configuration is used as Q_0, e. g., for calculating the electronic structure of H_2O or NH_3. On the other hand, the real low-symmetry form of a molecule can be treated as a distortion of its high-symmetry prototype caused by various reasons (Jahn–Teller effect or pseudo-Jahn–Teller effect, see Section 2.2). It is then expedient to select the high-symmetry nuclear configuration of a molecule as Q_0, and this is how we are going to proceed in the sequel:

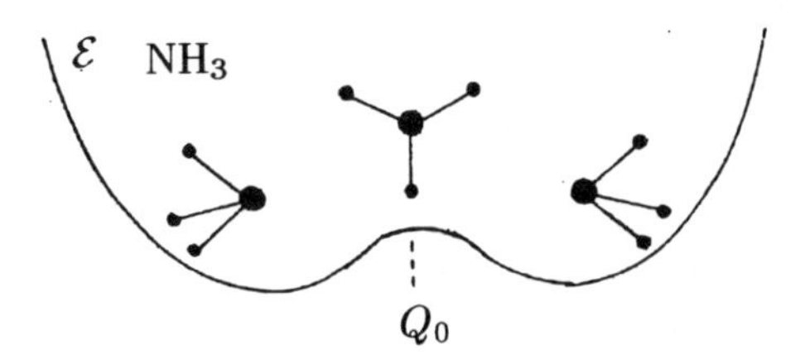

No matter how selected, Q_0 will satisfy the requirement of the optimum size of a molecule; it will be assumed that Q_0 is optimized with respect to all totally symmetric deformations changing all interatomic distances by the same factor. Energy $\mathcal{E}$ of a molecule in the Q_0 configuration should be minimum with respect to all totally symmetric deformations:

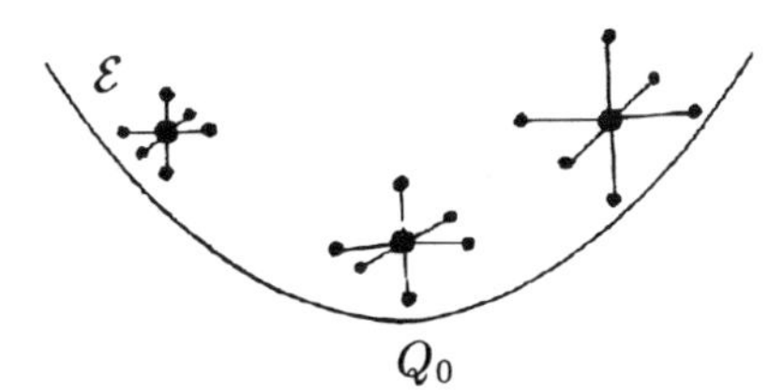

Physically, it is obvious that such a Q_0 configuration should exist for any molecular system, if only this system does not tend to collapse to a point or dissociate to its components.

2. The determination of the electronic terms corresponding to point Q_0 through solving the Schrödinger equation with the $\mathcal{H}_0$ operator obtained by substituting $Q = Q_0 = 0$ into (1.8),

$$\mathcal{H}_0\Psi = \mathcal{E}\Psi; \quad \mathcal{H}_0 = T_e + V_{ee} + V_{en}(q, 0) + V_{nn}(0), \tag{1.9}$$

is the key problem of the theory of the electronic structure of molecules. Its solution yields the wave functions of the terms, Ψ_j, as the eigenfunctions of operator $\mathcal{H}_0$ and the corresponding energies of the terms, $\mathcal{E}_j$, as the eigenvalues of the same operator. Both contain information about the electronic subsystem of the molecule at point Q_0.

For completeness, two important properties of Ψ_j functions should be mentioned. First, the Ψ_j eigenfunctions of Hamiltonian $\mathcal{H}_0$ given by (1.9) are real or can be selected real. The second property derives from the Hermitian character of operator $\mathcal{H}_0$ (note in passing that operator $\mathcal{H}_{\text{tot}}$ defined by (1.1) is also Hermitian). For this reason, any two functions Ψ_i and Ψ_j corresponding to different eigenvalues $\mathcal{E}_i$ and $\mathcal{E}_j$ are mutually orthogonal, that is, the $\int \Psi_i \Psi_j \, dq_1 \, dq_2 \ldots dq_N$ integral taken over the set of the coordinates of all electrons is zero. Here, the integral symbol implies the integration with respect to the spatial electron coordinates and the summation with respect to the spin variables. It is also possible that different Ψ_j functions will correspond to the same Hamiltonian $\mathcal{H}_0$ eigenvalue; such Ψ_j functions may also be considered orthogonal, or they can be orthogonalized by replacing the initially nonorthogonal functions by their appropriately chosen linear combinations. This means that we can (and will) treat all different eigenfunctions of operator (1.9) as mutually orthogonal.

In addition, Ψ_j can conveniently be normalized to unity through dividing each Ψ_j function by its norm, that is, the $\left(\int \Psi_j^2 \, dq_1 \ldots dq_n\right)^{1/2}$ value. Eigenfunctions Ψ_j of Hamiltonian (1.9) then satisfy the orthonormalization conditions

$$\int \Psi_i \Psi_j \, dq_1 dq_2 \ldots dq_N = \delta_{ij}, \tag{1.10}$$

where δ_{ij} is the Kronecker symbol,

$$\delta_{ij} = \begin{cases} 0, & \text{for } i \neq j \\ 1, & \text{for } i = j. \end{cases}$$

In the sequel, all term wave functions Ψ_j will be considered orthonormalized.

3. An important property of the eigenfunctions of Hamiltonian (1.9) is their relation to molecular symmetry at point Q_0: wave functions Ψ_j of each term transform according to one of the irreducible representations of the point symmetry group of a molecule in spatial configuration Q_0. (Almost every textbook on quantum chemistry contains an introduction to the theory of the representations of groups as applied to the electronic structure of molecules, e. g., [6, 7, 9]. Also see monographs on the subject [22–25].) In short, term wave functions Ψ_j show characteristic behaviors under various transformations, such as rotations and reflections, that constitute the molecular point symmetry group G. Every function Ψ_j of a certain term transforms into functions of the same term under symmetry operations from G.

In the simplest situation when the term has a single Ψ_j function, this function remains invariant within a factor equal to one in absolute value and unimportant, because generally, wave functions can only be considered definite within factors of this kind. Such terms are said to correspond to one-dimensional irreducible representations of the molecular symmetry group. In more complex situations, a term is formed by several linearly independent Ψ_j functions. Generally, any term function then transforms into a linear combination of linearly independent functions of the same term under symmetry operations from G. The term and the corresponding wave functions are then said to correspond to a non-one-dimensional representation.

For the symmetry groups that will most frequently be considered below, namely, O_h (octahedron), D_{4h} (square), and T_d (tetrahedron), irreducible representations are one-dimensional (A_i, A_{ig}, B_{ig}, A_{iu}, and B_{iu}), two-dimensional (E, E_g, and E_u), or three-dimensional (T_i, T_{ig}, and T_{iu}). Representations of types A and B (of the groups mentioned above, the latter are only present among the D_{4h} group representations) differ in the behavior of the corresponding wave functions under rotations about the principal molecular axis of symmetry. For instance, for molecules of D_{4h} symmetry, the functions corresponding to representation A remain unchanged, and those of representation B are multiplied by -1 under rotations through 90°. Indices g and u are only used when the molecule has a center of symmetry. The term functions corresponding to the representations with index g remain unchanged under reflection in the center of symmetry (gerade, i. e., even functions), and the functions with index u are multiplied by -1 (ungerade, i. e., odd functions). In addition to indices g and u, term symbols are assigned indices $i = 1$ or 2, because representations of the same dimensionality and parity may be different.

Lastly, molecules can have different spin states. Accordingly, terms are assigned indices that indicate multiplicity of spin degeneracy of the system. For instance, the ^2A, ^2E, or ^2T terms are doubly degenerate with respect to spin. In these states, a molecule has the total spin equal to 1/2, that is, contains one unpaired electron or one hole.

1.3. Molecular Orbitals and the MO LCAO Method

1. Because of its generality, the concept of molecular terms as solutions to equation (1.9) is fairly formal and abstract. If we only know the arrangement and charges of atomic nuclei and the number of electrons in a molecule, it is difficult to identify the term that corresponds to its ground state and the terms that correspond to excited states or at least qualitatively arrange the excited terms according to their excitation energies. Solving such problems requires specifying the form of the wave functions corresponding to the terms by relating them to more

detailed characteristics of the electronic structure of the molecule. For this purpose, the concept of one-electron molecular orbitals is used. These orbitals are then employed to construct many-electron functions Ψ_j.

It is assumed that separate electrons in a molecule are characterized by wave functions of their own, or molecular spin-orbitals. (Because of quantum indistinguishability of electrons in a system, we can specify spin-orbitals but cannot say which electron of the N present in the molecule occupies one or another spin-orbital.) Each spin-orbital represents a function of the spatial coordinates of the electron and its spin. It is assumed in the simplest approach that spin-orbitals can be written as products of spatial and spin functions; the spatial function ($\psi = \psi(x, y, z)$) is called molecular orbital (MO). Each MO corresponds to two molecular spin-orbitals, $\psi\alpha$ and $\psi\beta$, where α and β are the spin functions of electrons with spins "up" and "down," respectively. A study of the electronic structure of a molecule therefore reduces to a study of its MOs. These MOs are functions defined in three-dimensional space and are therefore easy to visualize. A large number of chemical problems can be handled with the use of pictures of MOs (pictorial quantum chemistry [26]).

For each molecular system, MOs ψ_j together with the corresponding one-electron levels ε_j are found by solving the one-electron Schrödinger equation

$$H_0\psi = \varepsilon\psi \tag{1.11}$$

with the effective one-electron Hamiltonian (as previously, corresponding to point Q_0)

$$H_0 = -\frac{1}{2}\left(\frac{\partial^2}{\partial x^2} + \frac{\partial^2}{\partial y^2} + \frac{\partial^2}{\partial z^2}\right) + U_0 \equiv -\frac{1}{2}\Delta + U_0. \tag{1.12}$$

This Hamiltonian contains the electron kinetic energy operator, $-(1/2)\Delta$, and the operator U_0 describing the summed action on the electron under consideration of all other electrons of the system and all its atomic nuclei (index 0 in U_0 indicates that the configuration of atomic nuclei corresponds to coordinates Q_0). The term describing repulsion of nuclei does not appear in Hamiltonian (1.12). Note that U_0 does not necessarily have the simple form $U_0 = U_0(x, y, z)$, that is, does not necessarily implies mere multiplication by a function of three spatial coordinates of the electron. For instance, in the Hartree–Fock approximation [7], U_0 is a so-called nonlocal operator. Finding U_0 is a nontrivial task, which is differently solved by different quantum-chemical methods, *ab initio* or semiempirical. We do not need to go into details, because quantitative calculations of MOs are not discussed in this book and are only included as literature data. A description of these methods can be found in, e. g., [8, 12, 16–21]. In addition, the form of the effective one-electron potential is of little importance for the further presentation, because instead of the explicit form (1.12), we will only use Hamiltonian matrix elements in the basis of atomic orbitals. Such an approach to π electrons in organic compounds is known as the Hückel method. In application to arbitrary systems, it is called the Hoffmann or extended Hückel method. We will therefore refer to Hamiltonian (1.12) as a Hückel-type Hamiltonian (although we will not apply the technique for calculating matrix elements developed in the Hückel and Hoffmann methods; see also Section 1.4).

2. The concept of MOs as solutions to (1.11) *per se* does not predetermine in what way these one-electron functions should be determined, and there exist methods for finding MOs that are not directly related to taking into account the structure of constituent atoms. For instance, for diatomic molecules, equation (1.11) can be solved by separation of variables in spheroidal coordinates [7, 27], and for arbitrary systems, by the method of scattered waves [8, 11, 20]. However when we deal with chemical problems, it is convenient that 'genetic' relations between MOs and one-electron atomic functions, or atomic orbitals (AOs), be retained. This requirement is met by the "method of molecular orbitals – linear combinations of atomic orbitals" (MO LCAO), which is most popular in quantum chemistry. According to this method, MOs are written in the basis set of AOs of the atoms constituting the molecule (in the valence approximation, in the basis set of valence AOs), that is,

$$\psi_j = \sum_{\mu=1}^{n} c_{\mu j}\chi_\mu, \quad j = 1, 2, \ldots, n, \tag{1.13}$$

where n is the total number of AOs in the basis set. Here, $c_{\mu j}$ are the coefficients determining the contribution of the μth AO to the jth MO.

Coefficients $c_{\mu j}$ and orbital energies ε_j can most easily be found if we rewrite (1.11) as $(H_0 - \varepsilon)\psi = 0$ and substitute the representation of ψ in terms of χ_μ, $\psi = \sum_{\mu=1}^{n} c_\mu \chi_\mu$, into this equation. The sum obtained should be multiplied from the left first by χ_1, next by χ_2, etc., and lastly, by χ_n. The resulting equalities should each be integrated over the whole space. The c_μ coefficients in the expression for MOs can then be calculated by solving the system of linear homogeneous equations

$$\begin{aligned} &(H^0_{11} - \varepsilon)c_1 + H^0_{12}c_2 + \cdots + H^0_{1n}c_n = 0 \\ &H^0_{21}c_1 + (H^0_{22} - \varepsilon)c_2 + \cdots + H^0_{2n}c_n = 0 \\ &\ldots\ldots\ldots\ldots\ldots\ldots\ldots\ldots\ldots\ldots\ldots \\ &H^0_{n1}c_1 + H^0_{n2}c_2 + \cdots + (H^0_{nn} - \varepsilon)c_n = 0, \end{aligned} \tag{1.14}$$

where $H^0_{\mu\nu} = H^0_{\nu\mu}$ are the matrix elements of operator (1.12) in the basis set of AOs, that is, the integrals of the form (all AOs are used as real)

$$H^0_{\mu\nu} = \langle\chi_\mu|H_0|\chi_\nu\rangle = \int \chi_\mu H_0 \chi_\nu \, dv. \tag{1.15}$$

Note that here and below, we use the approximation of "zero differential overlap" [12, 17, 11, 16] very popular in quantum chemistry. According to this approximation, AOs of different atoms are considered orthogonal, that is, the corresponding overlap integrals for such functions, $\langle\chi^A_\mu|\chi^B_\nu\rangle = \int \chi^A_\mu \chi^B_\nu \, dv$, are set equal to zero (for different AOs of the same atom, these integrals are strictly zero).

To determine c_μ from (1.14), we must know not only the $H^0_{\mu\nu}$ matrix elements but also the ε values, that is, the energies of MOs (orbital energies), which are, of course, also of interest in themselves. The technique for finding them is simple. Nontrivial solutions to (1.14) exist if its determinant is zero, that is,

$$\mathrm{Det}||H_{\mu\nu} - \varepsilon\delta_{\mu\nu}|| = \mathrm{Det}\begin{Vmatrix} H^0_{11} - \varepsilon & H^0_{12} & \ldots & H^0_{1n} \\ H^0_{21} & H^0_{22} - \varepsilon & \ldots & H^0_{2n} \\ \ldots & \ldots & \ldots & \ldots \\ H^0_{n1} & H^0_{n2} & \ldots & H^0_{nn} - \varepsilon \end{Vmatrix} = 0. \tag{1.16}$$

It follows that the sought MO energy values are obtained as the roots $\varepsilon = \varepsilon_j$ $(j = 1, 2, \ldots, n)$ of secular equation (1.16). Sequentially substituting the obtained ε_j values into (1.14) yields a set of $c_{\mu j}$ $(j = 1, 2, \ldots, n)$ coefficients for each ε. By (1.13), this set determines the MOs of a given energy level. As with term wave functions (see Section 1.2), the eigenfunctions of Hamiltonian (1.11), or MOs, can be considered real. And again, like term wave functions, MOs corresponding to different energy levels ε_i and ε_j are mutually orthogonal, that is,

$$\langle\psi_i|\psi_j\rangle = \int \psi_i\psi_j \, dv = 0.$$

As concerns different Hamiltonian (1.11) eigenfunctions of the same degenerate level (solutions of this kind appear when the secular determinant (1.16) has multiple roots), these functions can, if necessary, be replaced by their mutually orthogonal linear combinations. Therefore without the loss of generality, all the different MOs obtained from (1.14) and (1.16) can be considered mutually orthogonal. In addition, MOs can be normalized to unity through dividing each MO by its norm $\langle\psi_i|\psi_i\rangle^{1/2}$, where

$$\langle\psi_i|\psi_i\rangle = \int \psi_i^2\,dv = \sum_\mu c_{\mu i}^2.$$

After this operation, all MOs of a given molecular system will be orthonormalized, that is,

$$\langle\psi_i|\psi_j\rangle = \delta_{ij}. \tag{1.17}$$

In the sequel, all MOs of any molecular system will be assumed to be orthonormalized.

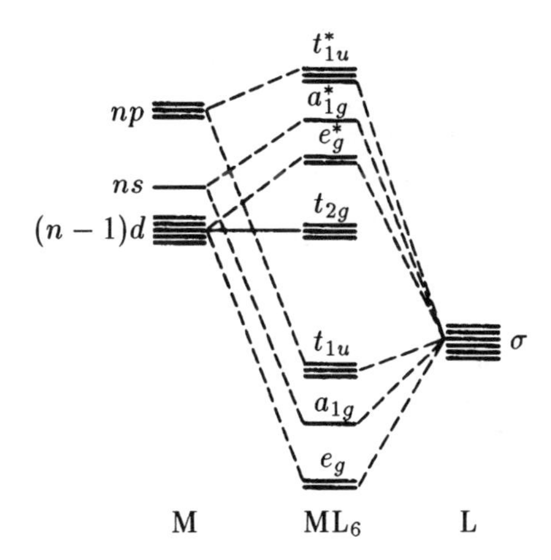

FIG. 1. TYPICAL ENERGY LEVEL SCHEME OF MOS FOR AN OCTAHEDRAL σ-BONDED TRANSITION METAL COMPLEX.

3. As with molecular terms, MOs are classified according to the irreducible representations of the point symmetry group of a molecule. Molecular orbitals and one-electron levels are denoted by small letters to distinguish them from terms, which are denoted by capital letters. The relation between MOs and irreducible representations is much easier to perceive than with terms, because the results of the action of group symmetry operations on an MO are often quite obvious, which can not always be said of transformations of many-electron Ψ_j functions. For instance, a cursory glance at the a_{1g} MO of an octahedral, say, transition metal complex (Fig. 1 and Table 1) is sufficient

to confirm that under all O_h group operations, the a_{1g} orbital remains unchanged. Similarly, t_{1u} MOs do not change (as the $t_{1u,z}$ MO under rotations about axis z), change sign, or transform into each other (e. g., the $t_{1u,x}$ orbital transforms into $t_{1u,y}$ when rotated through 90° counterclockwise about the z axis). Similarly, we see that the e_g MOs remain unchanged or are multiplied by -1 under O_h group rotations about the z axis. It is not immediately obvious what becomes of the e_g MOs under O_h group rotations, say, about the x and y axes. However then too, using the representation of the MOs in terms of AOs, we easily find that each MO of the e_g type (e_{g,z^2} or e_{g,x^2-y^2}) transforms into a linear combination of the same two orbitals. (Indices x, y, z, z^2, and $x^2 - y^2$ mean that the corresponding MOs transform like the central atom AOs that they contain, *viz.*, p_x, p_y, p_z, d_{z^2}, and $d_{x^2-y^2}$.)

As with many-electron states, or terms, the multiplicity of one-electron state degeneracy equals the dimension of the corresponding irreducible representation. Molecules of O_h symmetry have one-electron levels with degeneracy multiplicities of one, two, and three. The types of these levels possible in principle are determined by the irreducible representations of the point symmetry group of the molecule (for each point group comprising a finite number of elements, there exists a finite number of irreducible representations and, therefore, a finite number of different types of levels). However the set of levels of a particular system does not include levels of all possible symmetry types. The number and types of actual levels depend on the type of the molecule and the basis set of AOs used to describe this molecule. For instance, for an octahedral ML_6 transition metal complex, the scheme of MOs constructed from the valence ns, np, and $(n-1)d$ AOs of the central atom and the σ AOs of the ligands has the form shown in Fig. 1. If the basis set is extended by including the p_π ligand orbitals, new levels appear (see Fig. 33 in Chapter 6).

The relation between MOs and molecular symmetry is not only fundamentally important but also used in techniques for calculating MOs. Because each MO corresponds to some irreducible representation (Γ), it can be treated as composed of properly symmetrized combinations of AOs selected in such a way that all these combinations undergo identical transformations under symmetry operations, that is, according to the same representation Γ. In particular, for ML_n molecules, symmetrized functions will be pure AOs of the central M atom on the one hand and ligand "group orbitals" of the corresponding symmetry types on the other.

The symmetrized ligand group orbitals of ML_n molecules with most frequently encountered symmetries are given in Table 1 [11] together

Table 1. Atomic functions of central atom χ and ligand σ- and π-type group orbitals φ_σ and φ_π for complexes of O_h, D_{4h}, and T_d symmetry

Symmetry type	χ	φ_σ	φ_π
		Octahedral complex ML_6	
A_{1g}	s	$(\sigma_1+\sigma_2+\sigma_3+\sigma_4+\sigma_5+\sigma_6)/\sqrt{6}$	—
A_{2u}	f_{xyz}	—	—
T_{1u}	p_x, f_{x^3}	$(\sigma_3-\sigma_4)/\sqrt{2}$	$(\pi_{1x}+\pi_{2x}+\pi_{5x}+\pi_{6x})/2$
	p_y, f_{y^3}	$(\sigma_5-\sigma_6)/\sqrt{2}$	$(\pi_{1y}+\pi_{2y}+\pi_{3y}+\pi_{4y})/2$
	p_z, f_{z^3}	$(\sigma_1-\sigma_2)/\sqrt{2}$	$(\pi_{3z}+\pi_{4z}+\pi_{5z}+\pi_{6z})/2$
E_g	$d_{x^2-y^2}$	$(\sigma_3+\sigma_4-\sigma_5-\sigma_6)/2$	—
	d_{z^2}	$(2\sigma_1+2\sigma_2-\sigma_3-\sigma_4-\sigma_5-\sigma_6)/2/\sqrt{3}$	—
T_{2g}	d_{xy}	—	$(\pi_{3y}-\pi_{4y}+\pi_{5x}-\pi_{6x})/2$
	d_{xz}	—	$(\pi_{1x}-\pi_{2x}+\pi_{3z}-\pi_{4z})/2$
	d_{yz}	—	$(\pi_{1y}-\pi_{2y}+\pi_{5z}-\pi_{6z})/2$
T_{1g}	—	—	$(\pi_{3y}-\pi_{4y}-\pi_{5x}+\pi_{6x})/2$
	—	—	$(\pi_{1x}-\pi_{2x}-\pi_{3z}+\pi_{4z})/2$
	—	—	$(-\pi_{1y}+\pi_{2y}+\pi_{5z}-\pi_{6z})/2$
T_{2u}	$f_{x(z^2-y^2)}$	—	$(\pi_{1x}+\pi_{2x}-\pi_{5x}-\pi_{6x})/2$
	$f_{y(x^2-z^2)}$	—	$(-\pi_{1y}-\pi_{2y}+\pi_{3y}+\pi_{4y})/2$
	$f_{z(x^2-y^2)}$	—	$(\pi_{3z}+\pi_{4z}-\pi_{5z}-\pi_{6z})/2$

Table 1. (continued from the previous page)

Symmetry type	χ	φ_σ	φ_π
		Square-planar complex ML_4	
A_{1g}	s, d_{z^2}	$(\sigma_1+\sigma_2+\sigma_3+\sigma_4)/2$	—
B_{1g}	$d_{x^2-y^2}$	$(\sigma_1+\sigma_2-\sigma_3-\sigma_4)/2$	—
E_u	p_x	$(\sigma_1-\sigma_2)/\sqrt{2}$	—
	p_y	$(\sigma_3-\sigma_4)/\sqrt{2}$	—
		Tetrahedral complex ML_4	
A_1	s	$(\sigma_1+\sigma_2+\sigma_3+\sigma_4)/2$	—
T_2	p_x, d_{yz}	$(\sigma_1-\sigma_2+\sigma_3-\sigma_4)/2$	—
	p_y, d_{xz}	$(\sigma_1-\sigma_2-\sigma_3+\sigma_4)/2$	—
	p_z, d_{xy}	$(\sigma_1+\sigma_2-\sigma_3-\sigma_4)/2$	—

Note: Group σ AOs are given in local coordinates fixed on ligands, and π AOs, in the coordinate system fixed on the central atom. For the square-planar and the tetrahedral complexes, π orbitals are not given, because they are not discussed in text. The coordinate systems used are shown in Figs. 4, 14, and 19.

with the central atom AOs of the corresponding symmetry types. Symmetry considerations therefore allow the atomic composition of an MO to be determined to a substantial degree without any calculations. The $c_{\mu j}$ coefficients of the symmetrized functions, which remain to be found, are then determined for MOs of each symmetry type Γ by solving the corresponding secular equation. As previously, this equation is of form (1.16), but the matrix elements that it contains are calculated in the basis of symmetrized functions rather than AOs. For this reason, the degree of this secular equation equals not the total number of AOs but n_Γ, which is the number of different symmetrized functions that undergo identical transformations (the number of irreducible representations of type Γ contained in the reducible representation corresponding to the selected set of AOs). This secular equation is therefore said to be reduced by symmetry. As the roots of a secular equation give MO energies, the number of levels with different energies equals n_Γ for each Γ. Each of these levels has degeneracy equal to the dimension of representation Γ.

1.4. The Electronic Structure of High-Symmetry Molecules and Complexes

Molecules and complexes ML_n considered in this book almost always (except in Chapter 7) have O_h, D_{4h}, or T_d symmetry or are derivatives of systems with such symmetries. Reading the principal chapters of the book therefore only requires knowledge of the schemes of MOs for systems with octahedral, square-planar, and tetrahedral geometries. Octahedral ML_6 systems with O_h symmetry are characterized by the simplest scheme of MOs, because the valence s, p, and d central atom AOs then correspond to different irreducible representations, and the σ ligand AOs are almost fully separated by symmetry from their π AOs (Table 1). We already mentioned the scheme of the MOs of an octahedral σ-bonded ML_6 transition metal complex in the basis of the $(n-1)d$, ns, and np AOs of the central M atom and ligand σ AOs (Fig. 1). A similar typical scheme of MOs of a σ-bonded nontransition metal ML_6 complex or molecule is shown in Fig. 2. The basis set then includes the valence ns and np AOs of atom M and the σ AOs of ligands L. This basis is augmented by the unoccupied nd AOs of the metal, because these AOs appear to participate in bonding to some degree [20].

It often suffices to include ligand σ orbitals, because typical ligands such as halogen atoms, NH_3 group, and methyl CH_3 group are usually approximately treated as systems only forming σ bonds with the central atom. In other circumstances or when undertaking a more detailed

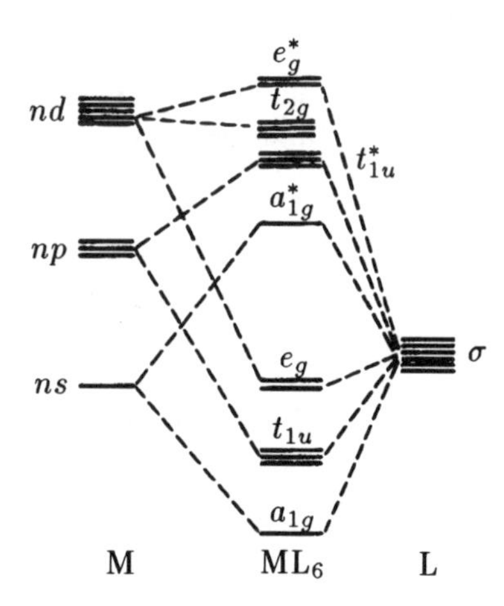

FIG. 2. TYPICAL ENERGY LEVEL SCHEME OF MOS FOR OCTAHEDRAL σ-BONDED NONTRANSITION ELEMENT COMPLEXES.

inquiry into the electronic structure of σ-bonded complexes, the p_π ligand orbitals should be included. The corresponding schemes of MOs are given in Fig. 33, Chapter 6; they are also typical of transition metals (Fig. 33a) and nontransition elements (Fig. 33b). A part of the scheme of MOs for ML_6 complexes of actinides is shown in Figs. 34 and 35, Chapter 6. In constructing these schemes, not only ligand AOs and the s, p, and d central atom AOs but also the valence f metal AOs were included in the basis set. Similar schemes of MOs for ML_4 molecules and complexes with D_{4h} and T_d point symmetry groups are also well known. They are shown in Figs. 13, 15, and 18, Chapter 4, for systems with σ-bonded ligands; species of these symmetry types with multiply bonded ligands will not be considered.

2. The schemes of MOs shown in these figures are in qualitative agreement with the calculated data, some of which were obtained by computations. They are substantiated by experimental data such as photoelectron spectra for occupied MOs and optical spectra for unoccupied MOs. They can therefore be regarded as based on calculations and experiment. Nevertheless, to gain insight into the patterns of the electronic structure of molecular systems, we must elucidate the genetic relation between the MOs and the AOs of atoms constituting the molecule (complex). To do this, it suffices to use group theory plus certain assumptions on the matrix elements of one-electron Hamiltonian (1.12). The diagonal matrix elements in the basis of AOs called Coulomb integrals,

$$\alpha_\mu = H^0_{\mu\mu} = \langle\chi_\mu|H_0|\chi_\mu\rangle, \tag{1.18}$$

are assumed to have values close to the corresponding one-electron levels of separate atoms, that is, atomic orbital ionization potentials taken with the opposite sign. Their values can be determined from atomic spectra [29] or calculations of the electronic structure of atoms [30, 31]. In addition to matrix elements (1.18), the energies of MOs depend on the off-diagonal matrix elements of Hamiltonian (1.12),

$$\beta_{\mu\nu} = H^0_{\mu\nu} = \langle \chi_\mu | H_0 | \chi_\nu \rangle, \tag{1.19}$$

which describe interaction between AOs of different atoms and are called resonance integrals. These integrals can be of different types depending on the form and mutual orientation of AOs χ_μ and χ_ν, for instance,

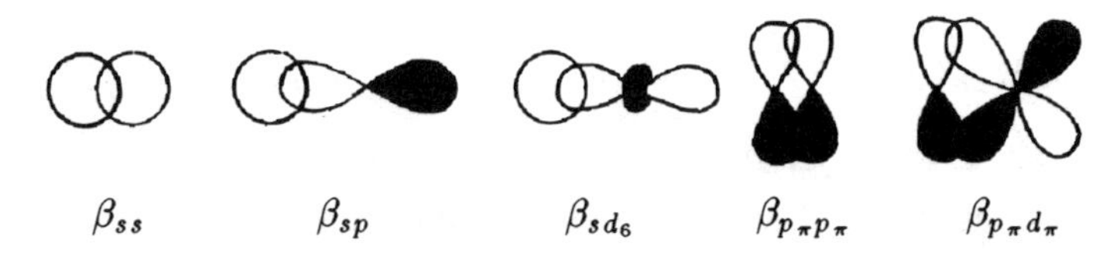

It is usually believed that $\beta_{\mu\nu}$ integrals vary in parallel with overlap between the χ_μ and χ_ν AOs, more exactly, are proportional in magnitude and opposite in sign to the $\langle \chi_\mu | \chi_\nu \rangle$ overlap integrals between these AOs. It follows that resonance integrals rapidly decrease in absolute value as the interatomic distance increases; therefore qualitative conclusions can be drawn from consideration of resonance integrals only between AOs of bonded atoms. For ML_n molecules, this means that interactions between the AOs of the central atom and ligand AOs are taken into account, whereas interactions between ligand AOs are not.

Note the difference between the concept of one- (α) and two-center (β) molecular integrals that we use from that inherent in the Hückel and Hoffmann methods [12]. As in these methods, Coulomb integrals α are qualitatively associated with AO energies, but no quantitative estimates of α will be used in this book. Similarly, resonance integrals β are treated at a qualitative level as values proportional to the overlaps between AOs. But we do not use quantitative estimates of the type $\beta_{\mu\nu} = F\langle \mu | \nu \rangle (\alpha_\mu + \alpha_\nu)$, where F is a constant, inherent in the Hückel and Hoffmann approaches.

3. The method for constructing qualitative schemes of MOs is easiest to demonstrate for the example of σ-bonded octahedral ML_6 complexes (Figs. 1 and 2). According to the group theory, the MOs of ML_6 built of the s, p, and d central atom and ligand σ AOs transform according to the A_{1g}, T_{1u}, E_g, and T_{2g} irreducible representations (Table 1).

Consider the a_{1g} and a_{1g}^* MOs (A_{1g} representation). It follows from the table that there is only one central atom AO, its ns AO, and only one symmetrized combination of ligand AOs of this symmetry type. The corresponding secular equation will therefore be of degree two. The same refers to the E_g and T_{1u} non-one-dimensional representations, because each of two (d_{z^2} and $d_{x^2-y^2}$) and three (p_x, p_y, and p_z) central atom AOs has only one counterpart among the symmetrized combinations of ligand AOs that transforms under O_h group symmetry operations like the corresponding central atom AO. To sum up, the secular equation for each of the three (a_{1g}, e_g, and t_{1u}) levels will have the form

$$\mathrm{Det}\left\| \begin{matrix} \langle\chi_\mu|H_0|\chi_\mu\rangle - \varepsilon & \langle\chi_\mu|H_0|\varphi_\mu\rangle \\ \langle\chi_\mu|H_0|\varphi_\mu\rangle & \langle\varphi_\mu|H_0|\varphi_\mu\rangle - \varepsilon \end{matrix} \right\| \equiv \begin{vmatrix} \alpha_\mu - \varepsilon & m_\mu\beta_{\mu\sigma} \\ m_\mu\beta_{\mu\sigma} & \alpha_\sigma - \varepsilon \end{vmatrix} = 0. \tag{1.20}$$

Here, χ_μ is the central atom AO, φ_μ is the corresponding symmetrized combination of ligand AOs, α_μ is the Coulomb integral for the s, p, or d central atom AO, and α_σ is the Coulomb integral for the ligand σ AO. The resonance integrals between central atom AOs and ligand σ AOs for the standard orientation of both orbitals are denoted as $\beta_{\mu\nu}$. The m_μ numbers appear when ligand group orbitals are replaced in the off-diagonal matrix elements by their expressions through individual ligand AOs. They can easily be obtained from the data given in Table 1 and, for the a_{1g}, e_g, and t_{1u} MOs that we consider, equal $\sqrt{6}$ (a_{1g}), $\sqrt{3}$ (e_g), and $\sqrt{2}$ (t_{1u}). The roots of (1.20) are

$$\varepsilon = (\alpha_\mu + \alpha_\sigma)/2 - \sqrt{(\alpha_\mu - \alpha_\sigma)^2/4 + m_\mu^2\beta_{\mu\sigma}^2} \tag{1.21a}$$

$$\varepsilon^* = (\alpha_\mu + \alpha_\sigma)/2 + \sqrt{(\alpha_\mu - \alpha_\sigma)^2/4 + m_\mu^2\beta_{\mu\sigma}^2} \tag{1.21b}$$

It follows from (1.21) that, in complete agreement with Figs. 1 and 2, two levels correspond to each of the A_{1g}, E_g, and T_{1u} representations, one bonding and one antibonding (labeled with an asterisk). The levels of the a_{1g}, e_g, and t_{1u} bonding MOs have lower energies ε than both the corresponding atomic levels of the central M atom and the σ orbital levels of ligand L. Conversely, every antibonding level is higher in energy (ε^*) than both AOs constituting it. The MOs can schematically be drawn as

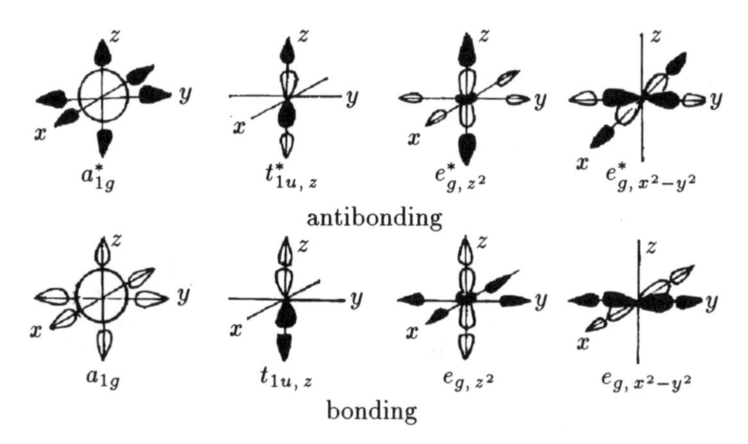

Here, the MOs are given in the style of pictorial quantum chemistry. The regions where the wave functions are negative are given in black, and the regions of their positive values are left white. Of the t_{1u} MOs, only one is shown; the others have the same shape but are oriented along the x ($t_{1u,x}$) and y ($t_{1u,y}$) axes. The antibonding MOs are labeled by asterisks. Generally, the order in which various bonding and various antibonding orbitals are arranged on the energy scale depends on resonance integral values. This order is, however, the same as that of atomic levels if the distances between them are fairly large.

Apart from the σ MOs discussed above, ML_6 complexes have t_{2g} MOs corresponding to the T_{2g} representation. These MOs derive from the d_{xy}, d_{xz}, and d_{yz} central atom AOs, which have no counterparts among ligand σ AOs. In a σ-bonded complex, the t_{2g} MOs therefore remain "atomic."

4. The arrangement of the MO levels in Figs. 1 and 2 is easy to explain in terms of the MO LCAO method. The ns, np, and vacant nd levels of nontransition elements form the sequence $s < p < d$, and the distance between the atomic levels in this sequence is of the order of 10 eV (Fig. 3). Typical resonance integral $\beta_{\mu\sigma}$ values (for all rather than only nontransition elements and for all orbitals χ_μ) are about -1 eV. Therefore in the nontransition element complexes under consideration, MO energy levels are ordered as central atom levels, that is, $a_{1g} < t_{1u} < e_g$ and $a_{1g}^* < t_{1u}^* < e_g^*$ (Fig. 2). In transition metal complexes, the situation is not so simple. The np transition metal levels lie several eV higher than the ns and $(n-1)d$ levels. Accordingly, the t_{1u} MOs have higher energies than those of a_{1g} and e_g. The energy difference between the ns and $(n-1)d$ atomic levels is, however, of the same order of magnitude as resonance integral values. It

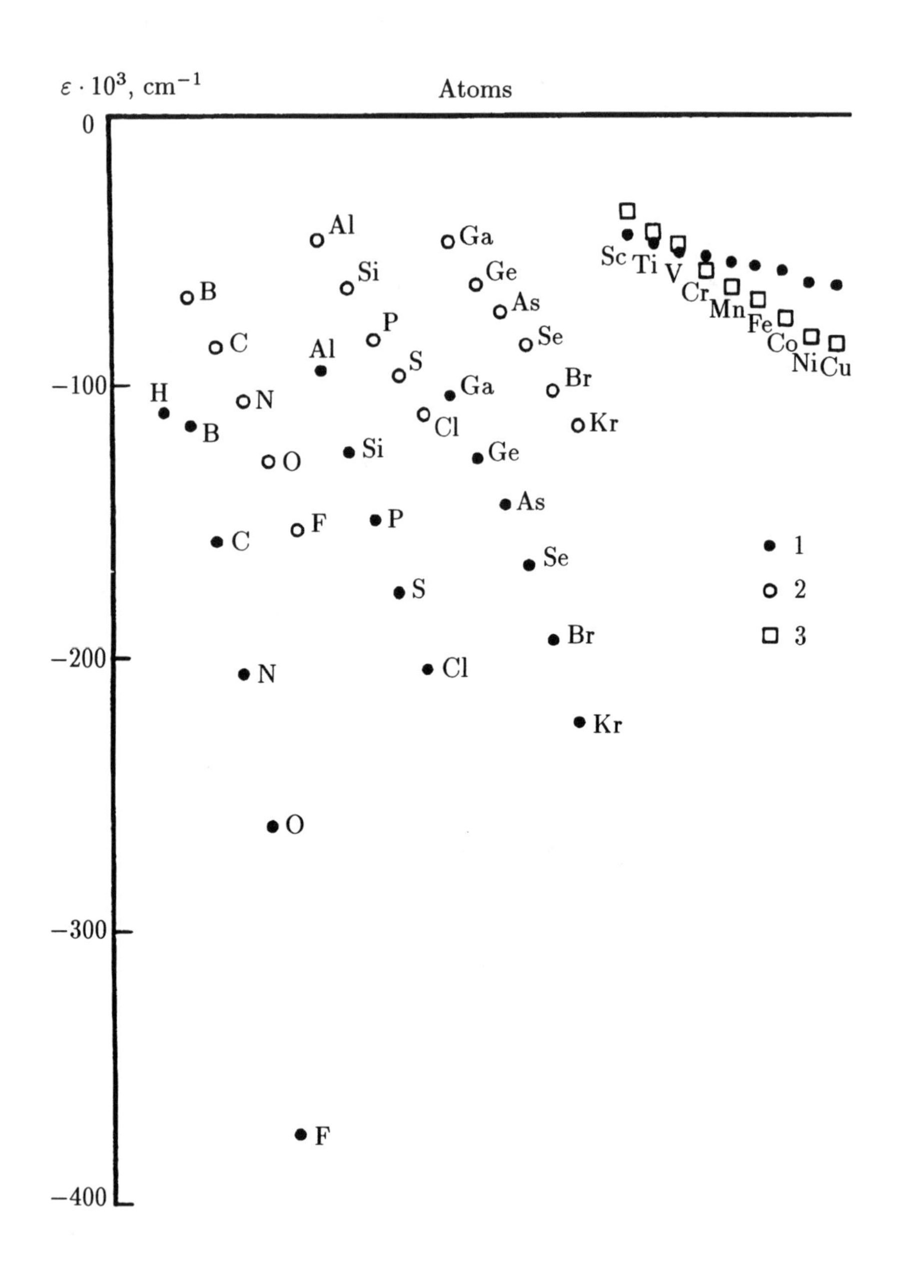

FIG. 3. VALENCE AO LEVELS OF NONTRANSITION ELEMENT AND TRANSITION METAL ATOMS. LEVELS: *1*, ns; *2*, np; AND *3*, $(n-1)d$.

is therefore difficult to predict the relative arrangement of the a_{1g} and e_g MOs. Experiments and calculations show that the $e_g \leqslant a_{1g} < t_{1u}$ and $e_g^* < a_{1g}^* < t_{1u}^*$ sequences (Fig. 1) are typical. Lastly, the t_{2g} MOs in σ-bonded complexes are purely atomic; in qualitative schemes, they are assigned the same energy as the free central atom d AOs (Figs. 1 and 2).

Qualitative schemes of the MOs of square-planar and tetrahedral σ-bonded ML_4 complexes (Figs. 13, 15, and 18) and of octahedral complexes with simultaneously taking into account the σ and p_π ligand AOs (Fig. 33) are constructed in a similar way. Complications that we encounter with σ-bonded complexes are largely due to reduction in symmetry. As a consequence, one irreducible representation can correspond to more than one or two independent sets of basis functions that transform in the same way under symmetry operations. For instance, in σ-bonded square-planar ML_4 complexes, three basis functions (the s and d_{z^2} AOs of atom M and one ligand group orbital) transform according to the A_{1g} representation. This results in the appearance of MOs of three types, *viz.*, bonding, nonbonding, and antibonding. The bonding orbital is lower and the antibonding one higher in energy than the constituent AOs, whereas the nonbonding orbital lies in the region of intermediate energies (in transition metal complexes, near the s and d_{z^2} levels of atom M, Fig. 13).

Similar complications are also characteristic of tetrahedral σ-bonded ML_4 complexes. In the basis of the s and p central atom and σ ligand AOs, two independent symmetrized sets of basis functions correspond to each of the irreducible representations describing transformations of the wave functions of the complex, A_1 and T_2. Therefore there are bonding and antibonding MOs of two types (a_1 and a_1^* and t_2 and t_2^*, respectively) (Fig. 18). However if the basis set includes the d AOs of M, three sets of basis functions correspond to the T_2 representation; these are the p and d_{xy}, d_{xz}, and d_{yz} central atom AOs and the ligand t_2-group σ AOs. In addition, there appear purely atomic MOs derivative from the central atom d_{z^2} and $d_{x^2-y^2}$ AOs. Their symmetry properties are described by the E representation absent for the ligand σ AOs.

The situation becomes even more complicated if the number of basis function is increased by including the p_π ligand AOs alongside the σ ones. Some of the central atom AOs then find counterparts among ligand p_π-group AOs. A typical example is the d_{xy}, d_{xz}, and d_{yz} AOs in octahedral ML_6 complexes; these AOs generate bonding t_{2g} and antibonding t_{2g}^* π MOs. The corresponding energy levels are assigned relative positions (relative to the σ MOs) based on the assumption that

π-interactions are weaker than σ bonds. In octahedral ML_6 complexes, the e_g σ-bonding MOs are therefore believed to be situated lower in energy than the t_{2g} π-bonding MOs, and the e_g^* σ-antibonding MOs, higher than the t_{2g}^* π-antibonding ones. The inclusion of ligand π AOs also results in the appearance of new orbitals of a purely ligand character. Examples are the t_{1g} and t_{2g} ML_6 orbitals; in qualitative schemes of MOs, they are drawn at the level of the corresponding ligand orbitals. For a more detailed description of the schemes of MOs for ML_n complexes of the O_h, T_d, and D_{4h} symmetry types, the reader is referred to book-length monographs [9, 11, 29].

5. Along with energies, MOs are characterized by their atomic compositions, that is, the coefficients of the AOs constituting them. For ML_n molecules and complexes, these are the coefficients of the central atom AOs and ligand group AOs. The square secular equation (1.20) is easy to use for explicitly finding not only the energies of the MOs, but also the coefficients, which are given by ($c^* = l$, $l^* = -c$)

$$
\begin{aligned}
c^2 &= \frac{1}{2} - \frac{\alpha_\mu - \alpha_\sigma}{2\sqrt{(\alpha_\mu - \alpha_\sigma)^2 + m_\mu^2 \beta_{\mu\sigma}^2}}, \\
l^2 &= \frac{1}{2} + \frac{\alpha_\mu - \alpha_\sigma}{2\sqrt{(\alpha_\mu - \alpha_\sigma)^2 + m_\mu^2 \beta_{\mu\sigma}^2}}.
\end{aligned}
\tag{1.22}
$$

Here, c and l (c^* and l^*) are the coefficients of the central atom AOs and the ligand group AOs, respectively, in the bonding (antibonding) MOs (c = central and l = ligand). The c and l coefficients are often indexed by MO symbols. For instance, for σ-bonded octahedral complexes, the a_{1g}, e_g, and t_{1u} MOs are written in terms of c_a, l_a, etc. coefficients. The normalization conditions give $c^2 + l^2 = 1$ and $(c^*)^2 + (l^*)^2 = 1$.

The explicit expressions (1.22) for the coefficients of MOs in terms of Coulomb and resonance integrals [47] can be used to analyze the dependence of the coefficients on atomic levels. In ML_n molecules, ligand valence levels often have energies noticeably different from those of central atom levels. (For instance, in transition metal complexes, the AO levels of typical ligands such as halogens, oxygen, etc. lie substantially lower than central atom AO levels, see Fig. 3.) The atomic level lowest in energy then makes the largest contribution to the bonding MO, and the atomic level highest in energy, to the antibonding MO. In transition metal complexes, bonding MOs are largely localized on ligands, and antibonding MOs, on the central atom. In nontransition element compounds, the situation is more complex. The valence AOs of typical ligands then have lower energies than those of the vacant nd

AOs of the central atom. The relative arrangement of the AO levels of typical ligands L and the valence *ns* and *np* AOs of the central M atom, however, depends on the position of M in the Periodic Table of the Elements. If M is situated in the left half of the table (Group III and IV element halides), the valence *p* AOs of ligands are close in energy to the valence *ns* AO of the central atom and noticeably lower than its valence *np* AOs. For instance, in such σ-bonded ML_6 octahedral complexes, the ns(M) AO and the a_{1g} group orbital of ligands make comparable contributions to the a_{1g} (and a_{1g}^*) MO. At the same time, the contribution of the central atom to the t_{1u} MOs is smaller than that of the group t_{1u} orbital of ligands. Still smaller is the contribution of the central atom to the e_g MOs. In contrast, the central atom makes the major contribution to the t_{1u}^* and, especially, e_g^* MOs.

6. A description of a molecular system in terms of the MO LCAO approach requires specifying not only the energies and the atomic compositions of MOs, but also their populations, that is, the numbers of electrons in various one-electron orbitals. Below, among octahedral σ-bonded ML_6 molecules and complexes, we will largely concentrate on transition metal compounds with 12 σ MO electrons and nontransition element compounds with 12 and 14 σ MO electrons. Figures 1 and 2 show that in 12-electronic transition metal complexes, all MOs to t_{1u} inclusive are occupied. The t_{1u} orbitals are the highest occupied MOs (HOMO), whereas the lowest unoccupied MOs (LUMO) are the e_g^* ones. In ML_6 nontransition metal 12-electronic complexes, all MOs up to the e_g HOMO are occupied, and the LUMO is the a_{1g}^* MO. In ML_6 nontransition metal 14-electronic complexes, the HOMO and the LUMO are the a_{1g}^* and the t_{1u}^* MOs, respectively.

The MO populations in complexes of other types will be considered in the corresponding sections. Note only that the type of the HOMO and the LUMO (together, they are called frontier orbitals) to a great extent determines the properties of molecular systems including their steric behaviors in ligand substitution reactions and reactivities [32, 33].

The populations of MOs also determine the distribution of charges in the system, that is, the effective atomic charges. Suppose that MO ψ_j contains n_j electrons ($n_j = 0$, 1, or 2). The $n_j c_{\mu j}^2$ value, where $c_{\mu j}$ is the coefficient of AO χ_μ in MO ψ_j, then, in the approximation of zero differential overlap, equals the partial electronic population of the corresponding AO. The (total) population of the AO is obtained as the sum of its partial populations over all MOs including this AO, and the electronic population of an atom in a molecule (atomic population) is the sum of the populations of all its AOs. The effective atomic charge is then calculated as the difference of the positive charge of its nucleus and its atomic population (in the valence approximation, as the difference of the charge of the core and the population of the valence AOs). This can be written as

$$P_{\mu j} = n_j c_{\mu j}^2, \tag{1.23}$$

$$P_\mu = \sum_j^{\text{occupied MOs}} n_j c_{\mu j}^2, \tag{1.24}$$

$$P_\text{A} = \sum_\mu^{\text{AOs of atom A}} \text{P}_\mu, \tag{1.25}$$

$$q_\text{A} = \zeta_\text{A} - P_\text{A}, \tag{1.26}$$

where $P_{\mu j}$ is the partial population of AO χ_μ, P_μ is the total population of AO χ_μ, P_A is the atomic population, and q_A is the effective charge of atom A. Formulas (1.23)–(1.26) describe the characteristics of individual AOs and separate atoms in a molecular system. In addition, the MO LCAO method considers the characteristics of various possible pairs of AOs and various possible pairs of atoms, such as

$$P_{\mu\nu} = \sum_j^{\text{occupied MOs}} n_j c_{\mu j} c_{\nu j}, \tag{1.27}$$

$$P_\text{AB} = \sum_\mu^{\text{AOs of atom A}} \sum_\nu^{\text{AOs of atom B}} P_{\mu\nu}, \tag{1.28}$$

where $P_{\mu\nu}$ characterizes the bond order between AOs χ_μ and χ_ν, and P_AB is the bond order between atoms A and B. It is usually assumed that the $P_{\mu\nu}$ value determines the contribution of AOs χ_μ and χ_ν to stability of the molecule, and the P_AB value, bond strength between atoms A and B, for instance, between the central M atom and some ligand L. For the M—L bond and for valence-bonded atoms in general, the P_AB value is treated in terms of the classical theory of the structure of molecules: it is assumed that the calculated P_AB value is directly related to the actual strength of the bond between the corresponding atoms. Instead of bond orders $P_{\mu\nu}$ and P_AB, other characteristics having approximately the same physical meaning are often used, for instance, overlap populations, which take into account overlap integrals $S_{\mu\nu}$ between AOs in addition to the $c_{\mu j}$ and $c_{\nu j}$ coefficients of these AOs in an MO. Wiberg indexes are yet another characteristic of this type.

1.5. Molecular Terms in the MO Method

1. In constructing term wave functions from one-electron MOs, or, more exactly, spin-orbitals, the guiding principle is that the many-electron wave function of an arbitrary many-electron system must have the property of being antisymmetric, which directly corresponds with the Pauli principle. The function should change sign under permutations of the space and spin coordinates of any two electrons. This requirement is fulfilled no matter in what form many-electron wave functions are written. When such wave functions are constructed from one-electron spin-orbitals, there is, however, an approximate method for automatically meeting the condition specified above. For this purpose, a many-electron wave function is written in the form of a determinant composed of one-electron spin-orbitals (Slater determinant) or as a linear combination of such Slater determinants. The antisymmetric property of the many-electron wave function is then a trivial consequence of the properties of any determinant, which changes sign under permutation of any two rows of columns.

In the simplest situation, electrons in a molecule are fully distributed over a certain number of one-electron levels in such a way that each k-degenerate level is occupied by exactly $2k$ electrons. Such systems with filled (closed) electronic shells include the majority of molecules and complexes in the ground state, for instance, CH_4 (Fig. 18), SF_6, and $[Co(NH_3)_6]^{3+}$. For the last two species, we have

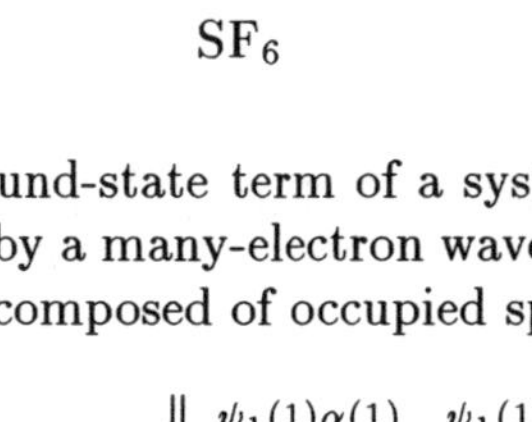

e_g^*, t_{1u}^*, a_{1g}^*, e_g, t_{1u}, a_{1g}

SF_6

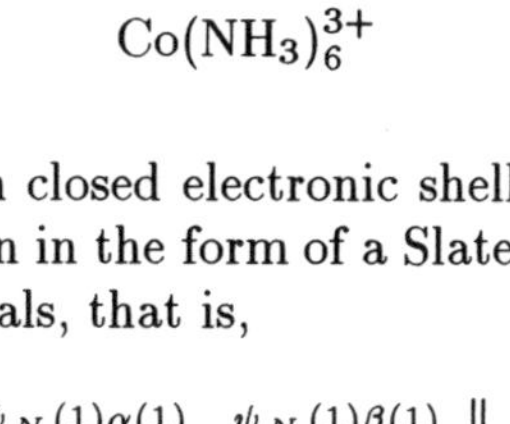

t_{1u}^*, a_{1g}^*, e_g^*, t_{2g}, t_{1u}, a_{1g}, e_g

$Co(NH_3)_6^{3+}$

The ground-state term of a system with closed electronic shells is described by a many-electron wave function in the form of a Slater determinant composed of occupied spin-orbitals, that is,

$$\Phi = \frac{1}{\sqrt{N!}} \mathrm{Det} \left\| \begin{matrix} \psi_1(1)\alpha(1) & \psi_1(1)\beta(1) \ldots & \psi_{\frac{N}{2}}(1)\alpha(1) & \psi_{\frac{N}{2}}(1)\beta(1) \\ \psi_1(2)\alpha(2) & \psi_1(2)\beta(2) \ldots & \psi_{\frac{N}{2}}(2)\alpha(2) & \psi_{\frac{N}{2}}(2)\beta(2) \\ \cdots & \cdots & \cdots & \cdots \\ \psi_1(N)\alpha(N) & \psi_1(N)\beta(N) \ldots & \psi_{\frac{N}{2}}(N)\alpha(N) & \psi_{\frac{N}{2}}(N)\beta(N) \end{matrix} \right\| . \quad (1.29)$$

This expression is often replaced by its short form with only the main diagonal of the determinant written explicitly,

$$\Phi = \frac{1}{\sqrt{N!}}\mathrm{Det}||\psi_1(1)\alpha(1)\ \psi_1(2)\beta(2)\ldots \\ \psi_{\frac{N}{2}}(N-1)\alpha(N-1)\ \psi_{\frac{N}{2}}(N)\beta(N)|| \tag{1.30}$$

Here, ψ_j are the MOs; α and β are the spin functions; numbers $1, 2, \ldots,$ N refer to electrons, and we write, e. g., $\psi_1(1)$ instead of the cumbersome denotation $\psi_1(x_1, y_1, z_1)$; and $1/\sqrt{N!}$ is the normalization factor.

Closed-shell systems are the only ones whose electronic configuration can be described by one Slater determinant. It is easy to see that under molecular symmetry operations, determinant (1.29) transforms according to the totally symmetric representation (A_{1g} for O_h and D_{4h} and A_1 for T_d). The electronic configuration of an open-shell system can only be represented by several Slater determinants. The simplest systems of this type contain one electron in the nondegenerate HOMO. This electron can have either spin α or spin β, and the two corresponding Slater determinants only differ in the type of the spin function for the specified spin-orbital. This fairly trivial situation of a spin-degenerate but orbitally nondegenerate state is of little importance for the further presentation.

2. Of greater importance are systems with k-fold degenerate one-electron levels containing less than $2k$ electrons. A typical example is $Cu^{II}L_6$ octahedral complexes with three electrons in the e_g^* HOMO (the e_g^{*3} electronic configuration, Fig. 21). In this configuration, electrons can be distributed over MOs differently, e. g., in such a way that occupied one-electron states will not only differ in spin but also be characterized by different space-dependent one-electron functions. For instance, for the already mentioned $Cu^{II}L_6$ complex, there exist two variants of the arrangement of electrons on the e_g^* MOs corresponding to spin α of the unpaired electron, *viz.*, with two electrons on the $\psi_2 = \psi_{x^2-y^2}$ MO and one electron on the $\psi_1 = \psi_{z^2}$ one and with two electrons on the $\psi_1 = \psi_{z^2}$ MO and one electron on the $\psi_2 = \psi_{x^2-y^2}$ one (see Figs. 1 and 21). Similar two variants also correspond to unpaired electron spin β. With spin and orbital degeneracy taken into account, we therefore obtain four variants of the distribution of electrons over the e_g^* MOs, each with a Slater determinant of its own. Two of these corresponding to unpaired electron spin α (the other two only differ by the spin direction for this electron) are

$$\begin{aligned}\Phi_1 &= \frac{1}{\sqrt{N!}}\mathrm{Det}\ ||\mathrm{A}\psi_2\alpha\psi_2\beta\psi_1\alpha||, \\ \Phi_2 &= \frac{1}{\sqrt{N!}}\mathrm{Det}\ ||\mathrm{A}\psi_1\alpha\psi_1\beta\psi_2\alpha||.\end{aligned} \tag{1.31}$$

In (1.31), only partly occupied e_g^* MOs are written explicitly; the remaining part of the determinant matrix corresponding to the electrons on the fully occupied MOs is denoted by A.

The symmetry properties of the systems with open shells deserve special notice. The simplest systems are those with a nondegenerate MO containing one unpaired electron or a k-fold degenerate MO with one unpaired electron or one hole (that is, $(2k-1)$ electrons). No matter how many closed shells, such electronic systems are described by the irreducible representation governing transformations of the corresponding nondegenerate or degenerate MO. The wave function of a molecule with a half-filled nondegenerate MO of the b_1 type in addition to filled MOs transforms according to the B_1 representation. Similarly, functions (1.31) written above for the $Cu^{II}L_6$ complex with three e_g^* electrons are of the E_g symmetry type.

These are the spatial symmetry properties of many-electron functions. As concerns spin multiplicity, the systems with one unpaired electron or one hole are spin doublets.

The symmetry properties of a system with more than one electron or hole on a degenerate orbital are of a more complex character. For such a molecule, we can write several Slater determinants for all distributions of electrons over MOs allowed by the Pauli principle, and generally, their behavior under symmetry operations does not correspond to any of the irreducible representations. Certain linear combinations of these determinants, however, transform under various molecular group irreducible representations, and the corresponding electronic configuration therefore corresponds to several molecular terms. The standard example is the $e_g^{*\,2}$ configuration of an octahedral transition metal complex. For this configuration, six different determinants can be constructed, and it generates three terms, *viz.*, $^3A_{2g}$, ^1E_g, and $^1A_{1g}$ [11].

Clearly, such complications arise not only with molecules containing one open shell with two or several unpaired electrons. They are also characteristic of systems with two or more open shells with one unpaired electron (or one hole) in each. Such systems are, e. g., formed in excitation of molecules with closed shells. For instance, excitation of an electron from the t_{2g}^* MO to the e_g^* one in octahedral ML_6 complexes with the d^6 configuration, e. g., Co(III) and Rh(III) complexes, causes the appearance of two open shells. To the resulting complex with the $t_{2g}^{*\,5}e_g^{*\,1}$ electronic configuration, there correspond several terms.

3. Consider the method for approximately replacing term wave functions by simpler combinations of Slater determinants; this is the technique that we will often apply. Let $\Phi_1, \ldots, \Phi_m$ be the determinants

whose linear combinations are the sought term wave functions Ψ_j,

$$\Psi_j = \sum_{\mu=1}^{m} B_{\mu j} \Phi_\mu. \tag{1.32}$$

These wave functions are the eigenfunctions of electronic Hamiltonian (1.9) and, together with term energies, can therefore be found in the $\{\Phi\}$ basis by solving the system of equations

$$\begin{aligned}
&(\mathcal{H}^0_{11} - \mathcal{E})\mathrm{B}_1 + \mathcal{H}^0_{12}\mathrm{B}_2 + \ldots + \mathcal{H}^0_{1m}\mathrm{B}_m = 0 \\
&\mathcal{H}^0_{21}\mathrm{B}_1 + (\mathcal{H}^0_{22} - \mathcal{E})\mathrm{B}_2 + \ldots + \mathcal{H}^0_{2m}\mathrm{B}_m = 0 \\
&\cdots\cdots\cdots\cdots\cdots\cdots\cdots\cdots\cdots\cdots\cdots \\
&\mathcal{H}^0_{m1}\mathrm{B}_1 + \mathcal{H}^0_{m2}\mathrm{B}_2 + \ldots + (\mathcal{H}^0_{mm} - \mathcal{E})\mathrm{B}_m = 0
\end{aligned} \tag{1.33}$$

The corresponding secular equation is

$$\mathrm{Det}\,||\mathcal{H}^0_{\mu\nu} - \mathcal{E}\delta_{\mu\nu}|| = 0. \tag{1.34}$$

Equations (1.33) and (1.34) differ from (1.14) and (1.16) in that calculations of their matrix elements involve many-electron functions Φ_μ and many-electron Hamiltonian $\mathcal{H}_0$. The determinant form of the Φ_μ functions can be used to express $\mathcal{H}^0_{\mu\nu}$ through matrix elements written in terms of one-electron MOs. Indeed, operator (1.9) is, within an unimportant constant $V_{nn}(0)$, the sum of one- and two-electron operators

$$\begin{aligned}
&\mathcal{H}_0 - V_{nn}(0) = \mathcal{H}_0^{(1)} + \mathcal{H}_0^{(2)}; \\
&\mathcal{H}_0^{(1)} = \sum_{i=1}^{N} H_0(i); \quad \mathcal{H}_0^{(2)} = \frac{1}{2}\sum_{i\neq j}^{N} G(i,j).
\end{aligned} \tag{1.35}$$

Here, all one-electron operators $H_0(i)$ have the same form and act on the coordinates of different electrons; two-electron operators also have the same form and act on the coordinates of different pairs of electrons. The validity of (1.35) follows from (1.9). In (1.35), $H_0(i)$ is the kinetic energy operator of the ith electron plus the potential energy of its attraction to all nuclei in the molecule, and $G(i,j)$ represents electrostatic repulsion between electrons i and j. Representation (1.35) is not, however, the only possible form of $\mathcal{H}_0$. For our purposes, it is convenient to assume (in the spirit of self-consistent field theory) that the term corresponding to potential energy in every one-electron operator $H_0(i)$ describes attraction of an electron to nuclei partially screened by the other electrons including valence ones rather than naked nuclei or atomic cores. The $G(i,j)$ operator will then describe residual interelectronic interaction not included in $H_0(i)$.

Next, let the $\mathcal{H}^0_{\mu\nu} = \langle\Phi_\mu|\mathcal{H}_0|\Phi_\nu\rangle$ matrix element be written as the sum of two matrix elements with operators $\mathcal{H}_0^{(1)}$ and $\mathcal{H}_0^{(2)}$, and let in the first one, determinants Φ_μ and Φ_ν be written in the usual form as the sums of the products of spin-orbitals. It is easy to see that the resulting expression will only contain nonzero elements when determinants Φ_μ and Φ_ν either coincide or differ from each other by a single spin-orbital. We then have [5, 34]

$$\langle\Phi_\mu|\mathcal{H}_0^{(1)}|\Phi_\mu\rangle = \sum_\lambda \langle\varphi_\lambda|H_0|\varphi_\lambda\rangle, \tag{1.36}$$

$$\langle\Phi_\mu|\mathcal{H}_0^{(1)}|\Phi_\nu\rangle = \langle\varphi_\lambda|H_0|\varphi_\kappa\rangle \ (\mu \neq \nu). \tag{1.37}$$

In (1.36), the summation is over all spin-orbitals (φ_λ) of which the Φ_μ determinant is built, and in (1.37), φ_λ and φ_κ are the spin-orbitals that make Φ_μ and Φ_ν different. As each spin-orbital is the product of an MO and a spin function ($\xi = \alpha$ or β), we have

$$\langle\varphi_\lambda|H_0|\varphi_\kappa\rangle = \begin{cases} \langle\psi_\lambda|H_0|\psi_\kappa\rangle, & \text{for } \xi_\lambda = \xi_\kappa \\ 0, & \text{for } \xi_\lambda \neq \xi_\kappa \end{cases} \tag{1.38}$$

that is, matrix element (1.38) can only be nonzero if the only difference between φ_λ and φ_κ is in their spatial components (in MOs that constitute them); their spin components should coincide. Formulas (1.36)–(1.38) express the $\langle\Phi_\mu|\mathcal{H}_0^{(1)}|\Phi_\nu\rangle$ matrix element through the matrix elements of one-electron operator H_0 in the basis of MOs. It is also easy to obtain formulas for the $\langle\Phi_\mu|\mathcal{H}_0^{(2)}|\Phi_\nu\rangle$ matrix element in terms of the matrix elements of two-electron operator $G = G(1,2)$ in the same basis, that is, in terms of integrals of the form

$$\int \psi_\lambda(1)\psi_\kappa(2)G(1,2)\psi_\mu(1)\psi_\nu(2)\, dv_1 dv_2. \tag{1.39}$$

(We do not write out these formulas explicitly, for we will not need them in what follows.)

For MOs, we can select the eigenfunctions of one-electron operator H_0 introduced in (1.35), that is, we take this operator to be the effective one-electron Hamiltonian (1.12). The $\mathcal{H}_0^{(1)}$ operator should then be chosen such that the $\mathcal{H}_0^{(2)}$ operator be as small as possible. If the latter operator is ignored, all off-diagonal $\langle\Phi_\mu|\mathcal{H}_0^{(1)}|\Phi_\nu\rangle$ $(\mu \neq \nu)$ matrix elements vanish; clearly, all operator $\mathcal{H}_0^{(2)}$ matrix elements will also be absent. It follows that when the $\mathcal{H}_0 \approx \mathcal{H}_0^{(1)}$ approximate operator is used, secular equation (1.34) only contains diagonal terms. This means that for each electronic configuration, different Slater determinants are *per se* the eigenfunctions of operator $\mathcal{H}_0$ and correspond to the same energy.

In other words, in the approximation $\mathcal{H}_0^{(2)} = 0$, there is no difference between the terms corresponding to a given electronic configuration. The energy of a molecule at fixed nuclear positions is then unambiguously determined by the scheme of its MOs and MO populations, whereas the eigenfunctions of electronic Hamiltonian $\mathcal{H}_0$ can be selected in the form of separate Slater determinants or their arbitrary orthonormalized linear combinations, which can be constructed such that the new eigenfunctions of $\mathcal{H}_0$ will correspond to definite spin states (singlet, triplet, etc.). In this respect, these eigenfunctions are akin to the wave functions of the electronic terms of a molecule. The basis set of such eigenfunctions of a certain spin multiplicity (in the approximation $\mathcal{H}_0^{(2)} = 0$) is equivalent to the basis set composed of the wave functions of terms of the same multiplicity. Such states are, however, easier to find than terms, because it is not necessary to determine whether or not that these eigenfunctions correspond to the irreducible representations of the point symmetry group of the molecule.

4. Excited states of molecules and complexes formed from closed-shell ground states by promoting an electron from a doubly occupied MO (ψ_n) to a vacant MO (ψ_m) are of importance for the further presentation. One electron on MO ψ_n and one electron on MO ψ_m may have spins α and β in any combination, which means that the resulting excited state is described by the Slater determinants

$$\begin{aligned}
\Phi_1 &= (1/\sqrt{N!})\mathrm{Det}\|A\psi_n\alpha\psi_m\alpha\|,\\
\Phi_2 &= (1/\sqrt{N!})\mathrm{Det}\|A\psi_n\alpha\psi_m\beta\|,\\
\Phi_3 &= (1/\sqrt{N!})\mathrm{Det}\|A\psi_n\beta\psi_m\alpha\|,\\
\Phi_4 &= (1/\sqrt{N!})\mathrm{Det}\|A\psi_n\beta\psi_m\beta\|.
\end{aligned} \tag{1.40}$$

Their structure can be illustrated by the scheme

Φ_1	Φ_2	Φ_3	Φ_4
ψ_m ↑	ψ_m ↓	ψ_m ↑	ψ_m ↓
ψ_n ↑	ψ_n ↑	ψ_n ↓	ψ_n ↓

In the approximation $\mathcal{H}_0^{(2)} = 0$, all these determinants correspond to the same energy, although they do not correspond to a definite total electron spin of the system. Functions (1.40) are, however, easy to combine to obtain four new functions with definite total spin values,

$$\begin{aligned}
{}^{(1)}\Psi &= \left(\frac{1}{\sqrt{2}}\right)\{\Phi_2 - \Phi_3\} \\
&= \frac{1}{\sqrt{2(N!)}}\{\,\mathrm{Det}\|A\psi_n\alpha\psi_m\beta\| - \mathrm{Det}\|A\psi_n\beta\psi_m\alpha\|\,\}, \\
{}^{(3)}\Psi_1 &= \Phi_1 = (1/\sqrt{N!})\mathrm{Det}\|A\psi_n\alpha\psi_m\alpha\|, \\
{}^{(3)}\Psi_2 &= \Phi_4 = (1/\sqrt{N!})\mathrm{Det}\|A\psi_n\beta\psi_m\beta\|, \\
{}^{(3)}\Psi_3 &= \left(\frac{1}{\sqrt{2}}\right)\{\Phi_2 + \Phi_3\} \\
&= \frac{1}{\sqrt{2(N!)}}\{\,\mathrm{Det}\|A\psi_n\alpha\psi_m\beta\| + \mathrm{Det}\|A\psi_n\beta\psi_m\alpha\|\,\},
\end{aligned} \tag{1.41}$$

Here, the ${}^{(1)}\Psi$ state, like ground state (1.30), is singlet (both have zero total spin), whereas the ${}^{(3)}\Psi_1$, ${}^{(3)}\Psi_2$, and ${}^{(3)}\Psi_3$ states are triplet (have total spins equal to one).

As mentioned, partitioning the set (1.41) into singlet and triplet states is indicative of a relation between functions (1.41) and electronic terms. If MOs ψ_n and ψ_m are nondegenerate, then functions (1.41) coincide with the wave functions of singlet and triplet terms. Generally, when ψ_n belongs to a degenerate set of MOs $\{\psi_n', \psi_n'', \ldots\}$, and ψ_m, to another set $\{\psi_m', \psi_m'', \ldots\}$, there is no such coincidence. However then also, we can easily obtain a basis set of functions (1.41) equivalent to the basis set of the wave functions of singlet terms, and another basis set equivalent to the basis set of the wave functions of triplet terms. The required basis set of singlet functions (1.41) comprises all ${}^{(1)}\Psi$ functions with all possible ψ_n', ψ_m', ψ_n', ψ_m'', ψ_n'', ψ_m'', etc. pairs taken as ψ_n and ψ_m. The basis set of triplet functions (1.41) can be obtained in a similar way.

By way of example, consider an octahedral transition metal d^6 complex of the $[Co(NH_3)_6]^{3+}$ type in the $e_g^4 a_{1g}^2 t_{1u}^6 t_{2g}^5 e_g^{*\,1}$ state (a diagram of the MOs of such a complex is shown in Fig. 1 and in the scheme given above, Section 1.5.1). Singlet states (1.41) are obtained by successively substituting various $(n, m) = (xy,\ z^2)$, $(xy,\ x^2-y^2)$, $(xz,\ z^2)$, etc. pairs into the expression for ${}^{(1)}\Psi$. (Subscripts n and m indicate central atom AOs contributing to a given MO.) In conformity with what has been said above, the basis set of the obtained six singlet functions (1.41) is equivalent to the basis set of six wave functions of the ${}^1T_{1g}$ and ${}^1T_{2g}$ terms originating from the $t_{2g}^5 e_g^{*1}$ configuration. In the same way, we obtain triplet functions (1.41) for the $t_{2g}^5 e_g^{*1}$ configuration; the basis set built of these functions is equivalent to the basis set of triplet term wave functions.

5. Note in conclusion that of all (1.41) functions, singlets are our major interest. Further, we will have to calculate matrix elements of the form $\langle\Psi_0|\mathcal{H}'|\Psi\rangle$, where Ψ_0 is the wave function of the singlet ground state (1.29) of a molecule, Ψ is the excited state wave function, and $\mathcal{H}'$ is the operator that does not act on electron spins. This matrix element is only nonzero when state Ψ as well as ground state Ψ_0 is singlet. Note also that in addition to one-electron excitations, Slater determinants or their suitable linear combinations can be written for transfer of two or more electrons from filled to vacant MOs. Such many-electron excitations are, however, of little interest to us for the following reason. Operator $\mathcal{H}'$, for which we calculate matrix elements between wave function (1.29) and excited state wave functions, has the form of the sum of one-electron operators. It follows from the analysis given above that for such an operator, only those matrix elements do not vanish that cause mixing of (1.29) with singly excited configurations.

CHAPTER 2

PERTURBATION THEORY AND ADIABATIC POTENTIAL

In the preceding chapter, we discussed the electronic structure of ML_n molecular systems with fixed high-symmetry nuclear configurations. We will, however, be interested in the dependence of the structure of molecules (largely, their ground-state energy) on an arbitrary arrangement of atomic nuclei, which may be devoid of some or even all symmetry elements. In addition, $ML_{n-k}XY...Z$ heteroligand complexes that we are interested in cannot in principle possess high symmetry of their homoligand counterparts.

There are two methods for studying low-symmetry molecules. In computational quantum chemistry, an ML_n molecule with a low-symmetry nuclear configuration or an $ML_{n-k}XY...Z$ heteroligand molecule is considered independent of its high-symmetry homoligand prototype ML_n. Conversely, in model approaches, e. g., in the theory of Jahn–Teller effects [11, 14, 15] or in the theory of heteroligand systems described below [35, 36], the electronic structure of a molecule with a low-symmetry nuclear configuration or a heteroligand molecule is treated as derivative from the electronic structure of its high-symmetry prototype. The corresponding mathematical formalism is provided by quantum-mechanical perturbation theory well suited for handling such problems. In the first section of this chapter, we give information pertaining to perturbation theory; further, we discuss two important examples that will be used in treatment of $ML_{n-k}XY...Z$ systems.

2.1. Perturbation Theory for Energies and Wave Functions

1. The main goal of quantum-mechanical perturbation theory [1–3] (time-independent Rayleigh–Schrödinger perturbation theory) is as follows. Let the energy levels $E_i^{(0)}$ and the corresponding wave functions $f_i^{(0)}$ of a quantum-mechanical system described by Hamiltonian $\widehat{H}_0$ be known. With molecules, these may be term energies $\mathcal{E}_i$ and many-electron wave functions Ψ_i, that is, the eigenvalues and eigenfunctions of Hamiltonian $\mathcal{H}_0$ (1.9), or one-electron levels ε_i and one-electron MOs

ψ_i, that is, the eigenvalues and eigenfunctions of H_0 (1.12). In this section, we use the denotations $\widehat{H}$, E_i, f_i without specifying them.

Consider a system whose Hamiltonian $\widehat{H}$ is close to $\widehat{H}_0$,

$$\widehat{H} = \widehat{H}_0 + \widehat{H}', \tag{2.1}$$

where $\widehat{H}'$ is the perturbation operator assumed to be small; a more exact meaning of this assumption is discussed below. The task is to approximately determine energy levels E_i and wave functions f_i of the perturbed system through levels $E_i^{(0)}$ and wave functions $f_i^{(0)}$ of the unperturbed system and the matrix elements of the perturbation operator in the basis of the wave functions of the unperturbed system. The method for solving this problem depends on the form of the energy spectrum $E_i^{(0)}$ of the unperturbed Hamiltonian. Usually, a Hamiltonian with all nondegenerate states and a Hamiltonian with a fully degenerate state are considered.

2. If all system states are nondegenerate, the unperturbed system is characterized by wave functions $f_i^{(0)}$ and level energies $E_i^{(0)}$, that is

$$f_1^{(0)},\ f_2^{(0)},\ \ldots, \tag{2.2}$$

$$E_1^{(0)},\ E_2^{(0)},\ \ldots, \tag{2.3}$$

where all energies are different. Under small perturbations $\widehat{H}'$, the energies and the wave functions change insignificantly, and each unperturbed system state can be put in correspondence with a perturbed system state. The energy of state f_i formed from $f_i^{(0)}$ under a perturbation is then given by the series in powers of small operator $\widehat{H}'$

$$E_i = E_i^{(0)} + \langle f_i^{(0)}|\widehat{H}'|f_i^{(0)}\rangle + \sum_{j\neq i} \frac{\langle f_i^{(0)}|\widehat{H}'|f_j^{(0)}\rangle^2}{E_i^{(0)} - E_j^{(0)}} + \ldots. \tag{2.4}$$

The corresponding perturbed wave function f_i will be

$$f_i = f_i^{(0)} + \sum_{j\neq i} \frac{\langle f_i^{(0)}|\widehat{H}'|f_j^{(0)}\rangle}{E_i^{(0)} - E_j^{(0)}} f_j^{(0)} + \ldots. \tag{2.5}$$

Series (2.4) is truncated after terms of the second order of smallness with respect to $\widehat{H}'$, and series (2.5) only contains zeroth-approximation and first-order (with respect to $\widehat{H}'$) terms. In model approaches, usually only these terms are taken into account.

According to (2.5), the perturbation operator may be considered small when corrections to functions $f_i^{(0)}$ are so, that is, when the $|\langle f_i^{(0)}|\widehat{H}'|f_j^{(0)}\rangle|$ value is small in comparison with the absolute value of the difference of $E_i^{(0)}$ and $E_j^{(0)}$ and the inequality $|H'_{ij}| \ll |E_i^{(0)} - E_j^{(0)}|$ holds. This inequality is an *a priori* criterion of the applicability of the method of perturbation, although in practice and in particular when studying molecules, it is far from being always stringently satisfied. The use of perturbation theory is then, strictly speaking, unwarranted. Nevertheless even then, the method gives a qualitatively correct description of the phenomena considered.

3. When the state of the unperturbed system is k-fold degenerate, the system is characterized by orthonormalized wave functions

$$f_1^{(0)},\ f_2^{(0)},\ \ldots,\ f_k^{(0)} \tag{2.6}$$

with the same energy value $E^{(0)}$. To a first approximation, the new energy values E_i of the perturbed system will be

$$E_i = E^{(0)} + E'_i, \tag{2.7}$$

where E'_i are the roots of the secular equation of degree k

$$\mathrm{Det}\begin{Vmatrix} \langle f_1^{(0)}|\widehat{H}'|f_1^{(0)}\rangle - E' & \langle f_1^{(0)}|\widehat{H}'|f_2^{(0)}\rangle & \cdots & \langle f_1^{(0)}|\widehat{H}'|f_k^{(0)}\rangle \\ \langle f_2^{(0)}|\widehat{H}'|f_1^{(0)}\rangle & \langle f_2^{(0)}|\widehat{H}'|f_2^{(0)}\rangle - E' & \cdots & \langle f_2^{(0)}|\widehat{H}'|f_k^{(0)}\rangle \\ \cdots & \cdots & \cdots & \cdots \\ \langle f_k^{(0)}|\widehat{H}'|f_1^{(0)}\rangle & \langle f_k^{(0)}|\widehat{H}'|f_2^{(0)}\rangle & \cdots & \langle f_k^{(0)}|\widehat{H}'|f_k^{(0)}\rangle - E' \end{Vmatrix} = 0. \tag{2.8}$$

Generally, for an arbitrary k, enrgy levels cannot be written in an explicit form similar to (2.4), although this is still possible, e. g., for $k = 2$. The system of linear equations corresponding to determinant (2.8),

$$\begin{array}{l} (\langle f_1^{(0)}|\widehat{H}'|f_1^{(0)}\rangle - E')c_1 + \langle f_1^{(0)}|\widehat{H}'|f_2^{(0)}\rangle c_2 + \ldots + \langle f_1^{(0)}|\widehat{H}'|f_k^{(0)}\rangle c_k = 0 \\ \langle f_2^{(0)}|\widehat{H}'|f_1^{(0)}\rangle c_1 + (\langle f_2^{(0)}|\widehat{H}'|f_2^{(0)}\rangle - E')c_2 + \ldots + \langle f_2^{(0)}|\widehat{H}'|f_k^{(0)}\rangle c_k = 0 \\ \cdots\cdots\cdots\cdots\cdots\cdots\cdots\cdots\cdots\cdots\cdots\cdots \\ \langle f_k^{(0)}|\widehat{H}'|f_1^{(0)}\rangle c_1 + \langle f_k^{(0)}|\widehat{H}'|f_2^{(0)}\rangle c_2 + \ldots + (\langle f_k^{(0)}|\widehat{H}'|f_k^{(0)}\rangle - E')c_k = 0, \end{array} \tag{2.9}$$

can be used to find zeroth-order approximations to the wave functions of the perturbed system as eigenfunctions of operator (2.1) in the form of linear combinations of functions (2.6).

4. Suppose the unperturbed system has one nondegenerate state with wave function $f_0^{(0)}$ and energy $E_0^{(0)}$ and a set of states with wave functions $f_i^{(0)}$ and energies $E_i^{(0)}$, $i \neq 0$:

$$f_0^{(0)},\ f_1^{(0)},\ f_2^{(0)},\ \ldots;\ E_0^{(0)},\ E_1^{(0)},\ E_2^{(0)},\ \ldots. \tag{2.10}$$

Suppose also that energy $E_0^{(0)}$ equals none of the $E_i^{(0)}$ energies, whereas some of the latter may be equal to each other. This is an important situation, which arises, e. g., when the $f_0^{(0)}$ function corresponds to the Ψ_0 ground state of a molecule with closed shells, and the $f_i^{(0)}$ functions correspond to its excited states Ψ_i.

The first- and second-order corrections to energy $E_0^{(0)}$ can then be calculated by the standard formula (2.4) for nondegenerate states,

$$E_0 \simeq E_0^{(0)} + \langle f_0^{(0)}|\hat{H}'|f_0^{(0)}\rangle + \sum_{j\neq 0} \frac{\langle f_0^{(0)}|\hat{H}'|f_j^{(0)}\rangle^2}{E_0^{(0)} - E_j^{(0)}}. \tag{2.11}$$

Equation (2.11) is easy to substantiate within the framework of the perturbation theory version suggested by Löwdin [37].

2.2. Vibronic Coupling as Perturbation

1. We can now turn to the problem of finding the adiabatic potential of an ML_n molecule; it will be assumed that the Schrödinger equation (1.9) is solved and, consequently, eigenfunctions Ψ_j and eigenvalues $\mathcal{E}_j$ of the electronic Hamiltonian at point Q_0 are known. As mentioned, in model approaches of the type of the accepted version for treatment of Jahn–Teller effects [11, 14, 15], this problem is solved by perturbation theory methods starting from a molecule or a complex with nuclear configuration Q_0 and eigenfunctions Ψ_j and eigenvalues $\mathcal{E}_j$ of Hamiltonian $\mathcal{H}_0$ as a zeroth-order approximation (unperturbed system). The perturbation operator then equals the difference $\mathcal{H} - \mathcal{H}_0 = W$ of operators (1.8) and (1.9), that is, the difference of the electronic Hamiltonians at an arbitrary point Q and at point Q_0. This difference is called the vibronic coupling operator. For simplicity, operator W is approximated by the first terms of its expansion in a series in powers of displacements $Q_\nu - Q_\nu^0$ (or Q_ν, if Q_0 is conventionally set equal to zero). Usually, the first or first two expansion terms are only taken into account. Operator $\mathcal{H} = \mathcal{H}_0 + W$ then takes the form (note that $\partial W/\partial Q_\nu = \partial \mathcal{H}/\partial Q_\nu$ etc.)

$$\mathcal{H} = \mathcal{H}_0 + \sum_\nu (\partial \mathcal{H}/\partial Q_\nu)_0 Q_\nu + (1/2) \sum_{\mu,\nu} (\partial^2 \mathcal{H}/\partial Q_\mu \partial Q_\nu)_0 Q_\mu Q_\nu. \tag{2.12}$$

Here, $\mathcal{H}_0$ includes an unimportant constant corresponding to internuclear repulsion energy $V_{nn}(Q_0)$ at point Q_0; this constant can easily be eliminated through replacing (1.8) and (1.9) by the operators

$$\mathcal{H}_0 = T_e + V_{ee} + V_{en}(q,\, Q_0), \tag{2.12a}$$

$$\mathcal{H} = T_e + V_{ee} + V_{en}(q,\, Q) + V_{nn}(Q) - V_{nn}(Q_0). \tag{2.12b}$$

This replacement does not affect the operator

$$W = V_{en}(q,Q) + V_{nn}(Q) - V_{en}(q,Q_0) - V_{nn}(Q_0). \tag{2.12c}$$

Note that the eigenfunctions of operators $\mathcal{H}$ and $\mathcal{H}_0$ also remain unchanged, while the eigenvalues of both operators shift by the same $V_{nn}(Q_0)$ value.

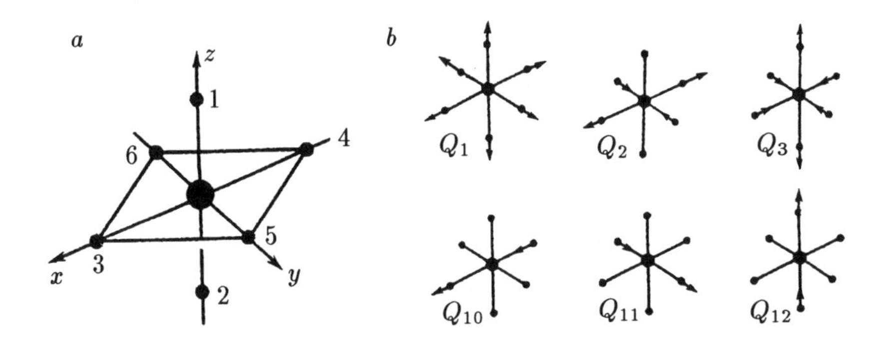

FIG. 4. (*a*) FRAME OF REFERENCE AND NUMBERING OF LIGANDS IN OCTAHEDRAL ML_6 COMPLEXES AND (*b*) NORMAL (SYMMETRIZED) STRETCHING MODES FOR OCTAHEDRAL ML_6 COMPLEXES (ALL $Q_\nu > 0$).

2. Coordinates Q_ν may be arbitrary variables describing the mutual arrangement of atomic nuclei in a molecular system. Especially convenient are the normal coordinates used in the theory of molecular vibrations [6, 11, 14, 15, 38, 39], because like term wave functions and MOs, normal coordinates are classified according to the irreducible representations of the point symmetry group corresponding to the Q_0 nuclear configuration (Table 2 and Fig. 4). Hamiltonian $\mathcal{H}$ derivatives with respect to normal coordinates have similar symmetry properties. In particular, the $(\partial\mathcal{H}/\partial Q_\nu)_0$ derivative as a function of the coordinates of electrons q transforms under the same irreducible representation as coordinate Q_ν [1, 14, 15]. This circumstance leads to important selection rules for $\langle\Psi_i|(\partial\mathcal{H}/\partial Q_\nu)_0|\Psi_j\rangle$ matrix elements (so-called linear vibronic constants), where wave functions Ψ_i and Ψ_j correspond to molecular electronic terms.

Let Γ_i, Γ_j, and Γ_ν be the irreducible representations that govern transformations of the Ψ_i and Ψ_j wave functions and the Q_ν normal coordinate, respectively. For the $\langle\Psi_i|(\partial\mathcal{H}/\partial Q_\nu)_0|\Psi_j\rangle$ matrix element

Table 2. Symmetrized displacements (normal coordinates) for the ML_6 and ML_4 molecular systems expressed through Cartesian displacement coordinates

Normal coordinate	Symmetry	Cartesian displacements
Octahedral ML_6 complex of O_h symmetry		
Q_1	A_{1g}	$(X_3 - X_4 + Y_5 - Y_6 + Z_1 - Z_2)/\sqrt{6}$
Q_2	E_g	$(X_3 - X_4 - Y_5 + Y_6)/2$
Q_3		$(X_3 + X_4 - Y_5 + Y_6 + 2Z_1 - 2Z_2)/2\sqrt{3}$
Q_7	T'_{1u}	$(X_1 + X_2 + X_5 + X_6)/2$
Q_8		$(Y_1 + Y_2 + Y_3 + Y_4)/2$
Q_9		$(Z_3 + Z_4 + Z_5 + Z_6)/2$
Q_{10}	T''_{1u}	$(X_3 - X_4)/\sqrt{2}$
Q_{11}		$(Y_5 + Y_6)/\sqrt{2}$
Q_{12}		$(Z_1 + Z_2)/\sqrt{2}$
Square-planar ML_4 complex of D_{4h} symmetry		
Q_1	A_{1g}	$(X_1 - X_2 + Y_3 - Y_4)/2$
Q_2	B_{1g}	$(X_1 - X_2 - Y_3 + Y_4)/2$
Q_3	E_u	$(X_1 + X_2)/\sqrt{2}$
Q_4		$(Y_3 + Y_4)/\sqrt{2}$

Table 2. (continued from the previous page)

Normal coordinate	Symmetry	Cartesian displacements
		Tetrahedral ML_4 complex of T_d symmetry
Q_1	A_1	$(Z_1 + Z_2 + Z_3 + Z_4)/2$
Q_4		$(Z_1 - Z_2 + Z_3 - Z_4)/2$
Q_5	T_2'	$(Z_1 - Z_2 - Z_3 + Z_4)/2$
Q_6		$(Z_1 + Z_2 - Z_3 - Z_4)/2$
Q_7		$(1/4)(-X_1 + X_2 - X_3 + X_4) + (\sqrt{3}/4)(-Y_1 + Y_2 - Y_3 + Y_4)$
Q_8	T_2''	$(1/4)(-X_1 + X_2 + X_3 - X_4) + (\sqrt{3}/4)(Y_1 - Y_2 - Y_3 + Y_4)$
Q_9		$(1/2)(X_1 + X_2 + X_3 + X_4)$

Note: (1) Displacements changing M—L bond lengths and some of the displacements changing L—M—L valence angles are given. (2) For octahedral and square-planar complexes, displacements are given in the general coordinate system, and for tetrahedral complexes, in the system of ligand-fixed local coordinates. (3) The Q_7–Q_{12} coordinates for ML_6 complexes, the Q_3 and Q_4 coordinates for ML_4 complexes of D_{4h} symmetry, and the Q_4–Q_9 coordinates for complexes of T_d symmetry are symmetry rather than normal coordinates. The coordinate systems used are shown in Figs. 4, 14, and 19.

to be nonzero, the following conditions should be satisfied:

$$\begin{cases} \Gamma_\nu \subset \Gamma_i \times \Gamma_j, & \text{if } \Psi_i \text{ and } \Psi_j \text{ correspond to} \\ & \text{different irreducible representations,} \\ \Gamma_\nu \subset [\Gamma_i^2], & \text{if } \Psi_i \text{ and } \Psi_j \text{ correspond to} \\ & \text{the same irreducible representation.} \end{cases} \tag{2.13}$$

here, $\gamma_i \times \gamma_j$ is the direct product of representations Γ_i and Γ_j, and $[\Gamma_i^2]$ is the symmetrical part of the direct product $\Gamma_i \times \Gamma_i$ (those who are interested in such group-theoretic constructions are referred to [1, 22–25]).

One point concerning Table 2 should be mentioned. Note that the Q_7–Q_{12} coordinates for octahedral complexes, the Q_3 and Q_4 coordinates for square-planar complexes, and the Q_4–Q_9 coordinates for tetrahedral complexes are mere "symmetry coordinates" [38]. They have the same symmetry properties as normal coordinates, which are combinations of symmetry coordinates of the same symmetry type. Symmetry coordinates can, however, be used instead of normal ones in consideration of static deformations of molecules in the "valence force field" approximation [39], according to which the potential energy of a molecule does not contain cross terms between bond stretches and valence angle deformations. We accept this model, because it qualitatively correctly describes the experimental data to be analyzed. Accordingly, all coordinates included in Table 2 will, for shortness, be called normal without specifying this every time that some of them are mere symmetry coordinates.

3. For the applications considered below, the major interest are molecules with closed shells such as $Pt^{II}L_4$ complexes and molecules with one open shell such as $Cu^{II}L_6$. Adiabatic potentials are convenient to discuss first for systems with open shells, which allows us to pass in a more natural way from standard perturbation theory to the idea of the adiabatic potential as a multidimensional and multisheeted potential energy surface.

In constructing adiabatic potentials for molecules with open shells, only the wave functions of the ground state (term) of a system are usually taken into account; let these be the functions

$$\Psi_1, \Psi_2, \ldots, \Psi_k, \tag{2.14}$$

where k is the multiplicity of orbital degeneracy of the electronic term at $Q_0 = 0$. We have to apply perturbation theory for degenerate levels considered in the preceding chapter. For simplicity, functions (2.14) are selected as many-electron functions only including open-shell MOs, for instance, the $\psi^*_{z^2}$ and $\psi^*_{x^2-y^2}$ MOs for $Cu^{II}L_6$. The other orbitals are not included in (2.14), and the electrons that they contain, both core and valence, are considered "internal" [11]; a more rigorous approach free of this simplification was taken in [14, 15]. As a zeroth-order perturbation theory approximation, the total molecular system energy as a function of coordinates Q_ν is described by the quadratic function $1/2\sum_\nu K_\nu Q_\nu^2$, where K_ν are the harmonic force constants. According to (2.7), the resulting adiabatic potential can then be written as [11]

$$\mathcal{E}(Q) = 1/2\sum_\nu K_\nu Q_\nu^2 + \mathcal{E}'(Q). \tag{2.15}$$

Here, the anharmonic vibronic part of adiabatic potential (2.15), $\mathcal{E}'$, is, according to (2.8), given by one of the roots of the secular equation

$$\mathrm{Det}\left\| \sum_\nu A^\nu_{ij} Q_\nu + \frac{1}{2}\sum_{\mu,\nu} A^{\mu\nu}_{ij} Q_\mu Q_\nu - \delta_{ij}\mathcal{E}' \right\| = 0. \tag{2.16}$$

This equation corresponds to the matrix of perturbation operator W from (2.12) in basis (2.14), and the values

$$A^\nu_{ij} = \langle \Psi_i | (\partial \mathcal{H}/\partial Q_\nu)_0 | \Psi_j \rangle, \tag{2.17}$$

$$A^{\mu\nu}_{ij} = \langle \Psi_i | (\partial^2 \mathcal{H}/\partial Q_\mu \partial Q_\nu)_0 | \Psi_j \rangle \tag{2.17a}$$

indicate the extent of mixing of ground-state wave functions Ψ_j under molecular deformations Q_ν. It follows from group-theory considerations (from the Wigner–Eckart theorem [11, 14, 15]) that not all (2.17) constants are independent. For the normal coordinates of the same symmetry type (associated with some irreducible representation of molecular symmetry group at point Q_0), all linear vibronic coupling constants can be expressed via one "reduced" linear vibronic constant. Matrix elements (2.17a) cal also be written in terms of a smaller number of "reduced" quadratic vibronic constants [11, 14, 15].

Consider the roots of equation (2.16) [11, 14, 15]. It has k roots at an arbitrary point $Q \neq 0$. The adiabatic potential surface in the multidimensional space $\{\mathcal{E}, Q_1, Q_2 \ldots, Q_\nu, \ldots\}$ is therefore a hypersurface comprising separate sheets, which coincide at point Q_0 because of k-fold term degeneracy at this point. The lowest sheet corresponds to the ground state, and the others, to excited states of the system. Further, we will only refer to the lowest sheet of the adiabatic potential energy surface; for shortness, it will often be referred to as the adiabatic potential surface.

In polyatomic molecules with a nonlinear spatial arrangement of nuclei and a degenerate term, the lowest sheet possesses the characteristic property that symmetrical configuration Q_0 can never be a minimum, even a local one, of the corresponding potential energy surface. This property rules out every possibility of a static high-symmetry nuclear configuration for a polyatomic nonlinear molecule with an orbitally degenerate ground state (the Jahn–Teller effect [11, 14, 15]). If the barriers separating various potential energy surface local minima are high enough to prevent nuclei from transitions between these minima, then we have the so-called static Jahn–Teller effect, which is virtually absent in isolated molecules and complexes but can be observed in crystals with Jahn–Teller centers, which interact with each other and the environment, in lattice sites. Another possibility is the dynamic Jahn–Teller effect, when atomic nuclei can penetrate barriers separating local minima. The dynamic effect is characteristic of isolated molecular systems, where no static distorted low-symmetry configuration can be observed, but motions of the atomic nuclei of the system have the character of continuous shifts between low-symmetry configurations.

4. Molecules of the second type have closed electronic shells, that is, nondegenerate ground-state terms. Adiabatic potentials are then constructed from ground-state wave function Ψ_0 and some number (p) of excited-state functions, among which there may be degenerate ones,

$$\Psi_0,\ \Psi_1,\ \ldots,\ \Psi_p. \tag{2.18}$$

To calculate the adiabatic potential, consider the action of (2.12b) on the basis functions (2.18) [11, 14, 15]. All adiabatic potential energy surface sheets are determined by the roots of the secular determinant $\mathrm{Det}\|\mathcal{H}_{ij} - \mathcal{E}\delta_{ij}\| = 0$, $i, j = 0, 1, \ldots, p$, and the lowest sheet of the surface corresponds to the smallest root. For our purposes, another technique [32, 35] based on Ψ_0 being nondegenerate is more convenient to apply. This technique directly yields an explicit expression for the lowest potential energy surface sheet, although, first, it is incapable of providing information about the other sheets, and, secondly, the expression obtained is only valid in a local neighborhood of initial configuration Q_0.

Let us apply perturbation theory with operator W as a perturbation operator to obtain corrections to the energy of the ground state of a molecule. We will only take into account terms not higher than second-order in variables Q_ν, that is, first- and second-order perturbation series terms. For the lowest adiabatic potential energy surface sheet, this yields $\mathcal{E}(Q) = \sum_\nu \langle\Psi_0|(\partial\mathcal{H}/\partial Q_\nu)_0|\Psi_0\rangle Q_\nu$ + terms quadratic in Q_ν to within an unimportant constant, and we can choose Q_0 in such a way that all terms linear in Q_ν will vanish. Indeed, for all modes except the totally symmetric one, the coefficients of Q_ν in linear terms are zero by virtue of conditions (2.13). As concerns the totally symmetric mode, the corresponding matrix elements can be reduced to zero by optimizing initial configuration Q_0 [32, 33] (as mentioned in Section 1.2, Q_0 can be chosen such that energy $\mathcal{E}(Q)$ will be minimum at Q_0, and, accordingly, its derivative with respect to totally symmetric deformations will be zero). It follows from the properties of normal coordinates Q_ν that the potential energy surface of a molecule is written as the sum of the squares of Q_ν with the corresponding coefficients. Eventually, the expression for the lowest adiabatic potential sheet will be

$$\mathcal{E}(Q) = 1/2 \sum_\nu K_\nu Q_\nu^2, \tag{2.19}$$

where K_ν are the force constants of the form

$$K_\nu = K_\nu^0 - \sum_{j\neq 0} 2(A_j^\nu)^2/\omega_j, \tag{2.20}$$

$$K_\nu^0 = \langle \Psi_0 | (\partial^2 \mathcal{H}/\partial Q_\nu^2)_0 | \Psi_0 \rangle, \tag{2.21}$$

$$A_j^\nu = \langle \Psi_0 | (\partial \mathcal{H}/\partial Q_\nu)_0 | \Psi_j \rangle. \tag{2.22}$$

The A_j^ν values are the off-diagonal linear vibronic coupling constants, which here show to what extent the ground state Ψ_0 and excited states Ψ_j are mixed by deformations Q_ν, and ω_j is the energy gap between the ground and jth excited states at point Q_0, that is, $\omega_j = \langle \Psi_j | \mathcal{H}_0 | \Psi_j \rangle - \langle \Psi_0 | \mathcal{H}_0 | \Psi_0 \rangle$.

Equations (2.19) and (2.20) show that the character of the lowest sheet of the adiabatic potential energy surface in the vicinity of Q_0 is determined for each normal coordinate by competition of two terms. The second one is negative and appears because of vibronic mixing of the ground and excited states, whereas the first term ("bare" force constant K^0) is believed to be positive [40a–40c]. Two situations may arise depending on the relative values of these two terms. If for all ν, we have $2\sum_{j\neq 0}(A_j^\nu)^2/\omega_j < K_\nu^0$, that is, if energy gaps are fairly broad and/or vibronic coupling constants A_j^ν are small, vibronic mixing only decreases bare force constants, and K_ν remain positive (weak pseudo-Jahn–Teller effect). The potential energy surface of a molecule then has a local minimum (I) at point Q_0, and vibronic coupling does not distort the form of the molecule. Conversely, if $2\sum_{j\neq 0}(A_j^\nu)^2/\omega_j > K_\nu^0$ for at least one ν value, vibronic coupling makes K_ν negative. The potential energy surface of the molecule then has the shape of a multidimensional saddle (II) in the neighborhood of Q_0, and the initial Q_0 configuration is then unstable along the Q_ν coordinate. This situation is known as a strong pseudo-Jahn–Teller effect:

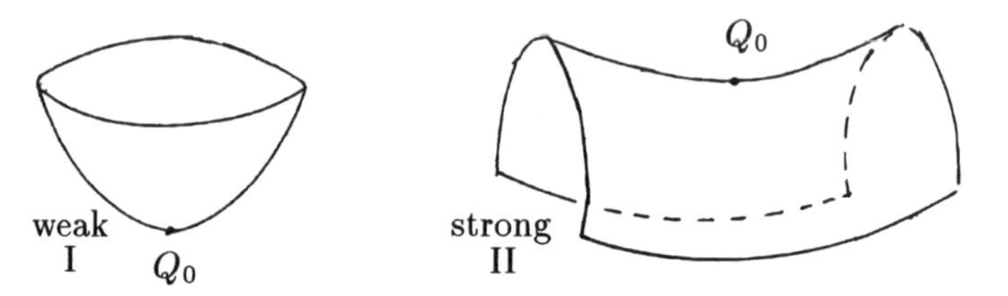

According to Pirson's criterion [32, 33], a strong pseudo-Jahn–Teller effect is usually observed when energy gap ω_j is minimal (narrower than 4 eV), although according to the calculations made in [40a, 40b], vibronic distortions are also possible when the minimum gap is much broader. As distinguished from the Jahn–Teller effect, a strong Jahn–Teller pseudoeffect may be present in isolated molecules and complexes. This effect explains distorted shapes of a number of molecular systems, for instance, the bent structure of water molecules and the nonplanar geometry of ammonia [40a, 40b]. (Strictly speaking, static geometry distortions only mean that the time of tunneling between different potential energy surface minima is substantially long. For instance, in the NH_3 molecule, the nitrogen nucleus does tunnel, although at a low frequency, through the barrier separating two minima which correspond to pyramidal structures. This tunneling should be taken into account in describing certain phenomena.)

An similar interesting example is instability of the $ZnCl_5^{3-}$ trigonal–bipyramidal complex, which tends to turn into a distorted $ZnCl_4^{2-}$ tetrahedron and an isolated Cl^- ion, as in the $[Co(NH_3)_6][ZnCl_4]Cl$ crystal [41].

5. There is another method for describing the potential energy surface of a molecule with closed shells. This method can be used to construct model adiabatic potentials not only in the neighborhood of Q_0 but in the whole configuration space. For instance, in the presence of a strong pseudo-Jahn–Teller effect, the method is capable of describing both equilibrium instability at point Q_0 and the minima that the system reaches after leaving point Q_0. This method is usually described for the simple example of a two-level system with nondegenerate ground (Ψ_0) and excited (Ψ_1) states of different symmetries which can be mixed by mode Q [14, 15] (for instance, Ψ_0 and Ψ_1 may be an even and an odd function, respectively, and Q, an odd mode). Taking into account only the term linear in Q in (2.12) and using perturbation theory, we then obtain the anharmonic part of the adiabatic potential

in the form of the secular equation

$$\text{Det}\begin{Vmatrix} -\frac{\omega_1}{2} - \mathcal{E}' & A_1 Q \\ A_1 Q & \frac{\omega_1}{2} - \mathcal{E}' \end{Vmatrix} = 0. \qquad (2.23)$$

Here, ω_1 is the energy gap between the ground and excited states, and A_1 is the linear vibronic coupling constant, $A_1 = \langle \Psi_0 | (\partial \mathcal{H}/\partial Q)_0 | \Psi_1 \rangle$. Solving the secular equation and using (2.15), we obtain the potential energy curve for the ground state in the explicit form

$$\mathcal{E}(Q) = (1/2)K^0 Q^2 - \sqrt{(\omega_1/2)^2 + A_1^2 Q^2}\,. \qquad (2.23a)$$

It follows from (2.23) that at $2A_1^2/\omega_1 < K^0$, the potential curve has one minimum at the origin $Q = 0$ (weak pseudo-Jahn–Teller effect). Conversely, at $2A_1^2/\omega_1 > K^0$, that is, in the presence of a strong pseudo-Jahn–Teller effect, the potential energy curve has a local maximum at $Q = 0$ and two minima situated symmetrically on both sides of point $Q = 0$:

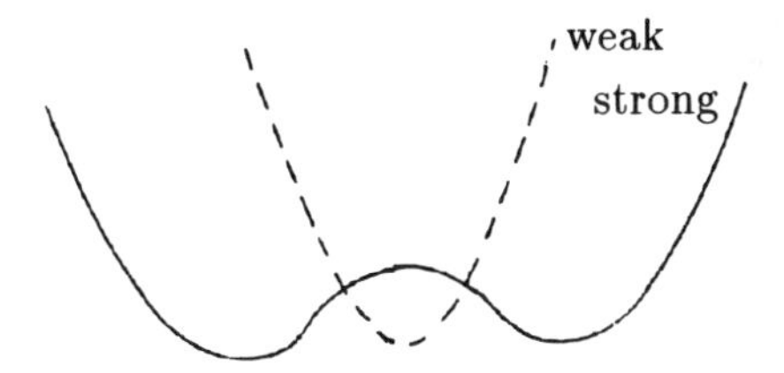

Equation (2.23) also gives the coordinates of the minima (from the standard condition $\partial\mathcal{E}/\partial Q = 0$),

$$Q = \pm\sqrt{(A_1/K^{(0)})^2 - (\omega_1/2A_1)^2}\,. \qquad (2.23b)$$

6. As mentioned, the method of perturbation is applied not only to many-electron states of molecular systems but also to their MOs. By analogy with the many-electron operator $\mathcal{H} - \mathcal{H}_0 = W$, we might introduce the one-electron vibronic coupling operator $\widetilde{H} - H_0 = U - U_0 = w_1$, that is, the difference between the one-electron Hamiltonian $\widetilde{H} = -(1/2)\Delta + U$ for an arbitrary nuclear configuration Q and operator (1.12) defined for configuration Q_0. Here, as in Section 1.3.1, U and U_0 are the operators describing the action on a given electron of the other electrons and all atomic nuclei of the system. It will, however, be more convenient to use a somewhat different one-electron perturbation

operator. Let us introduce operator w, which is the sum of w_1 and an additional term w_2, that is,

$$w = w_1 + w_2, \quad w_2 = (1/N)[V_{nn}(Q) - V_{nn}(Q_0)],$$

where N is the total number of electrons in the system, and $V_{nn}(Q)$ and $V_{nn}(Q_0)$ are the operators of mutual repulsion of atomic nuclei at points Q and Q_0, respectively. Operator w equals the difference of the one-electron operator

$$H = \widetilde{H} + w_2 = -(1/2)\Delta + U + (1/N)[V_{nn}(Q) - V_{nn}(Q_0)]$$

and operator (1.12). Like many-electron operator W (Section 2.2), w is approximated by the first terms of its expansion in powers of displacements $Q - Q_0$ (or, by virtue of the convention $Q_0 = 0$, in powers of Q). Using this approximation we can write

$$H = H_0 + \sum_{\nu}(\partial H/\partial Q_\nu)_0 Q_\nu + (1/2)\sum_{\mu\nu}(\partial^2 H/\partial^2 Q_\mu Q_\nu)_0 Q_\mu Q_\nu. \quad (2.24)$$

Here, the derivatives of one-electron Hamiltonian H with respect to normal coordinates possess the same symmetry properties as the derivatives of many-electron Hamiltonian $\mathcal{H}$. In particular, the $(\partial H/\partial Q_\nu)_0$ derivative transforms under the same irreducible representation of molecular point symmetry group as coordinate Q_ν [32].

This circumstance leads to important selection rules for the matrix elements $\langle\psi_n|(\partial H/\partial Q_\nu)_0|\psi_m\rangle$ calculated in the basis of MOs; here and below, we again consider operator H_0 eigenfunctions. Let Γ_n, Γ_m, and Γ_ν be the irreducible representations associated with the ψ_n and ψ_m MOs and the Q_ν normal coordinate. Then the $\langle\psi_n|(\partial H/\partial Q_\nu)_0|\psi_m\rangle$ matrix element may only be nonzero if the following conditions are met:

$$\begin{cases} \Gamma_\nu \subset \Gamma_n \times \Gamma_m, & \text{if } \psi_n \text{ and } \psi_m \text{ correspond to different irreducible representations,} \\ \Gamma_\nu \subset [\Gamma_n^2], & \text{if } \psi_n \text{ and } \psi_m \text{ correspond to the same irreducible representation.} \end{cases} \quad (2.25)$$

By analogy with the vibronic constants introduced earlier and calculated for term wave functions ("integral" vibronic coupling constants), similar constants for MOs are called orbital vibronic coupling constants [15]. For instance, linear orbital vibronic coupling constants can be written as

$$A_{nm}^{\nu} = \langle\psi_n|(\partial H/\partial Q_\nu)_0|\psi_m\rangle, \quad (2.26)$$

and (2.25) are therefore the selection rules for these constants. Orbital vibronic coupling constants are often written by small letters. We will not adhere to this tradition, for there will be no danger of confusing them with integral constants.

6. Orbital vibronic constants are easier to comprehend, in the same way as MOs are more visual than many-electron wave functions of molecules. Their physical meaning is simple. After a molecule is deformed along mode Q_ν, "new" MOs generally do not coincide with "old" ones but can be represented as linear combinations of "old" MOs. Off-diagonal ($m \neq n$) constants (2.26) describe the effectiveness of mixing of MOs ψ_n and ψ_m of the molecule with nuclear configuration Q_0 under deformation along mode Q_ν. Diagonal ($m = n$) constants (2.26) also have a simple physical meaning. They describe the "speed" at which the eigenvalues of operator H change under molecular deformations along mode Q_ν. We will see that calculations of constants (2.26) can be reduced to calculations of the matrix elements

$$\langle\psi_n\big|[\partial\widetilde{H}/\partial Q_\nu]_0\big|\psi_m\rangle = \langle\psi_n\big|\big[\partial(-(1/2)\Delta + U)/\partial Q_\nu\big]_0\big|\psi_m\rangle. \quad (2.26a)$$

In other words, in calculating such vibronic constants, it suffices to take into account the "electronic" component of operator H leaving out of consideration operator w_2.

7. Matrix elements (2.26) will be calculated using the approach suggested by perturbation MO (PMO) theory well known in organic chemistry [12], also see [42–44]. This theory deals with effective one-electron (Hückel-type) Hamiltonians of molecules with similar compositions and atomic spacings. One of these (e. g., benzene) is considered unperturbed, and the other (e. g., pyridine C_5NH_5) is treated as perturbed. It is assumed [12] that the basis set of the AOs of the unperturbed molecule can also be used to describe the perturbed one, but the matrix elements of the Hamiltonian of the perturbed system in this basis are assigned the values that they would have in the basis set of the AOs of the perturbed molecule. In particular, this approach is applied to obtain off-diagonal matrix elements, that is, resonance integrals β_{ij}, which characterize interaction between AOs χ_i and χ_j and can have different values in unperturbed and perturbed molecules, e. g., because of changes in interatomic distances.

Let us apply the same technique to take into account changes in interatomic distances under small deformations of molecules along vibrational modes. Deformation along mode Q_ν should then be described as changes in resonance integrals β_{ij} in the matrix of operator H_0 (1.12) calculated in the basis set of AOs of the unperturbed molecule, i. e., the molecule with the Q_0 nuclear configuration. (Clearly, Coulomb integrals α_i can be considered invariable under molecular deformations, because the natures of the atoms remain unchanged.) To sum up, "new" β_{ij} values are first calculated as resonance integral values in the basis of AOs of the deformed (perturbed) molecule,

$$\beta_{ij}(Q_\nu) = \beta_{ij}(Q_0) + (\partial\beta_{ij}/Q_\nu)_0 Q_\nu, \tag{2.26b}$$

where $(\partial\beta_{ij}/\partial Q_\nu)_0$ is understood to be the derivative at $Q = Q_0$ of function $\beta_{ij}(Q_\nu)$ describing β_{ij} variations under real deformations. On the other hand, the matrix elements of one-electron Hamiltonian $\widetilde{H}$ in the basis of AOs χ_i of the unperturbed molecule are

$$\begin{aligned} \langle\chi_i|\widetilde{H}|\chi_i\rangle &= \langle\chi_i|H_0 + (\partial\widetilde{H}/\partial Q_\nu)_0 Q_\nu|\chi_j\rangle \\ &= \langle\chi_i|H_0|\chi_i\rangle + \langle\chi_i|(\partial\widetilde{H}/\partial Q_\nu)_0|\chi_j\rangle Q_\nu. \end{aligned} \tag{2.26c}$$

According to the assumptions of PMO theory, the right-hand sides of (2.26b) and (2.26c) coincide. Therefore

$$\langle\chi_i|(\partial\widetilde{H}/\partial Q_\nu)_0|\chi_j\rangle = (\partial\beta_{ij}/\partial Q_\nu)_0, \quad i \neq j. \tag{2.27}$$

Performing the transformation to the basis set of MOs and taking into account (2.26) and the equality $H = \widetilde{H} + w_2$, we eventually obtain

$$\begin{aligned} A^\nu_{nm} &= \sum_{i\neq j} c_{in}c_{jm}\left(\frac{\partial\beta_{ij}}{\partial Q_\nu}\right)_0 + \langle\psi_n|(\partial w_2/\partial Q_\nu)_0|\psi_m\rangle \\ &= \sum_{i\neq j} c_{in}c_{jm}\left(\frac{\partial\beta_{ij}}{\partial Q_\nu}\right)_0 + \left(\frac{\partial w_2}{\partial Q_\nu}\right)_0 \delta_{nm}. \end{aligned} \tag{2.28}$$

It is taken into account here that w_2 is independent of the coordinates of electrons, and MOs are orthonormalized. It follows that off-diagonal constants (2.26) are expressed solely through the derivatives of resonance integrals,

$$A^\nu_{nm} = \sum_{i\neq j} c_{in}c_{jm}(\partial\beta_{ij}/\partial Q_\nu)_0, \quad m \neq n. \tag{2.28a}$$

Formula (2.28a) also applies to diagonal constants except when Q_ν is the totally symmetric deformation Q_1, because for the other modes, the $(\partial w_2/\partial Q_\nu)_0$ derivative vanishes. In many instances, this is easy to check by analyzing the behavior of $(\partial w_2/\partial Q_\nu)_0$ for various modes. More general considerations can also be given. Consider some symmetrical configuration of atomic nuclei Q_0 and suppose that not all $(\partial w_2/\partial Q_\nu)_0$ derivatives and, consequently (by the definition of w_2), not all $(\partial V_{nn}/\partial Q_\nu)_0$ derivatives reduce to zero for other than totally symmetric modes. For any molecule with nuclear configuration Q_0, there would then exist $Q_\nu \neq Q_1$ such that in the neighborhood of Q_0, the total energy (adiabatic potential) $\mathcal{E}$ of a molecule would be a linear function of Q_ν, i. e., $\mathcal{E} \sim Q_\nu$. Such a dependence characteristic of the Jahn–Teller effect would be caused solely by changes in the energy of internuclear repulsion no matter what the electronic structure of the molecule. The effect should then be observed for, e. g., nonlinear molecules with orbitally nondegenerate electronic configurations, which cannot exhibit such a Jahn–Teller behavior.

Lastly, consider formula (2.28) for diagonal orbital vibronic constants (2.26) and the totally symmetric mode $Q_\nu = Q_1$. The A^1_{nn} constant then contains a nonvanishing $(\partial w_2/\partial Q_1)_0$ derivative, which, for symmetrical ML_n molecules with octahedral, tetrahedral, and square-planar nuclear configurations, can easily be expressed through charges of atomic nuclei or cores and M — L interatomic distances. We can, however, again obviate the necessity of directly taking into account nuclear repulsion. Let Q_0 be an optimized configuration corresponding to a total energy (adiabatic potential, Section 1.2.1) minimum. The $(\partial w_2/\partial Q_1)_0$ derivative can then be written through values of the electronic origin as described below.

8. First consider the relation between integral vibronic coupling constants (2.17) and (2.22) and orbital vibronic constants (2.26), which is of interest *per se.* To determine this relation, we will use (1.35), where the $G(i, j)$ operators describe "residual" interelectronic interaction not included into the expressions for operators $H_0(i)$, see Section 1.5. Ignoring the corresponding $\mathcal{H}_0^{(2)}$ operator in $\mathcal{H}_0$ (2.12a), we have

$$\mathcal{H}_0 \approx \sum_i^{\text{electrons}} H_0(i). \tag{2.29}$$

A similar equality can be written for an arbitrary nuclear configuration Q with $\mathcal{H}$ taken in form (2.12b),

$$\mathcal{H} \approx \sum_i^{\text{electrons}} H(i). \tag{2.30}$$

Subtracting (2.29) from (2.30) and taking into account equations (2.12) and (2.24), we obtain

$$\begin{aligned}(\partial\mathcal{H}/\partial Q_\nu)_0 &\approx \sum_i^{\text{electrons}} \left(\frac{\partial H(i)}{\partial Q_\nu}\right)_0,\\ (\partial^2\mathcal{H}/\partial Q_\mu\partial Q_\nu)_0 &\approx \sum_i^{\text{electrons}} \left(\frac{\partial^2 H(i)}{\partial Q_\mu \partial Q_\nu}\right)_0.\end{aligned} \tag{2.31}$$

Next, consider vibronic coupling constants (2.17) and (2.22), where the many-electron functions will be expressed in terms of MOs. Using (2.31) then allows us to express integral vibronic constants (2.17) and (2.21) through orbital vibronic constants.

To conclude, we return to calculating diagonal orbital vibronic coupling constants for totally symmetric modes Q_1. Let us write the relation for the linear diagonal integral vibronic constant A_0^1 when Ψ_0 represents a single Slater determinant (e. g., for a molecule with closed shells),

$$\begin{aligned}A_0^1 = A_{00}^1 &= \sum_r n_r \left\langle \psi_r \middle| \left(\frac{\partial \widetilde{H}}{\partial Q_1}\right)_0 \middle| \psi_r \right\rangle + N\left(\frac{\partial w_2}{\partial Q_1}\right)_0\\ &= -\sum_r \sum_{i\neq j} n_r c_{ir} c_{jr} \left(\frac{\partial \beta_{ij}}{\partial Q_1}\right)_0 + N\left(\frac{\partial w_2}{\partial Q_1}\right)_0.\end{aligned} \tag{2.31a}$$

This relation directly follows from (2.31) and (1.36). Here, n_r is the population of the rth MO, and N is the total number of electrons in the system. Recall that the Q_0 configuration is optimized with respect to deformation Q_1, and at Q_0, the total energy of the molecule as a function of Q_1 has an extremum. By virtue of the definition of w_2, the right-hand side of (2.31a) then equals zero. Therefore,

$$(\partial w_2/\partial Q_1)_0 = -(1/N)\sum_r \sum_{i\neq j} n_r c_{ir} c_{jr} (\partial \beta_{ij}/\partial Q_1)_0. \tag{2.31b}$$

Substituting (2.31b) into (2.28) at $\nu = 1$ and $n = m$ yields the desired expression for the diagonal vibronic coupling constant under totally symmetric deformations Q_1 through the derivatives of resonance integrals. Of course, some resonance integrals in (2.28), (2.28a), (2.31a), and (2.31b) can be omitted in actual calculations. As is usual in chemistry, we will, for ML_n-type molecules, only take into account resonance integrals corresponding to interaction between the central M atom and ligands L.

2.3. Replacement of Atoms as Perturbation: Substitution Operator

1. Apart from the theory of vibronic coupling, the perturbation method was applied to study atomic (or atomic group) substitution effects in molecular systems. Even in prequantum chemistry, such molecules as, e. g., CH_3Hal and pyridine were treated as derivatives of simpler molecules (methane and benzene, respectively) which changed their properties (not radically, though) when a foreign atom was introduced. In quantum-chemical terms, such a "perturbative" treatment of substitution was developed within the framework of the already mentioned PMO theory [12, 44]. In this theory, substitution of atoms is described as changes of one-center and two-center matrix elements (Coulomb and resonance integrals) in the effective one-electron Hamiltonian of a molecule. It is accepted at a model level that the matrices of the one-electron Hamiltonians of both unperturbed and perturbed molecules can be written in the basis set of the AOs of the unperturbed molecule. In other words, substitution of atoms is treated as replacement of matrix element values in the matrix of the Hamiltonian of the unperturbed molecule by the values that these matrix elements would have in the basis set of the AOs of the substituted molecule. These changes of the Coulomb and resonance integrals are considered small, which invites applying standard perturbation theory (Section 2.1). As the eigenfunctions of the unperturbed problem, the MOs of the unsubstituted system are used, and the matrix elements of the perturbation operator are found in the basis set of these MOs. After this, the MOs of the perturbed system are calculated and used to determine changes in atomic charges, bond orders, and other changes in the electronic structure of the molecule caused by the substitution.

The described approach accepted in PMO theory is also used in inorganic chemistry. It was for the first time applied to the problem of ligand substitution $ML_n \longrightarrow ML_{n-k}XY \ldots Z$ in molecules and complexes in [45–47]; in somewhat more detail, these works are characterized in the beginning of Chapter 3. Further in this section, we will concentrate on methodology of applying perturbation theory to describe substitution of one or several ligands in molecules or complexes of the ML_n type. These problems will in essence be treated by concepts based on the PMO theory [12, 43], but with the use of the notation more convenient for our purposes and, in certain instances, of a more general character.

2. One of the key assumptions of the standard PMO theory and its version for treating ML_n molecules is that substitution of atoms (e. g., ligands) in a molecule only affects its electronic subsystem. It

is clear that in reality, substitution also causes changes in interatomic distances, atomic masses, etc. This assumption is, however, natural if the object of study is the electronic structure of molecules, that is, the energies and compositions of MOs, charge distribution, bond orders, and other similar characteristics. We will, therefore, in conformity with the PMO theory, assume that the replacement of one or several ligands $ML_n \longrightarrow ML_{n-k}XY\ldots Z$ in an ML_n complex with a Q_0 nuclear configuration,

can be considered a perturbation described by many-electron operator $\mathcal{H}_S$ acting on the electronic subsystem of the molecule [35, 36, 40, 48–52]. Electronic Hamiltonian $\mathcal{H}$ of a substituted molecule at point Q_0 therefore takes the form

$$\mathcal{H}(Q_0) = \mathcal{H}_0 + \mathcal{H}_S, \tag{2.32}$$

where $\mathcal{H}_0$ is the electronic Hamiltonian of unsubstituted system (1.9), and $\mathcal{H}_S$ is the "substitution operator." This operator will be specified later, but first, we will consider some formal properties of $\mathcal{H}_S$ independent of its actual form.

A comparison of (2.12) and (2.32) shows that these formulas are similar, except that (2.12) contains the term with second derivatives. The only difference is that the linear vibronic coupling operator is written as the sum whose terms are symmetrized with respect to the irreducible representations of symmetry group G of the molecule under consideration with nuclear configuration Q_0. The substitution operator can similarly be separated into symmetrized components, which is convenient for calculating the matrix elements of $\mathcal{H}_S$, because the wave functions of molecular terms are associated with the irreducible representations of group G.

To obtain the desired separation, we assume that the operator corresponding to the replacement of several ligands can be written as the sum across the separate ligands involved in substitution,

$$\mathcal{H}_S = \sum_l \mathcal{H}_S^l. \tag{2.33}$$

This representation is valid for different substitution operator versions, for instance, in the electrostatic crystal field model, where ligands are treated as spherically symmetrical charges or dipoles oriented along metal–ligand bonds [11, 53]. Representation (2.33) is also applicable in the "covalent" model that we will use, where ligands are characterized by their orbital ionization potentials and ligand–metal resonance integrals [35, 36, 45–50, 52], as is accepted in the PMO theory [12, 44].

Using the principle of additivity with respect to ligands (2.33), let us introduce group substitution operators symmetrized according to the irreducible representations of the symmetry group of the homoligand ML_n molecule [51, 55, 56] along with substitution operators for separate ligands. For instance, let the unsubstituted system be an octahedral ML_6 complex. For simplicity, we assume that the substituents X, Y, ..., Z are identical, and the substituted complex has the form $ML_{6-k}X_k$. For the O_h group, we can define six symmetrized substitution operators $\mathcal{H}_\Gamma^\gamma$ in the same way as for normal coordinates or group σ orbitals of ligands in MO theory (Fig. 4 and Tables 1 and 2):

$$\begin{aligned}
\mathcal{H}_{A_{1g}} &= (1/\sqrt{6})(\mathcal{H}_S^1 + \mathcal{H}_S^2 + \mathcal{H}_S^3 + \mathcal{H}_S^4 + \mathcal{H}_S^5 + \mathcal{H}_S^6),\\
\mathcal{H}_{E_g}^1 &= (1/2\sqrt{3})(2\mathcal{H}_S^1 + 2\mathcal{H}_S^2 - \mathcal{H}_S^3 - \mathcal{H}_S^4 - \mathcal{H}_S^5 - \mathcal{H}_S^6),\\
\mathcal{H}_{E_g}^2 &= (1/2)(\mathcal{H}_S^3 + \mathcal{H}_S^4 - \mathcal{H}_S^5 - \mathcal{H}_S^6),\\
\mathcal{H}_{T_{1u}}^x &= \frac{1}{\sqrt{2}}(\mathcal{H}_S^3 - \mathcal{H}_S^4), \qquad\qquad (2.34)\\
\mathcal{H}_{T_{1u}}^y &= \frac{1}{\sqrt{2}}(\mathcal{H}_S^5 - \mathcal{H}_S^6),\\
\mathcal{H}_{T_{1u}}^z &= \frac{1}{\sqrt{2}}(\mathcal{H}_S^1 - \mathcal{H}_S^2).
\end{aligned}$$

These formulas can be illustrated by the scheme that requires no special comments:

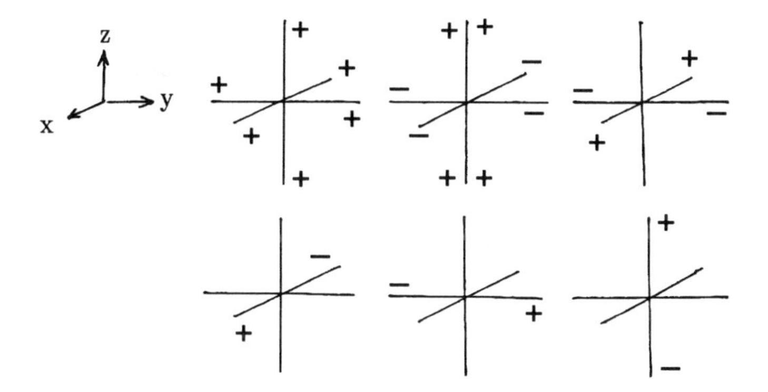

It can be seen from formulas (2.34) and the scheme that the $\mathcal{H}_{A_{1g}}$, ..., $\mathcal{H}^{z}_{T_{1u}}$ operators transform under O_h group symmetry operations according to the rows of the A_{1g}, E_g, and T_{1u} irreducible representations. Substitution operators $\mathcal{H}^{l}_{S}$ for separate ligands can be written in terms of operators (2.34) as

$$\begin{aligned}
\mathcal{H}^1_S &= (1/\sqrt{6})\mathcal{H}_{A_{1g}} + (1/\sqrt{3})\mathcal{H}^1_{E_g} + (1/\sqrt{2})\mathcal{H}^z_{T_{1u}},\\
\mathcal{H}^2_S &= (1/\sqrt{6})\mathcal{H}_{A_{1g}} + (1/\sqrt{3})\mathcal{H}^1_{E_g} - (1/\sqrt{2})\mathcal{H}^z_{T_{1u}},\\
\mathcal{H}^3_S &= (1/\sqrt{6})\mathcal{H}_{A_{1g}} - (1/2\sqrt{3})\mathcal{H}^1_{E_g} + (1/2)\mathcal{H}^2_{E_g}\\
&\quad + (1/\sqrt{2})\mathcal{H}^x_{T_{1u}},\\
\mathcal{H}^4_S &= (1/\sqrt{6})\mathcal{H}_{A_{1g}} - (1/2\sqrt{3})\mathcal{H}^1_{E_g} + (1/2)\mathcal{H}^2_{E_g}\\
&\quad - (1/\sqrt{2})\mathcal{H}^x_{T_{1u}},\\
\mathcal{H}^5_S &= (1/\sqrt{6})\mathcal{H}_{A_{1g}} - (1/2\sqrt{3})\mathcal{H}^1_{E_g} - (1/2)\mathcal{H}^2_{E_g}\\
&\quad + (1/\sqrt{2})\mathcal{H}^y_{T_{1u}},\\
\mathcal{H}^6_S &= (1/\sqrt{6})\mathcal{H}_{A_{1g}} - (1/2\sqrt{3})\mathcal{H}^1_{E_g} - (1/2)\mathcal{H}^2_{E_g}\\
&\quad - (1/\sqrt{2})\mathcal{H}^y_{T_{1u}}.
\end{aligned} \tag{2.35}$$

It follows from (2.35) that for an arbitrary $ML_{6-k}X_k$ molecule, its substitution operator $\mathcal{H}_S$ can be represented in the form of a decomposition in symmetrized operators $\mathcal{H}^{\gamma}_{\Gamma}$ (2.34) (Table 3). This result can be transferred to a more general case of an $ML_{n-k}X_k$ complex with arbitrary geometry and type of ligands, that is,

$$\mathcal{H}_S(ML_{n-k}X_k) = \sum_{\Gamma,\gamma} a_{\Gamma,\gamma}\mathcal{H}^{\gamma}_{\Gamma}, \tag{2.36}$$

where coefficients $a_{\Gamma,\gamma}$ depend on the number and mutual arrangement of substituents (structure coefficients) but are independent of their nature, which is fully determined by the set of operators $\mathcal{H}^{\gamma}_{\Gamma}$. Generalization to $ML_{n-k}XY...Z$ complexes is also possible.

The decomposition of substitution operator $\mathcal{H}_S$ into symmetrized components allows us to derive selection rules for its matrix elements. In the first place, for the same reasons as selection rules (2.13) apply to normal vibrations, similar selection rules apply to the $\langle\Psi_i|\mathcal{H}^{\gamma}_{\Gamma}|\Psi_j\rangle$ matrix elements, where Ψ_i and Ψ_j are the wave functions that transform under irreducible representations Γ_i and Γ_j. For these matrix elements to be nonzero, the following conditions must be met:

$$\begin{cases}
\Gamma \subset \Gamma_i \times \Gamma_j, & \text{if } \Psi_i \text{ and } \Psi_j \text{ correspond to different irreducible representations,}\\
\Gamma \subset [\Gamma_i^2], & \text{if } \Psi_i \text{ and } \Psi_j \text{ correspond to the same irreducible representation.}
\end{cases} \tag{2.37}$$

Table 3. Substitution operator separation into symmetrized components for quasi-octahedral $ML_{6-k}X_k$ complexes

Complex	Substituted ligands	$\mathcal{H}_S = \sum_{\Gamma,\gamma} a_{\Gamma,\gamma}\mathcal{H}_\Gamma^\gamma$
ML_5X	1	$(1/\sqrt{6})\mathcal{H}_{A_{1g}} + (1/\sqrt{3})\mathcal{H}_{E_g}^1$ $+(1/\sqrt{2})\mathcal{H}_{T_{1u}}^z$
cis-ML_4X_2	3, 5	$(2/\sqrt{6})\mathcal{H}_{A_{1g}} - (1/\sqrt{3})\mathcal{H}_{E_g}^1$ $+(1/\sqrt{2})(\mathcal{H}_{T_{1u}}^x + \mathcal{H}_{T_{1u}}^y)$
trans-ML_4X_2	1, 2	$(2/\sqrt{6})\mathcal{H}_{A_{1g}} + (2/\sqrt{3})\mathcal{H}_{E_g}^1$
cis-ML_3X_3	1, 3, 5	$(3/\sqrt{6})\mathcal{H}_{A_{1g}} + (1/\sqrt{2})(\mathcal{H}_{T_{1u}}^x$ $+\mathcal{H}_{T_{1u}}^y + \mathcal{H}_{T_{1u}}^z)$
trans-ML_3X_3	1–3	$(3/\sqrt{6})\mathcal{H}_{A_{1g}} + (1/2)\mathcal{H}_{E_g}^2$ $+(\sqrt{3}/2)\mathcal{H}_{E_g}^1 + (1/\sqrt{2})\mathcal{H}_{T_{1u}}^x$
cis-ML_2X_4	1, 2, 5, 6	$(4/\sqrt{6})\mathcal{H}_{A_{1g}} + (1/\sqrt{3})\mathcal{H}_{E_g}^1$ $-(1/\sqrt{2})(\mathcal{H}_{T_{1u}}^x + \mathcal{H}_{T_{1u}}^y)$
trans-ML_2X_4	3–6	$(4/\sqrt{6})\mathcal{H}_{A_{1g}} - (2/\sqrt{3})\mathcal{H}_{E_g}^1$
MLX_5	2–6	$(5/\sqrt{6})\mathcal{H}_{A_{1g}} - (1/\sqrt{3})\mathcal{H}_{E_g}^1$ $-(1/\sqrt{2})\mathcal{H}_{T_{1u}}^z$
MX_6	1–6	$\sqrt{6}\,\mathcal{H}_{A_{1g}}$

Note: See Fig. 4 for numbering of ligands.

The following rule is a corollary to (2.37): for the $\langle\Psi_i|\mathcal{H}_S|\Psi_j\rangle$ matrix element, where $\mathcal{H}_S$ is an arbitrary operator, to be nonzero it is necessary that the decomposition of $\mathcal{H}_S$ over symmetrized operators contain at least one $\mathcal{H}_\Gamma^\gamma$ operator satisfying conditions (2.37).

3. Although in a number of problems, it suffices to know the separation of $\mathcal{H}_S$ into $\mathcal{H}_\Gamma^\gamma$ operators, in many instances, explicit expressions for the matrix elements of $\mathcal{H}_S$ and, sometimes, their numerical estimates are required. For the purpose of determining this information, let us introduce one-electron operator H_S in addition to the many-electron $\mathcal{H}_S$. Operators of this type are used in the classical perturbation MO theory [12, 44], where they are given by their matrix elements in the basis set of AOs. As concerns the systems under consideration, this means that the H_S operator is defined as the difference of the corresponding matrices for the $ML_{n-k}XY...Z$ and ML_n complexes. In the simplest approximation, the off-diagonal elements of the matrix of H_S are ignored, and only changes in the Coulomb integrals for the AOs of the ligands involved in substitution are taken into account. (In Chapter 3 in discussing angular distortions and in Chapter 6 in discussing the effects of multiply bonded ligands, the off-diagonal elements of the matrix of H_S are also taken into account.) For instance, when a σ-bonded ligand L is replaced by a σ-bonded ligand X having a different electronegativity, calculations of the matrix of H_S reduce to replacing the Coulomb integral α_L for the σ AO of L by the corresponding $\alpha_X = \alpha_L + \Delta\alpha$ value. It follows that in the basis set of AOs χ_i, the matrix elements of H_S are determined by the equality [48–50, 52]

$$\langle\chi_i|H_S|\chi_j\rangle = \delta_{ij}\Delta\alpha_i, \tag{2.38}$$

where $\Delta\alpha_i$ is the increment of the Coulomb integral of the ith basis AO due to the substitution $ML_n \longrightarrow ML_{n-k}XY\ldots Z$. It is implied that $\Delta\alpha_i$ values are only nonzero for those AOs χ_i that belong to the ligands involved in substitution. The S_{mn} matrix elements of operator H_S in the basis set of MOs $\psi_r = \sum_i c_{ir}\chi_i$ are expressed in terms of coefficients c_{ir} of AOs χ_i in MO ψ_r and matrix elements (2.38) as

$$S_{nm} = \langle\psi_n|H_S|\psi_m\rangle = \sum_i c_{in}c_{im}\Delta\alpha_i. \tag{2.39}$$

If the off-diagonal matrix elements of operator H_S are also taken into account, (2.38) transforms into

$$\langle\chi_i|H_S|\chi_j\rangle = \delta_{ij}\Delta\alpha_i + (1-\delta_{ij})\Delta\beta_{ij}, \tag{2.40}$$

and (2.39) is replaced by

$$S_{nm} = \langle\psi_n|H_S|\psi_m\rangle = \sum_i c_{in}c_{im}\Delta\alpha_i + \sum_{i\neq j} c_{in}c_{jm}\Delta\beta_{ij}. \tag{2.41}$$

Further, we will have to calculate the matrix elements of H_S in the basis set of MOs. It should therefore be borne in mind that like $\mathcal{H}_S$ and in the same way as $\mathcal{H}_S$, the operator H_S can be separated into components symmetrized with respect to the irreducible representations of the molecular symmetry group at point Q_0. For instance, by analogy with (2.36), we can write

$$H_S(\mathrm{ML}_{n-k}\mathrm{X}_k) = \sum_{\Gamma,\gamma} \alpha_{\Gamma,\gamma} H_\Gamma^\gamma. \tag{2.42}$$

The selection rules for symmetrized H_Γ^γ components are similar to those given by (2.37). That is, for the matrix element $\langle\psi_n|H_\Gamma^\gamma|\psi_m\rangle$, where MOs ψ_n and ψ_m transform according to representations Γ_n and Γ_m, to be nonzero, it is necessary that

$$\begin{cases} \Gamma \subset \Gamma_n \times \Gamma_m, & \text{if } \psi_n \text{ and } \psi_m \text{ correspond to different irreducible representations,} \\ \Gamma \subset [\Gamma_n^2], & \text{if } \psi_n \text{ and } \psi_m \text{ correspond to the same irreducible representation.} \end{cases} \tag{2.43a}$$

It seems pertinent to note here that for the one-electron operator H_S, symmetrized components can easily be visualized. For instance, operator $H_{A_{1g}}$ corresponding to the A_{1g} representation for an octahedral $\mathrm{ML_6}$ complex within a normalization factor of $\sqrt{6}$ corresponds to the replacement $\mathrm{ML_6} \longrightarrow \mathrm{MX_6}$ of all ligands L by ligands X with equal $\Delta\alpha$ values. Similarly, operator $H_{T_{1u}}^z$ corresponds to the substitution $\mathrm{ML_6} \longrightarrow$ *trans*-$\mathrm{ML_4XY}$, where substituents X and Y are characterized by $\Delta\alpha$ values equal in magnitude and opposite in sign, etc.:

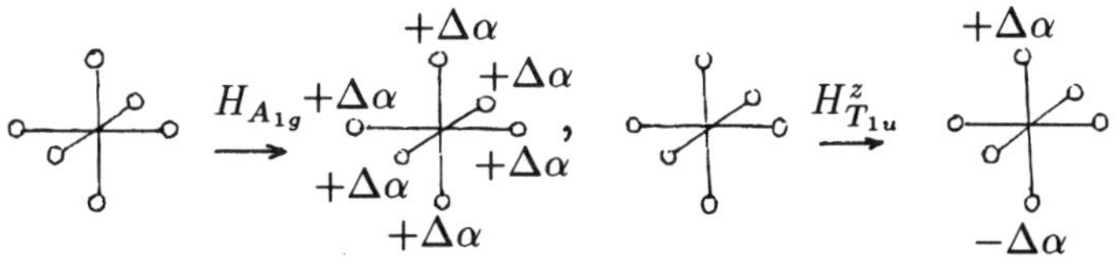

The visual character of operator H_S makes it possible to use one more selection rule. This rule is devoid of any group-theory meaning but is very simple and useful when operator H_S is given by (2.38) and (2.39). For $\langle\psi_n|H_S|\psi_m\rangle$ to be nonzero, it is necessary that

ψ_n and ψ_m simultaneously contain AOs of ligands undergoing replacement. (2.43b)

This condition is visually represented in the scheme where the ligand undergoing replacement is $\mathrm{L_1}$:

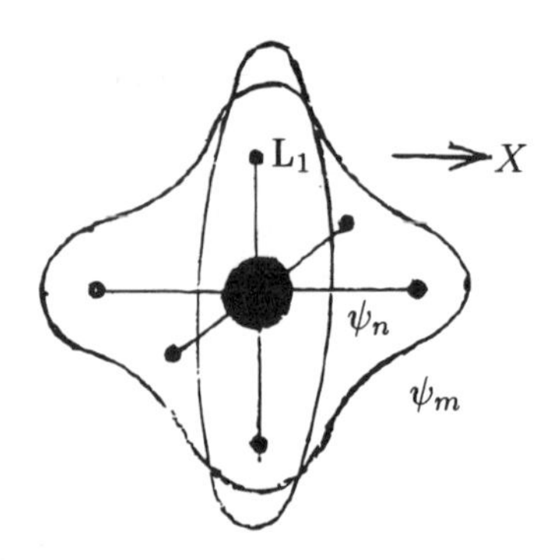

Rule (2.43b) can be violated when the matrix of operator H_S in the basis set of AOs of the initial system also contains off-diagonal elements. The $\langle\psi_n|H_S|\psi_m\rangle$ matrix element can then also be nonzero when AOs of the ligands undergoing replacement are only contained in one of two MOs, ψ_n or ψ_m.

Lastly, consider the problem of expressing the matrix elements of the operator $\mathcal{H}_S$ through the matrix elements of operator H_S. Note that from the point of view of the electronic subsystem, the replacement of an atom in a molecule reduces to changes in the potential acting on the electrons in the corresponding region of the molecule. Therefore $\mathcal{H}_S$ can approximately be written as the sum of operators H_S over all electrons (cf. formula (2.30)), that is,

$$\mathcal{H}_S = \sum_i^{\text{electrons}} H_S(i). \tag{2.44}$$

Using (2.44) and expressing many-electron wave functions of a molecule in terms of MOs, we can easily write the matrix elements of operator $\mathcal{H}_S$ in terms of the matrix elements of operator H_S.

4. To conclude this section, two points should be mentioned concerning perturbation theory applications to treatment of the replacement of atoms (ligands) in molecules and complexes. The first one refers to the conditions that should be met for perturbation theory to be applicable; these have already been specified. For treating the substitution $ML_6 \longrightarrow$ *trans*-ML_4XY as small perturbation, this substitution should not substantially affect the geometric structure and the electronic state of the first coordination sphere of the central atom. Ligand substitution can cause a distortion of the geometry of the environment of the central atom, but this distortion should not cause radical changes in the arrangement of the ligands (for instance, a square-planar complex should not transform into a tetrahedral one). Similarly, the substitution can cause some changes in the arrangement of MOs built of the AOs of the central atom and its immediate environment, but it should

not substantially change the MO energies in comparison with the energy gaps between them; the number of electrons on each MO and their spin states should remain unchanged.

The second remark refers to the already mentioned assumption of PMO theory. This assumption may cause confusion, especially with $ML_{n-k}XY...Z$ molecules. The fine point, usually left unstressed, is that the matrix of the perturbation operator is calculated as the difference of the matrices of effective one-electron Hamiltonians for the substituted and unsubstituted molecules, but these matrices are actually calculated using different basis sets, although conventionally, they are referred to the same basis. For instance, for the replacement of an atom in an aromatic ring, the Hückel Hamiltonian of the substituted system is constructed using the basis set including the AOs of the substituent which are not at all identical to the AOs of the initial C atom. Still more dramatically, this circumstance comes to the fore with ligand substitutions of the type $SF_6 \longrightarrow SF_5Cl$. Indeed, the AOs of Cl substantially differ from the AOs of F and are, in addition, centered at another point, because the atomic radius of fluorine is substantially different from that of chlorine. To answer these disturbing questions, it should be taken into consideration that in essence, perturbation MO theory considers pseudoatoms rather than real substituents. These pseudoatoms have the same properties (radii, forms of AOs, etc.) as unsubstituted system atoms but differ from the latter in Coulomb and/or resonance integral values. It is assumed that all characteristics of the electronic structure of molecules containing pseudoatoms (distribution of charges, bond orders, etc.) coincide with those of real molecules with real substituents. Note that this remark, although trivial, is of importance for correctly understanding the meaning of statements of the type "the $ML_6 \longrightarrow$ *trans*-ML_4XY substitution is considered at point Q_0 corresponding to the geometry of the ML_n molecule." It should be borne in mind that here, Q_0 refers to the geometry of a molecule with pseudoatoms rather than a real molecule with real substituents.

2.4. Adiabatic Potentials of Heteroligand Systems

1. The concept central to the further presentation is that of the adiabatic potentials of heteroligand molecules and complexes [35, 36, 50–52, 54–56]. At a model level, we should again use perturbation theory, and therefore, select an unperturbed Hamiltonian and a perturbation operator. Basically, we can directly apply formula (2.12) to heteroligand molecules. The $\mathcal{H}_0$ operator in (2.12) should then be understood as the Hamiltonian of the $ML_{n-k}XY...Z$ heteroligand system at a Q_0 nuclear

configuration, and the other terms in (2.12) should be treated as a perturbation operator. This variant of theory is, however, inconvenient, because generally, $ML_{n-k}XY...Z$ systems possess low symmetry, and their wave functions cannot be written in a simple form.

A more convenient approach is based on using the Hamiltonian and, consequently, energy levels and wave functions of a homoligand ML_n system as a zeroth-order approximation. Recall that ML_n molecules with closed shells are often high-symmetry systems (octahedral, square-planar, or tetrahedral), and their electronic structure has been studied in detail by diverse experimental and computational methods. Systems with open electronic shells are also treated as high-symmetry system in the initial approximation. Clearly, because of the Jahn–Teller effect, the minimum of the lowest sheet of their adiabatic potential energy surface does not correspond to a high-symmetry nuclear configuration. This circumstance is, however, treated as a result of perturbation described by the vibronic coupling operator.

These considerations lead us to formulate the approach to constructing adiabatic potentials of heteroligand systems as follows. Let Hamiltonian $\mathcal{H}$ for heteroligand molecule $ML_{n-k}XY...Z$ be written as

$$\mathcal{H} = \mathcal{H}_0 + \mathcal{H}_S + \sum_{\nu}(\partial\mathcal{H}/\partial Q_\nu)_0 Q_\nu + (1/2)\sum_{\mu,\nu}(\partial^2\mathcal{H}/\partial Q_\nu \partial Q_\mu)_0 Q_\nu Q_\mu, \qquad (2.45)$$

where zeroth-order Hamiltonian $\mathcal{H}_0$ refers to a high-symmetry ML_n system with nuclear configuration Q_0, and the perturbation operator comprises the substitution operator and the vibronic coupling operator. Note that here, it is assumed that the coordinates Q_ν and vibronic coupling operators of two systems, ML_n and $ML_{n-k}XY...Z$, are identical. This should not cause confusion because of the "pseudoatomic" treatment of substituents in PMO theory (see concluding remarks in Section 2.3).

2. As mentioned, we are interested in heteroligand derivatives of homoligand systems with closed as well as open electronic shells. For derivatives of both types, the method for finding adiabatic potential is in essence the same as for their homoligand counterparts. For heteroligand derivatives of molecules with open shells, wave functions (2.14) of the ground state should be used as a zeroth-order approximation. The same reasoning as in Section 2.2.3 leads to the adiabatic potential in form (2.15). However in this potential, $\mathcal{E}'$ are the roots of the secular

equation of a more general form [51, 54],

$$\mathrm{Det}\left\| \sum_{\nu} A_{ij}^{\nu} Q_{\nu} + \frac{1}{2} \sum_{\mu,\nu} A_{ij}^{\mu,\nu} Q_{\mu} Q_{\nu} + S_{ij} - \delta_{ij} \mathcal{E}' \right\| = 0, \qquad (2.46)$$

where S_{ij} are the matrix elements of the substitution operator in basis (2.14). Equation (2.46) makes it possible to calculate the adiabatic potential of a heteroligand molecule in terms of the wave functions of the homoligand system. The adiabatic potential can, however, be written in terms of the wave functions of the heteroligand system itself. For this purpose, basis (2.14) should be replaced by the equivalent basis $\widetilde{\Psi}_1$, $\widetilde{\Psi}_2$, ..., $\widetilde{\Psi}_k$. In this basis, operator $\mathcal{H}_S$ is given by the diagonal matrix $\widetilde{S}_{ii}\delta_{ij}$, and (2.46) takes the form

$$\mathrm{Det}\left\| \sum_{\nu} \widetilde{A}_{ij}^{\nu} Q_{\nu} + \frac{1}{2} \sum_{\mu,\nu} \widetilde{A}_{ij}^{\mu\nu} Q_{\mu} Q_{\nu} + \widetilde{S}_{ij}\delta_{ij} - \delta_{ij} \mathcal{E}' \right\| = 0. \qquad (2.46a)$$

Here, $\widetilde{S}_{ii}$ are the energies of the levels formed in splitting of degenerate state (2.14) under the action of $\mathcal{H}_S$ and counted from the energy of the degenerate state, and functions $\widetilde{\Psi}_j$ correspond to these levels. Equation (2.46a) therefore expresses the adiabatic potential of $ML_{n-k}XY...Z$ in terms of the wave functions of this heteroligand system (for configuration Q_0 of atomic nuclei).

Note that in some instances, the matrix of operator $\mathcal{H}_S$ is diagonal even in basis (2.14), that is, functions (2.14) are eigenfunctions of two Hamiltonians ($\mathcal{H}_0$ and $\mathcal{H}_0 + \mathcal{H}_S$) simultaneously. Equation (2.46) can then be interpreted in two ways. It may be thought of as written in terms of the wave functions of the ML_n system. All matrix elements in (2.46) then refer to perturbation. On the other hand, equation (2.46) can be taken to be written in terms of the wave functions of the $ML_{n-k}XY...Z$ system. Matrix elements $S_{ij} = S_{ii}\delta_{ij}$ then characterize the initial system, and vibronic coupling terms are a perturbation.

3. The adiabatic potential of heteroligand derivatives of ML_n molecules and complexes with closed shells requires a more detailed consideration, because for such systems, the potential energy surface for the ground state can be found in an explicit form. We will use basis (2.18) and perturbation theory (Section 2.2) and, in the expression obtained for the adiabatic potential, take into account terms of the first and second order of smallness with respect to two operators, $\mathcal{H}_S$ and the vibronic coupling operator. Next, using standard formula (2.11) with the perturbation operator from (2.45), we obtain [50, 55]

$$\mathcal{E}(Q) = S_0 - \sum_{j\neq 0} \frac{S_j^2}{\omega_j} + \sum_{\nu} \left\{ -2 \sum_{j\neq 0} \frac{S_j A_j^{\nu}}{\omega_j} Q_{\nu} + \frac{1}{2} K_{\nu} Q_{\nu}^2 \right\} \qquad (2.47)$$

for the lowest sheet of adiabatic potential $\mathcal{E}$ of a heteroligand molecule. Here, ω_j, K_ν, and A_j^ν are the energy gaps, force constants, and off-diagonal vibronic coupling constants (see (2.20)–(2.22)) for the initial ML_n system, and S_j are the matrix elements of the $\mathcal{H}_S$ operator in the basis of wave functions (2.18) of the ground and excited states of the homoligand system,

$$S_0 = \langle\Psi_0|\mathcal{H}_S|\Psi_0\rangle, \quad S_j = \langle\Psi_0|\mathcal{H}_S|\Psi_j\rangle. \tag{2.48}$$

It can be noticed that expression (2.47) for the adiabatic potential of a $ML_{n-k}XY...Z$ molecule is a generalization of (2.19); this expression turns into (2.19) if all S_j are zero. Explicit expression (2.47) for the adiabatic potential can be used to obtain explicit expressions for its two most important characteristics. The coordinates of the minimum of surface (2.47) correspond to equilibrium configuration Q_ν^f of the substituted molecule; they are found from the extremum condition $\partial\mathcal{E}/\partial Q_\nu = 0$. Applying this condition yields [50, 55]

$$Q_\nu^f = (2/K_\nu)\sum_{j\neq 0} S_j A_j^\nu/\omega_j. \tag{2.49}$$

Formula (2.49) describes geometric distortions of a molecule caused by the substitution $ML_n \longrightarrow ML_{n-k}XY\ldots Z$ (so-called static mutual influence of ligands). Another important characteristic of the adiabatic potential, the depth of its minimum, or, more exactly, the difference of the adiabatic potentials of the heteroligand and homoligand molecules in their equilibrium configurations, can be obtained by substituting expression (2.49) for Q^f (2.49) [55] into (2.47),

$$\mathcal{E}(Q^f) = S_0 - \sum_{j\neq 0}\frac{S_j^2}{\omega_j} - 2\sum_\nu \frac{1}{K_\nu}\Big\{\sum_{j\neq 0}\frac{S_j A_j^\nu}{\omega_j}\Big\}^2. \tag{2.50}$$

This formula describes the effect of the $ML_n \longrightarrow ML_{n-k}XY\ldots Z$ substitution on the total energy of the molecule.

Like (2.46), equation (2.47) and ensuing formulas (2.49) and (2.50) can be viewed from two angles. First, as mentioned above, this equation gives an expression for the potential energy surface of a heteroligand system in terms of the electronic structure of its homoligand counterpart, that is, through homoligand system wave functions (2.18) and energy gaps ω_j. The perturbation operator is then the sum $\mathcal{H}_S + W$, that is, three last terms in (2.45). On the other hand, equation (2.47) describes the same surface in terms of the electronic structure of the heteroligand system itself, which reduces the perturbation operator to only one term W. (This electronic structure of $ML_{n-k}XY...Z$ is, however, obtained within the framework of perturbation theory with $\mathcal{H}_S$ as a perturbation.) Indeed, consider the energy of a heteroligand molecule at point Q_0. To second-order terms inclusive, this energy equals the sum of two first terms of (2.47). Next, apply perturbation theory to find wave functions $\widetilde{\Psi}_j$, $j = 0, 1, 2, \ldots$ of the heteroligand system at the same point (Q_0). Using the $\widetilde{\Psi}_j$ basis in second-order perturbation theory with W as a perturbation then yields the two remaining terms in (2.47).

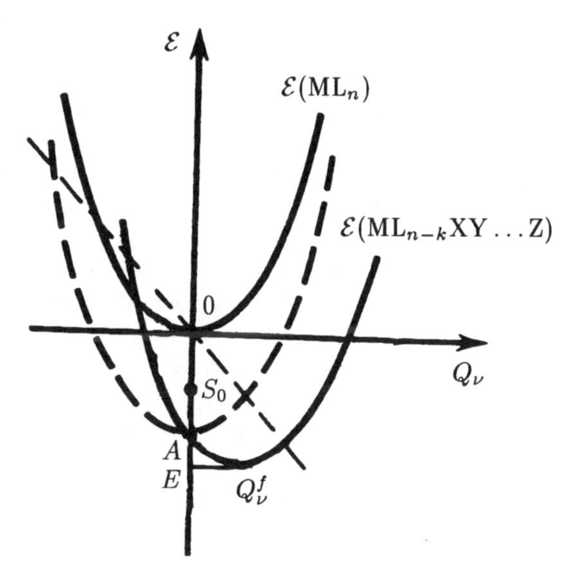

FIG. 5. CROSS-SECTION OF POTENTIAL ENERGY SURFACES OF HOMOLIGAND (ML_n) AND HETEROLIGAND ($ML_{n-k}XY...Z$) COMPLEXES ALONG THE Q_ν COORDINATE.

4. The results just obtained have a clear geometric meaning represented in Fig. 5. We assume that the ML_n molecule with configuration $Q_\nu = 0$ is stable with respect to the pseudo-Jahn–Teller effect. The cross-section of the lowest sheet of its adiabatic potential energy surface along an arbitrary normal coordinate Q_ν then has the form of a parabola passing through the origin and convex downward. The first-order correction S_0 for the $ML_n \longrightarrow ML_{n-k}XY\ldots Z$ substitution shifts the parabola along the energy axis by OS_0. The second term in (2.47) has a similar effect and shifts the parabola further by S_0A, that is, the total shift equals OA. In addition, the presence of substitution operator $\mathcal{H}_S$ results in the appearance of a component linear in Q_ν (the

dashed line passing through the origin) in the dependence of the energy of $ML_{n-k}XY...Z$ on coordinate Q_ν. The resultant cross section of the adiabatic potential energy surface for the $ML_{n-k}XY...Z$ molecule is obtained by summing two components, quadratic (dashed parabola) and linear (dashed straight line). The linear component shifts the minimum to point Q^f_ν, which makes one more contribution to the energy of the substituted system (segment AE, Fig. 5). As a result, the cross-section that we are considering has the form of a parabola shifted along the Q_ν and energy axes.

A shift of the adiabatic potential minimum along the Q_ν axis means that the equilibrium geometry of the $ML_{n-k}XY...Z$ molecule changes in comparison with the equilibrium geometry of ML_n as a result of ligand substitution. Analytically, this change is described by (2.49). A shift of the adiabatic potential along the $\mathcal{E}$ axis signifies a change in the total energy of the $ML_{n-k}XY...Z$ molecule in comparison with the total energy of ML_n. Figure 5 shows that the change in energy caused by the $ML_n \longrightarrow ML_{n-k}XY\ldots Z$ substitution is given by the sum of three contributions [56]. These contributions are described by three terms on the right-hand side of (2.50), and each of these contributions can be assigned a physical meaning of its own.

The first term S_0 on the right-hand side of (2.50), or the OS_0 segment in Fig. 5, describes the change in energy obtained without taking into account changes in the wave function of the ground state and the geometry of the molecule that accompany the $ML_n \longrightarrow ML_{n-k}XY\ldots Z$ substitution. For this reason, S_0 has an important property.

Consider the usual situation when all ligands in the initial homoligand ML_n molecule are equivalent by symmetry (as, e.g., in octahedral ML_6 and square-planar ML_4 molecules). Let operator $\mathcal{H}_S$ be represented by the sum over ligands undergoing replacement, that is, $\mathcal{H}_S = \sum_l \mathcal{H}^l_S$ (see Section 2.3). The S_0 value then only depends on the total numbers of ligands of each kind rather than the mutual arrangement of these ligands (see below) and can be ignored in calculating isomerization energies for $ML_{n-k}XY...Z$ molecules.

The second term on the right-hand side of (2.50), to which the S_0A segment corresponds (Fig. 5), describes the change in the ground-state energy of ML_n with configuration Q_0 caused by mixing of ground and excited states (electron density redistribution resulting from the $ML_n \longrightarrow ML_{n-k}XY\ldots Z$ substitution). This term has the meaning of the energy of electronic subsystem relaxation accompanying ligand substitution, however, as previously, without taking into account geometry changes.

Lastly, the third term in (2.50) (segment EA in Fig. 5) describes geometric relaxation effects, that is, the change in energy caused by a change in the geometry of the $ML_{n-k}XY...Z$ molecule in comparison with ML_n. The physical meaning of this term is made clear when equations (2.49) and (2.50) are compared; it then becomes obvious that this term can be written as $(1/2)\sum_\nu K_\nu (Q_\nu^f)^2$. It follows from (2.50) that both relaxation components have a stabilizing action on the substituted molecule. They depend not only on the number but also on the mutual arrangement of the substituents and, therefore, determine the relative stabilities of various $ML_{n-k}XY...Z$ molecule isomers.

Note that generally, ligand substitution is accompanied by the appearance of both relaxation components contributing to the total energy of the system, provided that the corresponding terms in (2.50) do not vanish for some special reason (e. g., because of the use of a limited set of MOs). This is obvious with electronic relaxation. So far as geometric relaxation is concerned, some comments appear to be necessary, because, e. g., pseudo-Jahn–Teller deformations in molecules occur not always but only under certain conditions (when vibronic coupling constants are fairly large and/or energy gaps ω_j and force constants K_ν are fairly small, see Section 2.2). The reason for geometric relaxation is the presence of terms linear in Q_ν in (2.47). When the substitution $ML_6 \longrightarrow$ *trans*-ML_4XY takes place, they always cause changes in the initial geometry of ML_6 no matter what the relative values of A_j^ν, K_ν, and ω_j, although these values can affect the extent to which ligand substitution distorts the geometry of the ML_6 molecule.

4. Let us show that the term S_0 in (2.50) depends, and depends linearly, only on the number of ligands X and is independent of the mutual arrangement of these ligands in the $ML_{n-k}X_k$ molecule [55]. To be more specific, consider the example of the substitution $ML_4 \longrightarrow ML_{4-k}X_k$ in square-planar complexes (Fig. 14). The symmetrized substitution operators corresponding to the A_{1g}, B_{1g}, and E_u representations of the D_{4h} molecular symmetry group [55] have the form

$$\begin{aligned}
\mathcal{H}_{A_{1g}} &= (1/2)(\mathcal{H}_S^1 + \mathcal{H}_S^2 + \mathcal{H}_S^3 + \mathcal{H}_S^4),\\
\mathcal{H}_{B_{1g}} &= (1/2)(\mathcal{H}_S^1 + \mathcal{H}_S^2 - \mathcal{H}_S^3 - \mathcal{H}_S^4),\\
\mathcal{H}_{E_u}^x &= (1/\sqrt{2})(\mathcal{H}_S^1 - \mathcal{H}_S^2),\\
\mathcal{H}_{E_u}^y &= (1/\sqrt{2})(\mathcal{H}_S^3 - \mathcal{H}_S^4).
\end{aligned} \tag{2.51}$$

Here, $\mathcal{H}_S^l$ is the partial substitution operator for the introduction of ligand X into position l. As is shown in Section 2.3, any $\mathcal{H}_S(ML_{4-k}X_k)$ operator can be represented as a linear combination of symmetrized substitution operators (2.51), that is, $\mathcal{H}_S = \sum_{\Gamma,\gamma} a_{\Gamma,\gamma}\mathcal{H}_\Gamma^\gamma$, where coefficients $a_{\Gamma,\gamma}$ are found by direct calculations (they are given in Table 4).

Table 4 shows that the $S_0 = \langle\Psi_0|\mathcal{H}_S|\Psi_0\rangle$ value for a totally symmetrical Ψ_0 wave function is only nonzero for the totally symmetrical substitution operator, which is represented in Table 4 by the totally symmetrical component $\mathcal{H}_{A_{1g}}$ of the substitution operator,

$$S_0 = \langle\Psi_0|\mathcal{H}_S|\Psi_0\rangle = a_{A_{1g}}\langle\Psi_0|\mathcal{H}_{A_{1g}}|\Psi_0\rangle. \tag{2.52}$$

The $a_{A_{1g}}$ coefficient of $\mathcal{H}_{A_{1g}}$ in the decomposition in symmetrized contributions is, however, the same (equals one) for the *cis* and *trans* isomers of ML_2X_2. Table 4 also shows that the $a_{A_{1g}}$ coefficient is proportional to the number of replaced ligands, which directly leads to a linear dependence of S_0 on k for $ML_{4-k}X_k$ complexes. Similarly, the data given in Table 3 evidence additivity of S_0 for quasi-octahedral complexes.

Table 4. Substitution operator separation into symmetrized components for square-planar $ML_{4-k}X_k$ complexes

Complex	Substituted ligands	$\mathcal{H}_S = \sum_{\Gamma,\gamma} a_{\Gamma,\gamma}\mathcal{H}_\Gamma^\gamma$
ML_3X	1	$(1/2)\mathcal{H}_{A_{1g}} + (1/2)\mathcal{H}_{B_{1g}} + (1/\sqrt{2})\mathcal{H}_{E_u}^x$
trans-ML_2X_2	1, 2	$\mathcal{H}_{A_{1g}} + \mathcal{H}_{B_{1g}}$
cis-ML_2X_2	1, 3	$\mathcal{H}_{A_{1g}} + (1/\sqrt{2})(\mathcal{H}_{E_u}^x + \mathcal{H}_{E_u}^y)$
MLX_3	2, 3, 4	$(3/2)\mathcal{H}_{A_{1g}} - (1/2)\mathcal{H}_{B_{1g}} - (1/\sqrt{2})\mathcal{H}_{E_u}^x$
MX_4	1–4	$2\mathcal{H}_{A_{1g}}$

Note: See Fig. 14 for numbering of ligands.

It is easy to prove that for an arbitrary $ML_{n-k}X_k$ heteroligand derivative of a closed-shell molecule, S_0 also possesses the properties specified above [36]. Let us divide ML_n molecule ligands into "transitive" sets each comprising ligands that transform into each other under molecular symmetry operations. A $ML_{n-k}X_k$ derivative can have isomers of two types depending on whether the replaced ligands are or are not equivalent by symmetry. Consider partitioning $\mathcal{H}_S = \sum_l \mathcal{H}_s^l$ of operator $\mathcal{H}_S(\mathrm{ML}_{n-k}\mathrm{X}_k)$ into the sum over the ligands undergoing replacement, and separate each $\mathcal{H}_S^l$ operator into a linear combination of symmetrized substitution operators (2.36),

$$\mathcal{H}_S^{(l)} = \sum_{\Gamma,\gamma} a_{\Gamma,\gamma}^l \mathcal{H}_\Gamma^{l\gamma}.$$

It is easy to see that for replaced ligands l equivalent by symmetry, the a_A^l coefficients of totally symmetric substitution operator $\mathcal{H}_A$ coincide, $a_A^l = a_A$, because these ligands transform into each other under symmetry operations. For any isomer $\mathrm{ML}_{n-k}\mathrm{X}_k$ in which replaced ligands belong to the same transitive set, we then have (Ψ_0 is the wave function of the ground state of the ML_n molecule)

$$\langle\Psi_0|\mathcal{H}_S|\Psi_0\rangle = \sum_l a_A^l \langle\Psi_0|\mathcal{H}_A|\Psi_0\rangle = k a_A \langle\Psi_0|\mathcal{H}_A|\Psi_0\rangle,$$

as was to be proved. The last equality can easily be generalized to $\mathrm{ML}_{n-k}\mathrm{XY}\ldots\mathrm{Z}$ molecules with substituents of different kinds, which shows that S_0 only depends on the chemical composition of a molecule rather than the mutual arrangement of ligands L, X, Y, ..., Z.

6. We can now turn to the problem of the isomerization energy of a heteroligand molecular system. Consider two heteroligand molecules that only differ in the mutual arrangement of the ligands. After writing expressions (2.50) for both isomers, we obtain the corresponding isomerization energy as their difference. For example, consider the energy released in the isomerization *trans*-$\mathrm{ML}_{n-k}\mathrm{X}_k \longrightarrow$ *cis*-$\mathrm{ML}_{n-k}\mathrm{X}_k$,

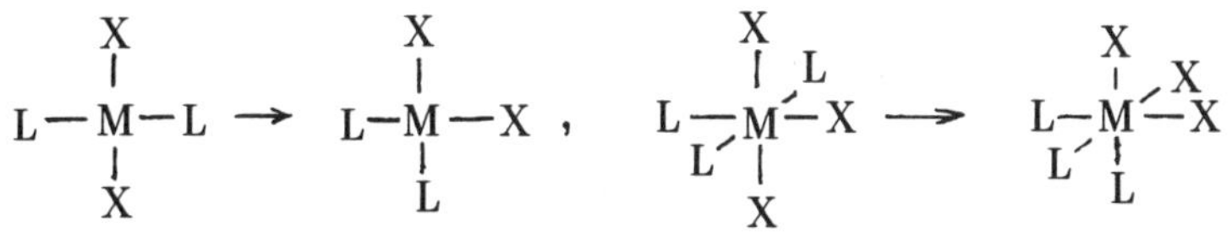

```
    X            X               X                X
    |            |               | L              | ,X
L—M—L  →   L—M—X ,       L—M—X   →       L—M—X
    |            |             L'|              L' |
    X            L               X                L
```

We have [56]

$$\Delta\mathcal{E}_{t\to c} = \sum_{j\neq 0} \frac{S_j^+ S_j^-}{\omega_j} + 2\sum_\nu \frac{1}{K_\nu}\Big(\sum_{j\neq 0}\frac{A_j^\nu S_j^+}{\omega_j}\Big)\Big(\sum_{j\neq 0}\frac{A_j^\nu S_j^-}{\omega_j}\Big). \quad (2.53)$$

Here, the notation

$$S_j^+ = S_j^c + S_j^t, \qquad S_j^- = S_j^c - S_j^t \quad (2.54)$$

is used, and indices c and t refer to the *cis* and *trans* isomers, respectively.

Note that (2.53) is derived without using any information about the type of the isomers. Equation (2.53) is therefore valid for the energy $\mathcal{E}_{1\to 2}$ of the transformation of an arbitrary isomer *1* of $ML_{n-k}X_k$ into any other isomer *2*. The left-hand side of (2.53) should then be written as $\mathcal{E}_{1\to 2}$, and (2.54) be replaced by

$$S_j^+ = S_j^{(2)} + S_j^{(1)}, \qquad S_j^- = S_j^{(2)} - S_j^{(1)}. \tag{2.54a}$$

7. Lastly, let us rewrite expressions (2.47), (2.49), and (2.50) in terms of the MOs [48, 56] of the ML_n molecule. The Ψ_0 wave function of the ground state of this molecule will be written in the form of Slater determinant (1.29) composed of doubly occupied MOs. As we do not take into account interelectronic interaction explicitly, the wave functions of excited states can be taken in form (1.41). Singlet states should only be considered, because the vibronic coupling operator and the substitution operator do not act on spin variables. For this reason, mixing between the ground and excited singlet states only occurs. States with more than one excited electron can be ignored, because both operators that we are considering are approximately represented as the sums of one-electron operators (see (2.29), (2.30), and (2.44)). After calculating the ω_j, S_j, and A_j^ν values with the functions specified above, we obtain the desired expression for the lowest sheet of the adiabatic potential energy surface of a heteroligand molecule in terms of MOs, that is,

$$\mathcal{E} = S_0 - 2\sum_n\sum_m \frac{S_{nm}^2}{\omega_{nm}} + \sum_\nu\Big\{-4\sum_n\sum_m \frac{S_{nm}A_{nm}^\nu}{\omega_{nm}}Q_\nu + \frac{1}{2}K_\nu Q_\nu^2\Big\}, \tag{2.55}$$

which gives

$$Q_\nu^f = (4/K_\nu)\sum_n\sum_m S_{nm}A_{nm}^\nu/\omega_{nm}. \tag{2.56}$$

Here, S_{nm} are the matrix elements of one-electron operator H_S, A_{nm}^ν are the orbital vibronic coupling constants, and ω_{nm} are the energies of transitions between occupied (n) and unoccupied (m) orbitals. From (2.55) and (2.56), we can also obtain the MO version of formula (2.50),

$$\mathcal{E}(ML_{n-k}XY\ldots Z) = S_0 - 2\sum_n\sum_m \frac{S_{nm}^2}{\omega_{nm}} - 8\sum_\nu \frac{1}{K_\nu}\Big(\sum_n\sum_m \frac{A_{nm}^\nu S_{nm}}{\omega_{nm}}\Big)^2. \tag{2.57}$$

Equation (2.57) makes it possible to easily calculate the isomerization energy in terms of MOs,

$$\Delta\mathcal{E}_{t\to c} = 2\sum_n\sum_m \frac{S^+_{nm}S^-_{nm}}{\omega_{nm}} + 8\sum_\nu \frac{1}{K_\nu}\Big(\sum_n\sum_m \frac{A^\nu_{nm}S^+_{nm}}{\omega_{nm}}\Big)\Big(\sum_n\sum_m \frac{A^\nu_{nm}S^-_{nm}}{\omega_{nm}}\Big), \quad (2.58)$$

$$S^+_{nm} = S^c_{nm} + S^t_{nm}, \qquad S^-_{nm} = S^c_{nm} - S^t_{nm}. \quad (2.59)$$

Obviously, like (2.53), equation (2.58) is valid for the energy $\mathcal{E}_{1\to 2}$ of an arbitrary $1 \to 2$ isomeric transformation if (2.59) is replaced by the more general expression

$$S^+_{nm} = S^{(2)}_{nm} + S^{(1)}_{nm}, \qquad S^-_{nm} = S^{(2)}_{nm} - S^{(1)}_{nm}. \quad (2.59a)$$

Note that like (2.47), (2.49), and (2.50), formulas (2.55)–(2.58) can be treated as expressed in terms of the MOs of the $ML_{n-k}XY...Z$ heteroligand molecule; these MOs are found by the perturbation method with H_S as a perturbation operator.

8. It remains to consider perturbation theory applicability to finding the potential energy surfaces of heteroligand systems, that is, the applicability of formulas (2.46)–(2.58), which are used in the further chapters.

Typically, minimum energy gaps ω_{nm} in closed-shell homoligand prototypes of heteroligand systems are of several eV. The difference of the Coulomb integrals of ligand-substituents X, Y, ..., Z and initial ligands L is usually of the order of 1–2 eV, and $\Delta\alpha/\omega_{nm} \approx 1/2$–$1/5$. In the S_{nm} matrix element, the $\Delta\alpha$ value is multiplied by the product of the coefficients of the AOs of ligand L (~1/3–1/10) and the number of replaced ligands (take it to be two). This yields a value of about 0.1 for the parameter of smallness $|S_{nm}/\omega_{nm}|$ determining the applicability of perturbation theory; this estimate can, in particular, be considered an *a priori* argument for the validity of the classical PMO theory and its modifications used with $ML_{n-k}XY...Z$ molecules. The criterion of perturbation theory applicability is also satisfied if ligands X, Y, ..., Z differ from L by resonance integral β(metal–ligand) values. Typically, these integrals are of about 1 eV, and the corresponding smallness parameter has approximately the same value as previously. Similar estimates can be made for the vibronic coupling operator. The derivative of the resonance integral with respect to distance is of the order of 1 eV/Å. Usually, deformations of closed-shell ML_n complexes caused by ligand substitution amount to several hundredths of an Ångstrom unit, and the smallness parameter has the same or even smaller value as for ligand substitution.

Heteroligand derivatives of ML_n systems with open shells are a more complex problem. Perturbation theory for degenerate states (Section 2.1) is applicable when the $|H'_{nm}/\omega_{nm}|$ ratio is small; here, the H'_{nm} element of the perturbation operator couples one of partially occupied degenerate MOs with the nearest unoccupied MO, and ω_{nm} is the energy gap between these MOs. The substitution operator poses no problem, because as previously, the $|S_{nm}/\omega_{nm}|$ ratio is of about 0.1. Deformations of Jahn–Teller systems counted from high-symmetry Q_0 configurations are, however, not small. They amount to several tenths of an Ångstrom unit, and the $|H'_{nm}/\omega_{nm}|$ ratio may increase severalfold. For instance, for Cu(II) complexes, this ratio is ~1/3 (see estimates given in Section 5.5). Clearly, it can hardly be claimed that $1/3 \ll 1$. The perturbative treatment of the Jahn–Teller effect [14, 15] nevertheless gives a correct description of the properties of complexes, or at least, a correct qualitative description.

It would be pertinent to give an *a posteriori* analysis of the problem of perturbation theory applicability. It will be shown in subsequent chapters that, almost always, the perturbation method qualitatively correctly reproduces the type of stereochemical distortions and relative isomer stabilities in heteroligand $ML_{n-k}XY...Z$ molecules and complexes. This method is also capable of estimating the magnitude of various effects. For instance, for closed-shell complexes, expected Q^f_ν distortions are of the order of $Q^f_\nu \approx 4SA/K\omega$, where S, A, K, and ω are the characteristic values of the S_{nm}, A^ν_{nm}, K_ν, and ω_{nm} parameters. Here, $S/\omega \approx 1/10$, K is of about 1 mdyn/Å, and A has the value of the order of 1 eV/Å (in conformity with (2.28), because the $\partial\beta/\partial Q_\nu$ value is of about several eV/Å). This gives the typical Q^f_ν value of several hundredths of Å, which is in agreement with experimental distortion values (see Chapters 3 and 4).

Typical isomerization energies of heteroligand derivatives of closed-shell ML_n systems are also easy to estimate, and these estimates also agree with experiment. Such estimates will be given for square-planar complexes of the type of platinum(II) complexes, for which thermochemical data are available (see Section 4.3). Similar estimates can be obtained for isomerization energies of heteroligand derivatives of open-shell systems (see Section 5.3).

CHAPTER 3

MUTUAL INFLUENCE OF LIGANDS AND RELATIVE ISOMER STABILITY IN σ-BONDED QUASI-OCTAHEDRAL COMPLEXES

In this Chapter, we begin to apply the approach developed in Chapter 2 (for definiteness, this approach will be called vibronic theory of heteroligand systems) to heteroligand molecules and complexes of various types. We will discuss heteroligand derivatives of σ-bonded octahedral molecules and complexes, which are the simplest systems for theoretical treatment because of high symmetry of their homoligand predecessors. The problem will be considered systematically starting with Section 3.2. To form a more complete idea of the matters involved, we will digress slightly into historical aspects of the problem. Section 3.2 will therefore be preceded by a short and, of necessity, only fragmentary description of the history of the theory of mutual influence of ligands and the related problem of isomer stability. For more historical details, the reader is referred to the book [57].

3.1. Mutual Influence of Ligands and Isomer Stability: a Short History

1. The problem of mutual influence of ligands, now traditionally regarded as one of the central problems in inorganic chemistry, dates back to the middle 1920s. The first phenomenon related to mutual influence of ligands was the *trans*-effect discovered by I. I. Chernyaev for square-planar Pt(II) complexes [58, 59]. First, this phenomenon was treated as a kinetic effect in substitution reactions of the type

$$\begin{array}{ccc} \mathrm{L} & & \mathrm{L} \\ & \mathrm{M} & \\ \mathrm{X} & & \mathrm{L} \end{array} \xrightarrow{[\mathrm{X}]} \begin{array}{ccc} \mathrm{L} & & \mathrm{X} \\ & \mathrm{M} & \\ \mathrm{X} & & \mathrm{L} \end{array}, \tag{3.1}$$

in which ligand X destabilized the bond between central atom M and the ligand L positioned *trans* with respect to X. In a number of X-ray structure studies that followed pioneering work [60], data on structural manifestations of *trans*-influence in platinum and other transition metal complexes were collected [61–63]. It was found that the $ML_n \longrightarrow ML_{n-1}X$ substitution predominantly affected the $M-L_{trans}$ bond length between central atom M and the ligand positioned *trans* with respect to the substituent (this phenomenon is called static *trans*-influence; to be terminologically precise, we must distinguish between the static *trans*-influence and the kinetic (discovered by Chernyaev) *trans*-effect).

In addition to X-ray structural studies, *trans*-influence was extensively investigated by other physical methods such as IR spectroscopy. The results obtained in these works were in conformity with the conclusions drawn based on X-ray data [61]. Still later, static mutual ligand influence was studied experimentally for nontransition element compounds such as square-planar Te(II) [64] and octahedral Sn(IV) [65, 66] complexes. It was found that in such complexes, mutual ligand influence manifestations were more diverse than in transition metal derivatives. In complexes with a central atom in a low oxidations state such as Te(II) compounds, mutual ligand influence was similar to that observed in transition metal complexes. On the other hand, in complexes with a central atom in its higher oxidation state such as Sn(IV) compounds, mutual influence of ligands of another type was observed. Starting with the work [65], this kind of phenomena became known as *cis*-influence, because substitution in such compounds had substantial influence on bond lengths between central atoms and ligands positioned *cis* to the substituent. It should, however, be borne in mind that the term "*cis*-effect" or "*cis*-influence" does not correctly describe the phenomenon, because simultaneously, *trans*-bond lengths change even more substantially, although in a direction opposite to changes in *cis*-bond lengths (more details are given in Section 3.5).

A little later, electron diffraction studies [67, 68] showed that nontransition element complexes in the higher oxidation states can also exhibit *trans*-influence similar to that characteristic of transition metals. Mutual influence of ligands of this type was discovered in, e. g., IOF_5 [68]. Lastly, in [69, 70], the singly substituted UCl_5O^- and $PaCl_5O^{2-}$ complexes were studied, which raised the question of mutual influence of ligands in actinide compounds. Work [71] was, we believe, the first to note that this influence was different from the *trans*-influence in transition metal complexes; rather, it was akin to *cis*-influence observed in Sn(IV) derivatives. Interestingly, this observation might have been made earlier, because the geometry of doubly substituted actinide complexes, e. g., uranyl compounds, had long been known.

2. Attempts at theoretically explaining *trans*-influence in platinum(II) complexes, which was chronologically the first observed phenomenon of mutual influence of ligands, began virtually immediately after Chernyaev's discovery, although until the work of Pidcock *et al.* [72], no clear-cut distinction between static *trans*-influence and kinetic *trans*-effect had been made.

From the mid 1920s, Chernyaev himself repeatedly tried to explain *trans*-influence either as a "jump of the chemical action" of the electron through the central atom of a complex [73a] or in terms of a collision between three atoms (the substituent, the central atom, and the *trans* ligand) [73b], or by a "vector electron cloud distribution" [73c]. Considering the state of theoretical inorganic chemistry at that time, it would hardly be right to criticize this author for the absence of a consistent physical view on the phenomenon. Rather, it should be mentioned that his main idea, that of an indirect mechanism of interaction between ligands, inevitably with the participation of the central atom, was fundamentally correct and is retained in all modern models based on perturbation theory.

In the 1930s–1940s, Grinberg and Nekrasov developed the "polarization" model of *trans*-influence, which laid emphasis on different ligand polarizabilities responsible for "a shift of the center of gravity of positive and negative charges" in heteroligand complex ions [73d] or "a shift of the atomic core" of the central atom with respect to the ligands [73e]. It is clear at present that the most important shortcoming of the Grinberg–Nekrasov approach is the absence of a quantum-chemical background. A comparison of the cited statements with the quantum-chemical treatment given in Section 2.4, however, shows them to contain an idea of electronic and geometric relaxation of substituted complexes, although of a purely descriptive character and expressed in "prequantum" language. Curiously, Grinberg notes [73f] that symmetry lowering in passing from a homoligand complex to a heteroligand system is one of the reasons for metal–ligand bond length changes. Of course, at that time (1928), symmetry considerations could not be effectively applied to studying the mutual influence of ligands, because doing so would require taking into account not only the point symmetry group of the complex but also symmetry of the wave functions in its ground and excited states. At present, group theory is an effective tool of the theory of heteroligand complexes (for instance, see Section 2.3).

The works considered above appeared at the first stage of the development of the theory of heteroligand complexes, when even very simple quantum-chemical concepts were not used. The beginning of the second stage dates to 1948, when Syrkin [74a] suggested an approach to *trans*-

influence in quasi-square-planar platinum complexes based on Pauling's ideas, popular at that time, of hybridization of the AOs of the central atom and superposition (resonance) of all valence structures that can be written for a molecule. Specifically, it was assumed in [74a] that the Pt atom with the $5d^9 6s$ electronic configuration formed covalent bonds with ligands by its $6s$ and one of the $5d$ AOs. These AOs gave two hybrid *sd* orbitals directed mutually orthogonally. For an ML_4 complex containing four equivalent ligands, superposition of the corresponding resonant structures gave four equivalent M—L bonds. In the presence of a different ligand X that formed stronger M—X bonds, the contribution of the structures including X and its *cis* neighbors increased at the expense of the contribution of the structures including the L ligand positioned *trans* with respect to X. Such a treatment, according to which the $M-L_{trans}$ bond in an ML_3X complex underwent seeming weakening against the background of strengthening of the $M-L_{cis}$ bonds, was at variance with experiment, as was noted as early as in [74b]. In addition, it is easy to see, especially from the point of view of the nowadays predominant MO theory, that the approach developed in [74a] was not free of conceptual drawbacks. It should, however, be emphasized that in [74a], the problem was for the first time formulated in quantum-chemical terms. In addition, this work raised the question of the relation between the character of mutual influence of ligands and the electronic structure of the central atom.

Starting with [74a], quantum-chemical concepts have become fundamental in all *trans*-influence and *trans*-effect theories. In the 1950s–1960s, works by Cardwell and Phil [74c], Chatt, Duncanson, and Venanzi [74d, 74e], Orgel [74f], Pearson [74g, h], and Langford and Gray [74i] appeared. In these works, ligand substitution in square-planar complexes with the formation of five-coordinate transition state complexes was considered and possible pathways of such reactions were discussed. For instance, Cardwell and Phil assumed that metal–ligand bonds in square-planar platinum complexes were formed by four dsp^2 hybrid orbitals, whereas the vacant $6p_z$ platinum orbital was considered responsible for bonding with a fifth ligand and the formation of an intermediate. Chatt and Orgel assigned significance to additional dative π bonds between the d_{xz} or d_{yz} platinum orbitals and a vacant π AO of one of the ligands. According to these authors, the formation of a dative bond decreased the electron cloud density of one of these d AOs and through this freed the room for a fifth ligand either above or below the xy plane of the square-planar complex. We will not discuss works [74c–74i] in more detail, because these works are concerned with the kinetic *trans*-effect rather than static mutual influence of ligands. There was, however, a fundamental shortcoming common to all of these works, namely, in all of them, the problem was treated in purely descriptive terms with no mathematical background, which, of course, made theoretical constructions less rigorous and theoretical conclusions less convincing.

The transition from a descriptive to a mathematical (perturbative) treatment of the problem marked the beginning of the third, modern stage in the development of the theory of heteroligand systems. This turn, which dates to the 1970s, was due to the work of many authors. The first mathematical model of *trans*-influence was based on analysis of the nodal structure of MOs in a linear X—M—L heteroligand fragment present in an $ML_{n-1}X$ complex and considered separately (Shustorovich and Buslaev [75a, b] and Shustorovich [75c]). Based on relations between the coefficients on the three-center MOs in the X—M—L fragment, the authors considered the distribution of wave function nodes in the X—M and M—L regions to study weakening of the *trans* M—L bond depending on the multiplicity and character (covalent or donor-acceptor) of both X—M and M—L bonds.

A year later, Baranovskii and Sizova [45], who studied *trans*-influence in quasi-square-planar Pt(II) complexes, applied perturbation theory to the MOs of the whole ML_3X complex rather than its fragments. In [45], the perturbation molecular orbital theory [12, 44] was used to determine the effective charges on the ligands positioned *cis* (L_{cis}) and *trans* (L_{trans}) with respect to substituent X. The difference of charges was treated as a measure of *trans*-influence of ligand X.

An important role in the development of the theory of mutual influence of ligands was played by the model elaborated by Popov for the example of quasi-octahedral transition metal and nontransition element ML_5X complexes [46, 47]. The influence of substituents was measured by changes in overlap populations between central atom AOs and the AOs of ligands positioned *cis* and *trans* to X. Further, we will consider this model together with work [45] in more detail. Note only that works [46, 47] performed independently of [45] were especially important because they revealed the potentialities of a thorough mathematical analysis of complexes of various types in terms of perturbation theory without actual calculations of the electronic structure of the systems.

In [45–47], the perturbation problem was treated in terms of delocalized canonical MOs. Nefedov and Buslaev [76] suggested another approach; further developments in this direction were summarized in [77]. In [77], an unsubstituted homoligand ML_n complex was described in terms of equivalent orbitals (EO LCAO), bonding (ϕ_i) and antibonding (ϕ_i^*), localized on separate $M-L_i$ bonds. The electronic structure of the derivative heteroligand complex was described by taking into account partial mixing between ϕ_i and ϕ_i^* orbitals as a result of perturbation caused by ligand substitution. The degree of mixing was considered a criterion of weakening of the $M-L_i$ bond.

All these perturbative approaches reduced the problem of mutual influence of ligands to calculations of atomic charges, overlap populations, and similar values, which were treated as criteria of ligand effects. From the point of view adopted in this book, such criteria of mutual influence of ligands are insufficient, because they indirectly describe the energy and geometric characteristics of molecules. A theory of the geometry and relative stability of isomers of molecules and complexes should relate ligand substitution $ML_n \rightarrow ML_{n-k}XY...Z$ to the position and depth of the minima of the adiabatic potential energy surface of the system, which implies the use of the vibronic theory of substitution effects.

In 1970, Tobias published a work that passed unnoticed, probably, because of the descriptive character of treatment, which was not substantiated mathematically [78]. *Trans*-influence in quasi-square-planar $AuHal_3(CH_3)^-$ complexes and *cis*-influence in quasi-octahedral $SnHal_4(CH_3)_2^{2-}$ complexes were treated in [78] in terms of a quite usual analysis of interactions between central-atom and ligand AOs. The work, however, contained an important idea of a relation between bonding in substituted complexes and their geometric distortions described by normal modes. This brought in the conception, although fairly vague, of the vibronic nature of mutual influence of ligands.

Porai-Koshits [66] used vibronic interactions in an explicit although still non-mathematical form to explain the influence of substituents X on the geometry of six-coordinate $SnL_2L'_2X_2$ complexes. Tetrahedral distortions of coordination quasi-octahedra observed in such complexes were treated as a "continuation of $O_h \rightarrow D_{4h}$ deformation" due to the pseudo-Jahn–Teller effect, which accompanied narrowing of the energy gap between the occupied and unoccupied MOs caused by $O_h \rightarrow D_{4h}$ distortions. Once more emphasis was laid on the relation between mutual influence of ligands and vibronic phenomena; subsequently, this idea, although in a form other than that suggested in [66], proved to be exceedingly fruitful.

Consistent vibronic treatment of all important stereochemical manifestations of mutual influence of ligands in diverse transition metal, nontransition element, and actinide complexes, which had much in common with the vibronic theory of catalysis developed by Bersuker [79, 15], was suggested and developed in [48–52, 55, 56]. The corresponding theory will be described in detail in this and subsequent chapters.

3. Chernyaev already clearly understood that the character of mutual influence of ligands was closely related to the preferred formation of certain geometric isomers. The very occurrence of reaction (3.1) means that one of the ML_2X_2 complex isomers should predominantly be formed, although for kinetic (*trans*-effect) rather than thermodynamic reasons. No wonder, this relation was repeatedly discussed in the literature. One of the latest, if not the last, reviews on the topic is Appleton's work [80a], which contains an analysis of the relation between *trans*-influence and *trans*-effect on the one hand and isomerism of diverse Pt(II) complexes on the other.

At an early stage of the development of the theory of mutual influence of ligands, the nature of the relation between isomerism and *trans*-effect in transition metal complexes was discussed by Chatt and Heaton [80b] and Pearson [80c] at a qualitative level, for instance, in terms of chemical "antisymbiosis" [80c]. The work by Tobias cited above [78] contained not only the idea of vibronic treatment of mutual influence of ligands but also its implications concerning geometric relaxation of M—L bonds in a heteroligand complex as the cause of stabilization of certain isomers (this idea is retained in our theory described below). In addition, the nature of differences in stability between *cis* and *trans* isomers was considered in [78], probably for the first time, not only for transition metal complexes but also for nontransition element compounds.

Subsequent rapid development of perturbative theories of mutual influence of ligands did not result in equally rapid progress in the theory of relative isomer stabilities because of the use of criteria of mutual influence of ligands which were indirectly related to the energy characteristics of complexes. Such a progress was only made possible after the introduction of the vibronic theory of heteroligand systems, which took into account isomer type effects on the adiabatic potential energy surface and, therefore, the total energy of a molecule or a complex. If both ML_n and $ML_{n-k}XY...Z$ systems have closed electron shells, a solution to the problem of the relative energetic stability of $ML_{n-k}XY...Z$ isomers immediately follows from the vibronic theory of mutual influence of ligands [55]. In more recent works [51, 52], this approach was extended to heteroligand derivatives of ML_n systems with open electron shells.

We considered the development of the theory of relative stabilities of isomers of heteroligand molecules and complexes in the context of the theory of mutual influence of ligands. Some additional remarks on the history of the problem will be made in Sections 3.9 and 5.5.

4. Consider works [45–47] in more detail. These works are of interest by themselves and in relation to the vibronic theory of mutual influence of ligands described in detail below. For uniqueness sake, we will, so far as is possible, use the notation accepted throughout this book. As mentioned, perturbation theory in the basis set of the MOs of the whole molecular system was for the first time applied to treat mutual influence of ligands in [45], where a quasi-square-planar $Pt^{II}L_3X$ complex with σ-bonded X and L ligands was considered. The strength of the $Pt-L_\mu$ bond was measured by population P_μ of the σ orbital of the μth ligand, because the more ionic bonds were believed to be weaker. The dependence of P_μ on the Coulomb integral of substituent X which replaced ligand L_1 was estimated using the expression for atom-atom polarizabilities ($\pi_{\mu 1}$)

$$\delta P_\mu = \Delta\alpha\pi_{\mu 1} = \Delta\alpha \sum_n^{\text{occ}} \sum_m^{\text{unocc}} \frac{c_{1n}c_{\mu m}c_{1m}c_{\mu n}}{\varepsilon_n - \varepsilon_m} \tag{3.2}$$

borrowed from PMO theory [12, 44]. Here, $c_{\lambda j}$ is the coefficient of the σ AO numbered λ in the jth MO. This expression was analyzed in [45] under various assumptions on the participation of the $6p$(Pt) orbitals in bonding to find that because of the diffuse character of these orbitals and their insignificant contribution to M—L bonding, they largely caused changes in the populations of the *trans* ligand σ AOs (in comparison with *cis* ligands). Increasing the contribution of the $6p$(Pt) AOs to bonding sharply weakened the *trans*-influence. These conclusions remained unaffected by additionally taking into account metal–ligand resonance integral changes caused by the $Pt^{II}L_4 \longrightarrow Pt^{II}L_3X$ substitution.

Whereas in [45], only the *trans*-influence in transition metal complexes was only considered, in [46], both *trans*-influence in transition metal complexes and *cis*-influence in nontransition element compounds were analyzed. The approach suggested in [46] was exemplified by applying it to octahedral σ-bonded complexes. The M—L central atom–ligand bond strength was measured by overlap populations N(M—L)

$$N(\text{M—L}) = 4\sum_\lambda \sum_n^{occ} c_{\lambda n} l_{\text{L}n} S_{\text{L}\lambda}, \tag{3.3}$$

where $c_{\lambda n}$ and $l_{\text{L}n}$ are the coefficients of the metal λ AO and the ligand σ orbital in the nth bonding MO, and $S_{L\lambda}$ is the overlap integral. Overlap population N(M—L) characterizes the electronic charge in the region between M and L, and N(M—L) should be expected to vary in parallel with metal–ligand bond strength. As previously, the replacement of ligand L_1 will be modeled by a change $\Delta\alpha$ in the Coulomb integral of its σ AO. The first-order perturbation theory equation for wave functions then gives change ΔN in overlap population $N(\mathrm{M-L})$ in the form

$$\Delta N = 4\Delta\alpha \sum_{\lambda}\sum_{n}^{occ}\sum_{m}^{unocc} \frac{c_{1n}c_{1m}(c_{\lambda n}l_{m\mathrm{L}} + c_{\lambda m}l_{n\mathrm{L}})S_{\lambda \mathrm{L}}}{\varepsilon_m - \varepsilon_n}. \tag{3.4}$$

Of interest for estimating mutual ligand effects are the ΔN_t and ΔN_c values, that is, the changes in the overlap populations of the M—L bonds *trans* and *cis* with respect to the new substituent, and the difference $\Delta N_t - \Delta N_c$ characterizing the difference of the *trans*- and *cis*-effects. According to [46], these values can be written explicitly as

$$\Delta N_t = \frac{\Delta\alpha}{36}\Bigg[(\eta_a + 2\eta_e - 3\eta_t)\Big(\frac{3c_t^2}{\omega_t} - \frac{2c_e^2}{\omega_e} - \frac{c_a^2}{\omega_a}\Big) + \Big(\frac{\eta_a}{\omega_a} + \frac{2\eta_e}{\omega_e} - \frac{3\eta_t}{\omega_t}\Big)(3c_t^2 - 2c_e^2 - c_a^2)\Bigg], \tag{3.5}$$

$$\Delta N_c = \frac{\Delta\alpha}{36}\Bigg[(\eta_a - \eta_e)\Big(\frac{c_e^2}{\omega_e} - \frac{c_a^2}{\omega_a}\Big) + \Big(\frac{\eta_a}{\omega_a} - \frac{\eta_e}{\omega_e}\Big)(c_e^2 - c_a^2)\Bigg], \tag{3.6}$$

$$\begin{aligned}\Delta N_t - \Delta N_c = \frac{\Delta\alpha}{12}\Bigg[&(2\eta_t - \eta_e - \eta_a)\Big(\frac{c_e^2}{\omega_e} - \frac{c_t^2}{\omega_t}\Big) \\ &+ \Big(\frac{\eta_t}{\omega_t} - \frac{\eta_e}{\omega_e}\Big)(c_a^2 + c_e^2 - 2c_t^2) + (\eta_t - \eta_e)\Big(\frac{c_a^2}{\omega_a} - \frac{c_t^2}{\omega_t}\Big) \\ &+ \Big(\frac{\eta_t}{\omega_t} - \frac{\eta_a}{\omega_a}\Big)(c_a^2 - 2c_t^2)\Bigg].\end{aligned} \tag{3.7}$$

Here, $\eta_i = 4S_i c_i l_i$ is the overlap population between the ith AO χ_i of atom M and the ligand group orbital that, together with χ_i, forms bonding MO ψ_i. The difference of the energies of the antibonding and bonding orbitals is denoted by $\omega_i = \varepsilon_i^* - \varepsilon_i$. Index $i = a, e, t$ corresponds to MO symmetry. Note that equations (3.5)–(3.7) were obtained in [46] from the general form of σ-MOs in octahedral complexes (Table 1) on the simplifying assumption that the differences of the energies of bonding σ MOs can be ignored in comparison with the energy differences between antibonding and bonding MOs. The expressions for ΔN_t etc. can, however, be written and analyzed without this assumption [47].

In [46, 47], the expressions for ΔN_t, ΔN_c, and $\Delta N_t - \Delta N_c$ were analyzed in detail for 12-electron transition metal complexes and 12- and 14-electron nontransition element compounds. This analysis showed that in transition metal σ-bonded octahedral complexes, the $ML_6 \longrightarrow ML_5X$ substitution largely caused weakening (or strengthening, depending on the sign of $\Delta\alpha$) of *trans* bonds. This gave an explanation, on some different premises than in [45], of the static *trans*-influence in transition metal complexes in terms of the MO LCAO method. A similar analysis performed for nontransition element 12-electron complexes [46, 47] explained *cis*-influence and predicted *trans*-influence in these compounds prior to its experimental observation [67, 68]. In addition, experimentally observed *trans*-influence in nontransition element complexes with central atoms in low oxidation states was given theoretical treatment [46, 47]. On the whole, the works [45–47] demonstrated the effectiveness of applying the method of perturbed molecular orbitals to the problem of mutual influence of ligands.

More recent similar calculations based on applying the perturbation method (for the latest works, see [81a, 81b]), however, added nothing conceptually new to the theory of mutual influence of ligands. We will see that in this respect, the theory of vibronic coupling offers much more promise.

5. In conclusion, consider attempts at theoretically studying mutual influence of ligands and relative stabilities of geometric isomers of heteroligand molecules and complexes by purely computational methods with the use of various quantum-chemical calculation procedures. Among the early works (the late 1960s and the mid 1970s), we should first and foremost mention the calculations performed by Zumdahl and Drago [82a] with the use of the extended Hückel method for several Pt(II) complexes with various ligands (H_2O, NH_3, Cl^-, H_2S, CH_3-, PH_3, and H^-) to study *trans*-influence and *trans*-effect. In addition, note the work by Burdett [82b] who applied the model of angular overlap to study relative stabilities of geometric isomers of quasi-square-planar, quasi-octahedral, and some other transition metal complexes. Since then, many computational works performed at various levels of approximation have been published. In these works, complexes with various central atoms and diverse ligands (polydentate ligands too) were studied; the most recent examples are [82c–82l]. The computer simulation results obtained in these works together with the data of diffraction and other experimental studies give important additional information about the behavior of real mixed-ligand molecules and complexes, although sometimes, even high-level calculations are at variance with experiment (see the discussion of works [196, 197] at the end of

Section 4.4). However on the whole, the contribution of computer simulations to the development of the theory of heteroligand systems was not very significant, because computations are mere numerical experiments, whose results not only do not give conceptual explainations of phenomena but require such explanations themselves.

3.2. Transition Metal Complexes: Bond Lengths

1. We will begin a systematic presentation of the vibronic theory of mutual influence of ligands with consideration of σ-bonded transition metal octahedral complexes with 12 electrons on σ MOs. First, the more visual approximation of frontier σ-MOs will be used. A more complete treatment taking into account all valence σ-MOs will be given in Section 3.5. As applied to the problem that we are considering, the approximation of frontier MOs ignores all terms in (2.49) and (2.56) except those which simultaneously contain the highest occupied MOs (HOMOs) and the lowest unoccupied ones (LUMOs). Usually, it is argued that for these terms, the energy differences in the denominators are the smallest, and they therefore make the largest contribution to the sum.

Table 5. Modes contributing to geometry distortions in σ-bonded ML_5X and ML_3X complexes in the approximation of frontier MOs

Complex	HOMO	LUMO	Modes satisfying selection rule (2.25)	MOs satisfying rule (2.43b)	Active mode
a	t_{1u}	e_g^*	T_{1u}: Q_{10}, Q_{11}, Q_{12}	ψ_z, $\psi_{z^2}^*$	Q_{12}
b	a_{1g}^*	t_{1u}^*	T_{1u}: Q_{10}, Q_{11}, Q_{12}	$\psi_{a_{1g}}^*$, ψ_z^*	Q_{12}
c	e_g	a_{1g}^*	E_g: Q_2, Q_3	ψ_{z^2}, $\psi_{a_{1g}}^*$	Q_3
d	e_u	b_{1g}^*	E_u: Q_3, Q_4	ψ_x, $\psi_{b_{1g}}^*$	Q_3
e	a_{1g}^*	e_u^*	E_u: Q_3, Q_4	$\psi_{a_{1g}}^*$, ψ_x^*	Q_3

Note: a, quasi-octahedral transition metal complexes with 12 electrons on σ MOs; b, quasi-octahedral 14-electron nontransition element complexes; c, quasi-octahedral 12-electron nontransition element complexes; d, quasi-square-planar transition metal d^8 complexes; and e, quasi-square-planar nontransition element complexes of the type of Te(II) complexes.

The frontier σ-MOs in the complexes under consideration are easy to identify. A qualitative consideration (see Section 1.4), calculations [20], and experimental data show that the frontier MOs of the complexes are the t_{1u} HOMO and the e_g^* LUMO. These MOs are given in the second and third columns of Table 5. It follows that the only terms that remain in equation (2.56) describing the coordinates of the minimum of the lower sheet of the adiabatic potential energy surface are those with the same ω_{nm} value equal to the energy gap between the t_{1u} HOMO and the e_g^* LUMO. At this stage, the symmetry characteristics of the frontier MOs can be used to draw conclusions on vibrational modes Q_ν^f active in mutual ligand effects. It immediately follows from rules (2.25) that T_{1u}-symmetry MOs are mixed with E_g ones only by T_{1u}-symmetry modes. Of the latter, only Q_{10}, Q_{11}, and Q_{12} (see Table 5) involve bond stretches. These are the modes that will be considered in this Section (the modes involving angle deformations will be discussed in Section 3.7).

Let us return to equation (2.56). Even in the approximation of frontier MOs, its right-hand side still contains many terms, because unless the problem is formulated more specifically, we must include couplings of each of three t_{1u} HOMOs with each of two e_g^* LUMOs. To continue the analysis based on reducing this set of terms to a single one, we should specify the type of the heteroligand complex. We consider an ML_5X complex and assume that substituent X is situated on the z axis at position 1 (Fig. 4). Then according to rules (2.43b), of all operator H_S matrix elements, only that between the $t_{1u,z}$ and e^*_{g,z^2} MOs does not vanish (indices z and z^2 label the MOs that transform as the p_z and d_{z^2} central atom AOs), and there only remain the terms of the form $(4S_{z,z^{2*}}A^\nu_{z,z^{2*}})/(K_{T_{1u}}\omega_{z,z^{2*}})$ in (2.56). Lastly, it is easy to see that the $t_{1u,z}$ and e^*_{g,z^2} MOs can only be coupled by mode Q_{12}; for other modes, the orbital vibronic coupling constant $A^\nu_{z,z^{2*}}$ vanishes (see Table 5). This leads us to conclude that in complexes under consideration undergoing the $ML_6 \longrightarrow ML_5X$ substitution at position 1, the adiabatic potential minimum in the space of modes possible for octahedral ML_6 molecules lies on the Q_{12} axis. Accordingly, sum (2.56) reduces to

$$Q_\nu^f = 4SA/K\omega, \tag{3.8}$$

where $Q_\nu^f = Q_{12}^f$, and S and A are the $S_{z,z^{2*}}$ and $A^{12}_{z,z^{2*}}$ matrix elements,

$$Q_{12}^f = 4S_{z,z^{2*}}A^{12}_{z,z^{2*}}/K_{T_{1u}}\omega_{z,z^{2*}}. \tag{3.8a}$$

2. To qualitatively characterize the deformation of the ML_6 complex caused by the $ML_6 \longrightarrow ML_5X$ substitution, it remains to find the sign of Q_{12}^f, that is, determine the signs of S and A in (3.8a), because the denominator cannot be negative. For this purpose, let us write S and A through the coefficients of AOs in the frontier MOs. Generally,

symmetry considerations and the normalization conditions for MOs allow us to write

$$\begin{aligned}&\text{HOMO } t_{1u}: \ \psi_z = c_t p_z + (l_t/\sqrt{2})(\sigma_1 - \sigma_2),\\&\text{LUMO } e_g^*: \ \psi_{z^2}^* = l_e d_{z^2} - (c_e/\sqrt{12})(2\sigma_1 + 2\sigma_2 - \sigma_3 - \sigma_4 - \sigma_5 - \sigma_6)\end{aligned} \tag{3.9}$$

for the complexes under consideration. In (3.9), as everywhere in this book, c_i (here, $i = t$, e) are the coefficients of valence AOs of the central atom in bonding MOs (here, t_{1u} and e_g), and l_i are the coefficients of the corresponding symmetrized group orbitals of ligands. Therefore, all $c_i > 0$ and all $l_i > 0$ and $c_i^2 + l_i^2 = 1$. (We use the approximation of zero differential overlap). It then follows from (3.9) and relations (2.27) and (2.39) that

$$S = -(1/\sqrt{6})c_e l_t \Delta\alpha, \tag{3.10}$$

$$A = l_e l_t [\partial\beta(\sigma, d)/\partial R]_0 - \sqrt{2/3}\, c_e c_t [\partial\beta(\sigma, p)/\partial R]_0. \tag{3.11}$$

In (3.11), the derivatives of resonance integrals describing metal–ligand interactions with respect to coordinate Q_{12} are expressed in terms of the derivatives with respect to the interatomic distance R. These derivatives are positive, because resonance integrals are negative and decrease in magnitude as the interatomic distance increases. Interactions between nonbonded atoms (ligands) are, as usual, ignored.

Consider (3.10) and (3.11) [48]. It follows from (3.10) that the sign of S is opposite to that of $\Delta\alpha$, that is, the difference of the Coulomb integrals of the σ AOs of substituent X and initial ligand L, $\Delta\alpha = \alpha(X) - \alpha(L)$. As concerns vibronic coupling constant A [48], its sign at first sight seems difficult to analyze, because (3.11) contains c_i and l_i coefficients present in the expressions for MOs that cannot be known beforehand. It therefore appears pertinent to stress here that such an analysis does not require detailed information about the atomic compositions of the MOs; it can be performed based on general considerations of the character of chemical bonds in the class of compounds that we are considering. (This circumstance, in particular, explains why a certain type of mutual ligand effects is not determined by the special features of a certain molecule, but is common to a class of systems.)

Clearly, the second term in (3.11) can be ignored for transition metal complexes, because it is small in comparison with the first one. Indeed, it follows from the relative arrangement of the valence levels of typical ligands and typical transition metals that bonding MOs should predominantly be localized on ligands (Section 1.4). Qualitative considerations therefore lead us to conclude that the c_e and c_t coefficients are smaller than l_e and l_t, and, consequently, the $c_e c_t$ product is smaller than $l_e l_t$. Experience of reliable calculations [20] lends support to this conclusion; according to [20], the contributions of vacant transition metal np AOs to bonding are far smaller than those of valence $(n-1)d$ AOs (for this reason, transition metal complexes are often calculated using the basis set of metal ns and $(n-1)d$ AOs plus ligand AOs; when such a basis set is used, the second term in coupling constant A is merely absent). Lastly, because of a greater diffusivity of metal vacant np AOs in comparison with $(n-1)d$ ones, the $\partial\beta(\sigma,p)/\partial R$ derivative should be relatively small in comparison with $\partial\beta(\sigma,d)/\partial R$, which may serve as an additional argument.

These considerations show that the sign of vibronic coupling constant A is determined by the first term in (3.11), in which all multipliers including the derivative of the resonance integral are positive. It follows that deformation of the ML_6 complex caused by the $ML_6 \longrightarrow ML_5X$ substitution is determined by the Q_{12}^f mode having the sign opposite to that of $\Delta\alpha$, that is, $Q_{12}^f < 0$ if $\Delta\alpha > 0$, and $Q_{12}^f > 0$ if $\Delta\alpha < 0$. If we recall the form of mode Q_{12} shown in Fig. 4, we can express the result obtained in familiar terms of structural chemistry: the substitution only affects the length of the bond positioned *trans* with respect to the substituent. The *trans* bond becomes longer (more labile) proportionally to the $\Delta\alpha$ value if ligand L is replaced by a more electropositive ligand X and shorter (more stable) if L is replaced by a more electronegative ligand X:

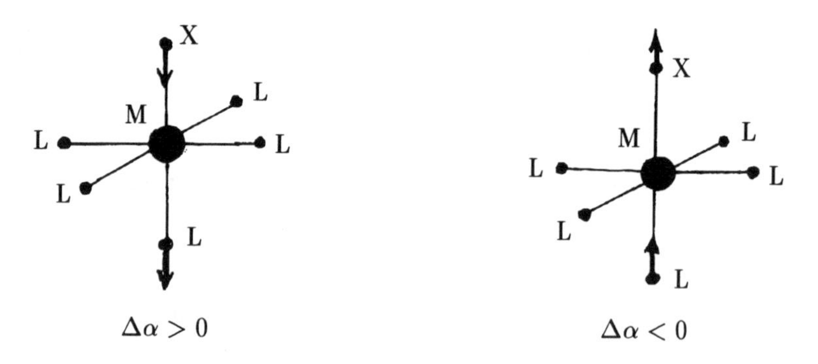

$\Delta\alpha > 0$ $\Delta\alpha < 0$

3. The formulated rule characterizes *trans*-influence typical of transition metal complexes (see reviews [61–63]). For completeness sake, recall data on quasi-octahedral σ-bonded transition metal complexes obtained by IR and diffraction methods. In the $Pt^{IV}Cl_2Me_2(PMe_2Ph)_2$ complex with two chlorine atoms positioned *trans* to each other, the

replacement of one Cl by CH_3 decreases the Pt—Cl vibration frequency from 332 to 265 or even 244 cm^{-1} [61],

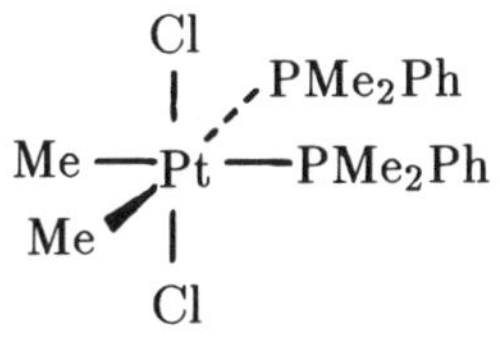

Cl
PMe$_2$Ph
Me—Pt—PMe$_2$Ph
Me
Cl

$\nu(\text{Pt–Cl}) = 332\ cm^{-1}$

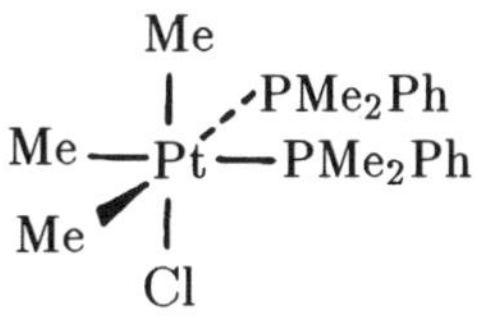

Me
PMe$_2$Ph
Me—Pt—PMe$_2$Ph
Me
Cl

$\nu(\text{Pt–Cl}) = 265 - 244\ cm^{-1}$

The Ir—H bond in Ir(III) complexes experiences similar weakening when Cl is replaced by a more electropositive ligand [61],

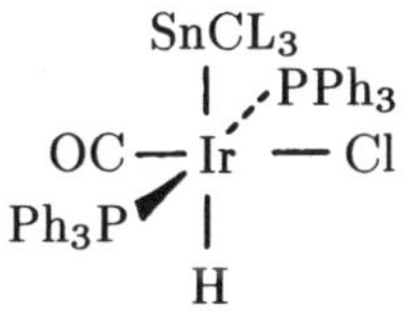

SnCL$_3$
PPh$_3$
OC—Ir—Cl
Ph$_3$P
H

$\nu(\text{Ir–H}) = 2198 - 2160\ cm^{-1}$

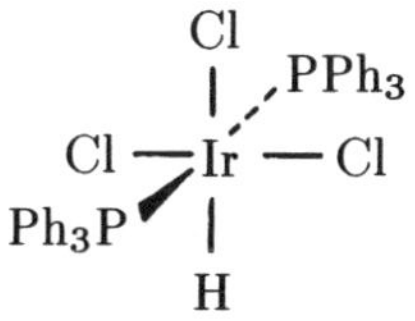

Cl
PPh$_3$
Cl—Ir—Cl
Ph$_3$P
H

$\nu(\text{Ir–H}) = 2239\ cm^{-1}$

or

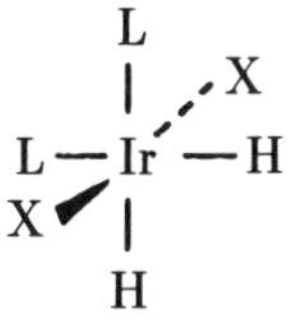

L
X
L—Ir—H
X
H

$\nu(\text{Ir–H}) = 2000 - 2100\ cm^{-1}$

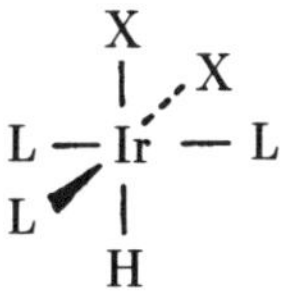

X
X
L—Ir—L
L
H

$\nu(\text{Ir–H}) = 2195 - 2220\ cm^{-1}$

(here, L are tertiary phosphines, and X are halogen atoms). The diffraction data [83a–83f] including those obtained recently provide an even clearer illustration of *trans*-influence patterns in the complexes under consideration (although strictly speaking, not all examples given in [83a–83f] can be considered relevant, because in addition to 12 σ electrons, some complexes contain electrons on quasi-atomic t_{2g} MOs). Note elongation of the bonds positioned *trans* with respect to the more electropositive ligands in $Rh(NH_3)_5Et^{2+}$ [83a] and $OsClF_5^{2-}$ [83b] (here and throughout, bond lengths are given in Å):

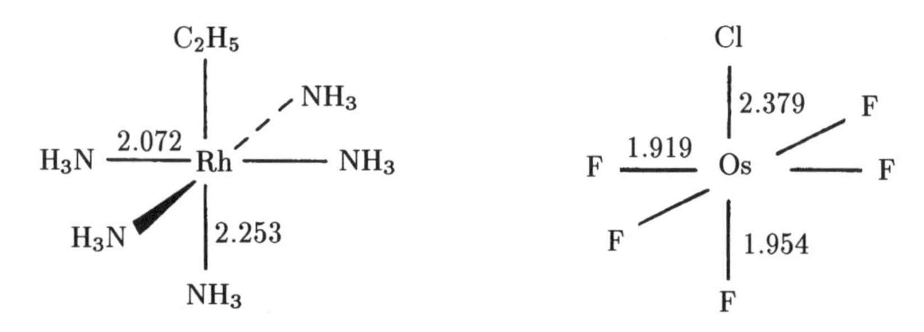

Note also the characteristic behavior of the Pt—Py bonds in the platinum(IV) complex:

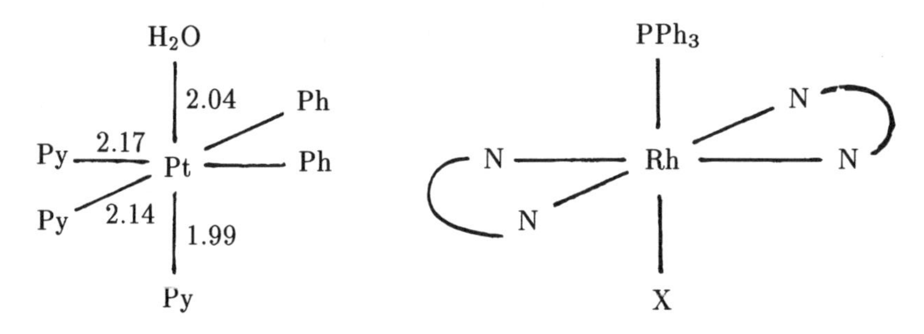

Here, elongation of Pt—Py bonds positioned *trans* to the more electropositive ligands (Ph) and shortening of this bond positioned *trans* to the most electronegative one (OH_2) is quite distinct [83b]. The structures of a series of Rh complexes containing the X—Rh—PPh_3 axial group [83b] are also instructive. In this group, the Rh—P bond heavily depends on X and is short (2.28–2.32 Å) for X=Cl and longer by 0.2 Å for X=Et. Among other complexes, whose structures were determined in recent years, $PyM(DH)_2Hal$, where M = Rh(III) and Co(III), DH is dimethylglyoxime, and Hal is Cl and I, deserve attention. Substituting I for Cl in this complex elongates the *trans* M—Py bond by ~0.2 Å [83d, e]. The $Co(OEP)(CH_3)Py$ complex exhibits a similar behavior; the axial CH_3 group causes elongation of the axial *trans* Co—N bond by 0.12 Å in comparison with the equatorial Co—N bonds [83f].

4. Many of the examples considered above may at first sight seem not to be exactly relevant, because the complexes involved are not monosubstituted, and it is not *a priori* clear to what extent the conclusions drawn for ML_5X complexes are valid for them. The point of view traditional for coordination chemistry and harmonizing with experiment is that in transition metal $ML_{n-k}XY...Z$ complexes, the substituent virtually only affects the bond positioned *trans*. It is easy to see that

according to (2.49) and (2.56), multiply substituted complexes should behave in precisely this way. Indeed, as mentioned in Section 2.3, the $\mathcal{H}_S$ operator corresponding to the substitution $ML_n \longrightarrow ML_{n-k}XY...Z$ can be represented as the sum of partial operators corresponding to substitutions of separate ligands. Each matrix element of the total substitution operator is then written as the sum of the matrix elements for separate substituents, that is, $S_j = \sum_l S_j^l$. It follows that any of the Q_ν^f coordinates describing deformation of the ML_n complex caused by the $ML_n \longrightarrow ML_{n-k}XY...Z$ substitution is also representable as

$$Q_\nu^f = \sum_l^{\text{ligands}} Q_\nu^f(l), \tag{3.12}$$

where $Q_\nu^f(l)$ is molecular deformation caused by the replacement of the lth ligand.

5. To conclude this section, consider a visual and rather general method for determining the signs of S and A in formulas of type (3.8a) described below for the example of quasi-octahedral transition metal complexes ML_5X. Here are the shapes of the LCAO forms of HOMO and LUMO (3.9) drawn according to the standards of "pictorial" quantum chemistry [26a, 26b]:

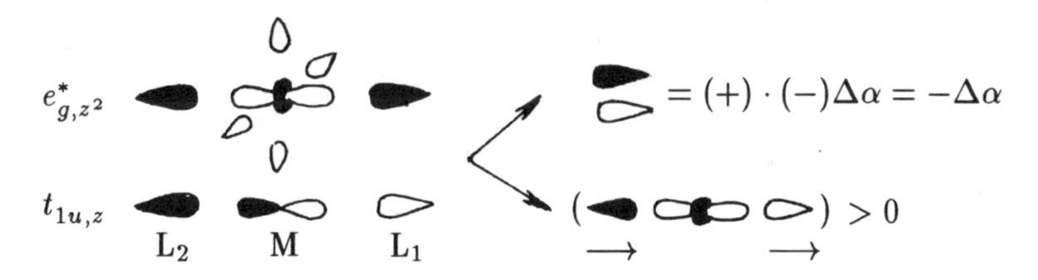

Positive MO regions are as usual drawn in white, and negative ones, in black. The z axis is directed horizontally from left to right. At the top right of the scheme, an estimate of S is given. The coefficients of the leaving ligand L_1 σ AO in the HOMO and LUMO have opposite signs. Therefore for S, which, according to (2.39), equals the product of the coefficients of σ_1 in both MOs by $\Delta\alpha$, we have $S \sim -\Delta\alpha$. At the bottom right of the scheme, we see an estimate of A. The contribution of the $np(M)$ AO to the t_{1u} MO is small. The corresponding off-diagonal matrix element of the one-electron Hamiltonian of the molecule in the basis set of $t_{1u,z}$ and e^*_{g,z^2} MOs is determined by the overlap between the $d_{z^2}(M)$ AO and the $\sigma_1 - \sigma_2$ combination. For ML_6 with the equilibrium configuration, this overlap equals zero, because the positive overlap with the σ_1 orbital is counterbalanced by the negative overlap with $-\sigma_2$. A displacement of ligands along the Q_{12} mode ($Q_{12} > 0$) shifts the balance to negative overlap values. The balance of the resonance integrals then shifts to the side of positive values, and the derivative of the sum of the corresponding resonance integrals with respect Q_{12}, or A according to (2.28a), is positive, that is $A > 0$. Therefore, $Q_\nu^f = Q_{12}^f \sim SA \sim -\Delta\alpha$, in complete agreement with the previous result.

3.3. 14-Electron Nontransition Element Complexes: Bond Lengths

1. To begin the presentation of the theory of mutual ligand effects in σ-bonded nontransition element molecules and complexes, let us stress once more that among these compounds, we distinguish between complexes with 12 electrons on σ MOs with central atoms in the higher oxidation states (e.g., SnF_6^{2-} and SF_6) and complexes with 14 electrons on σ MOs formed by atoms in low oxidation states (e.g., Se^{IV} and Te^{IV} derivatives). Complexes of these two types exhibit mutual ligand effects drastically different in character, and we will discuss them separately starting with 14-electron complexes, where these effects are similar to *trans*-influence in transition metal complexes. In view of this similarity, it seems pertinent to note that the behavior of 14-electron σ-bonded nontransition element compounds well illustrates the proposition that similar mutual influence of ligands may be characteristic of molecules with different electronic structures, because the same irreducible representation Γ_ν corresponding to modes Q_ν active in mutual ligand effects can be contained in the direct products of various pairs of irreducible representations Γ_n and Γ_m.

As in the preceding section, we will use the approximation of frontier MOs. According to Fig. 2, the frontier MOs of the compounds under consideration are the a_{1g}^* HOMO and the t_{1u}^* LUMO. Rules (2.25) asserts that these MOs can only be mixed by the Q_{10}, Q_{11}, and Q_{12} modes of T_{1u} symmetry; at this stage, we again only consider modes that change bond lengths. The further analysis is well illustrated by Table 5. Applying rules (2.43b) to ML_5X complexes with ligand X at position 1 (Fig. 4) leads us to conclude that the only nonvanishing matrix element of the H_S operator is S_{a^*, z^*}. The $t_{1u,z}^*$ and a_{1g}^* orbitals are in turn only mixed by the Q_{12} mode, and we eventually again obtain formula (3.8), where $Q_\nu^f = Q_{12}^f$, $S = S_{a^*, z^*}$, and $A = A_{a^*, z^*}^{12}$. To determine the signs of S and A, consider the shape of the frontier MOs of a homoligand ML_6 complex. In the basis set of the ns, np, and nd

central atom AOs and the ligand σ AOs, we have

$$\begin{aligned}&\text{LUMO } t_{1u}^{*}: \ \psi_z^{*} = l_t p_z - (c_t/\sqrt{2})(\sigma_1 - \sigma_2),\\ &\text{HOMO } a_{1g}^{*}: \ \psi_a^{*} = l_a s - (c_a/\sqrt{6})(\sigma_1 + \sigma_2 + \sigma_3 + \sigma_4 + \sigma_5 + \sigma_6),\end{aligned} \tag{3.13}$$

whence

$$S = (1/2\sqrt{3})c_a c_t \Delta\alpha, \tag{3.14}$$

$$A = -l_a c_t[\partial\beta(\sigma, s)/\partial R]_0 - (1/\sqrt{3})c_a l_t[\partial\beta(\sigma, p)/\partial R]_0. \tag{3.15}$$

Equations (3.14) and (3.15) show that the signs of S and A can be found without analyzing the electronic structure of the complexes. Clearly, the vibronic coupling constant $A < 0$, and the sign of matrix element S coincides with that of $\Delta\alpha$. This can be put more visually using the language of "pictorial" quantum chemistry as

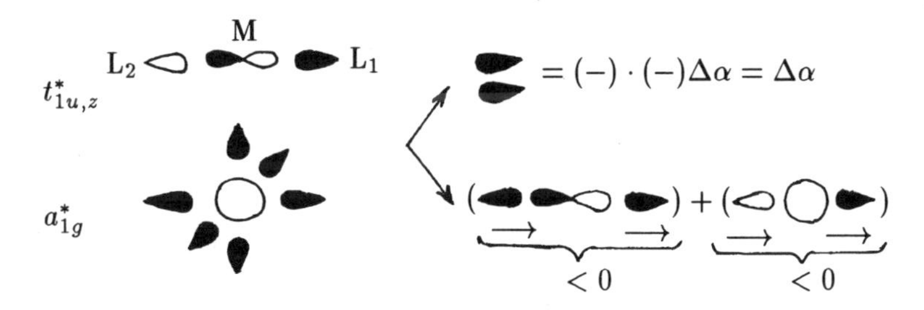

This scheme hardly requires comments because of its similarity to the scheme analyzed in detail in Section 3.2.

To summarize, formula (3.8) allows us to assert that 14-electron σ-bonded octahedral complexes of nontransition elements should exhibit the same substitution effects as transition metal complexes.

2. The mutual influence of ligands in 14-electron nontransition element complexes is usually illustrated by a considerable elongation of the Te—I bond positioned *trans* to the CH_3 group, which is more electropositive than iodine, in the six-coordinate Te(IV) complex. This is the $[CH_3TeI_4]^-$ anion in $[(CH_3)_3Te]^+[CH_3TeI_4]^-$, which is octahedral and has one free position involved in Te...I interaction of length 3.88 Å [84]. In the $[SeOCl_3^-]_n$ anion with six-coordinate Se(IV) as the central atom, the Se—Cl bond *trans* to oxygen experiences similar elongation (in the compound with 8-hydroxyquinolinium, $SeOCl_3^-$ anions form chains containing five-coordinate Se, which participates in weak (3.38 Å) Se...Cl intermolecular interactions along the sixth direction) [84]:

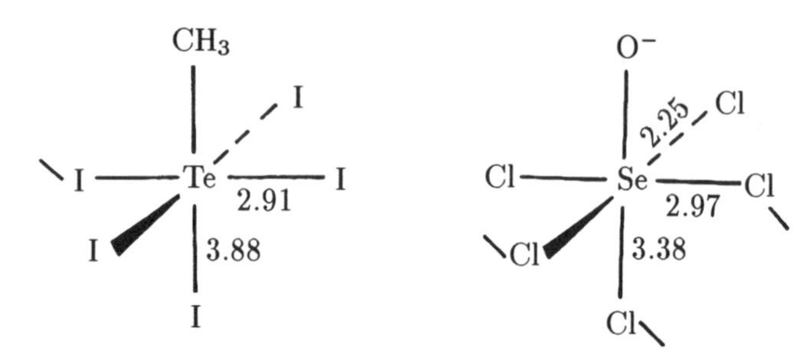

It should, however, be borne in mind that such long M$-$L bonds are akin to intermolecular contacts, and the complexes discussed above are not isolated. For this reason, discussing quasi-octahedral complexes of nontransition elements in low oxidation states is rather of a methodological interest. Nevertheless, there exist examples of "pure" *trans*-influence in isolated nontransition element complexes with central atoms in low oxidation states. These are square-planar complexes of Te(II), which will be considered in Chapter 4.

3.4. 12-Electron Nontransition Element Complexes: Bond Lengths

1. As mentioned, mutual influence of ligands in σ-bonded quasi-octahedral molecules and complexes of nontransition elements with central atoms in the higher oxidation states is of a more involved character than in transition metal complexes or 14-electronic complexes of nontransition elements. Nevertheless, an analysis of mutual influence of ligands is performed following the same standard scheme with the use of Table 5 as in Sections 3.3 and 3.4.

Consider a typical scheme of MOs (Fig. 2) paying attention to the HOMO and LUMO. In 12-electronic complexes, these will be the e_g and a_{1g}^* MOs. Rules (2.25) lead us to conclude that of all ML_6 vibrational modes, the frontier orbitals can be mixed only by the Q_2 and Q_3 ones of E_g symmetry (Fig. 4). Assuming that the ligand in position 1 is replaced (Fig. 4) and applying selection rule (2.43b), we see that the only nonvanishing matrix element of the substitution operator is $S_{z^2,\,a^*}$, and of all frontier orbitals, the H_S operator only mixes the $e_{g,\,z^2}$ and a_{1g}^* MOs. Of two vibrational modes of E_g symmetry, only Q_3 contributes to the mixing. This again gives equation (3.8), now with $S = S_{z^2,\,a^*}$ and $A = A^3_{z^2,\,a^*}$. To determine the sign of $Q_\nu^f = Q_3^f$, it remains to obtain explicit expressions for S and A. Considering the form of the frontier MOs

$$\begin{aligned} &\text{HOMO } e_g: \ \psi_{z^2} = c_e d_{z^2} + (l_e/\sqrt{12})(2\sigma_1 + 2\sigma_2 - \sigma_3 - \sigma_4 - \sigma_5 - \sigma_6), \\ &\text{LUMO } a^*_{1g}: \ \psi^*_a = l_a s - (c_a/\sqrt{6})(\sigma_1 + \sigma_2 + \sigma_3 + \sigma_4 + \sigma_5 + \sigma_6) \end{aligned} \tag{3.16}$$

we easily find that

$$S = -(1/3\sqrt{2})c_a l_e \Delta\alpha, \tag{3.17}$$

$$A = l_a l_e[\partial\beta(\sigma, s)/\partial R]_0 - (1/\sqrt{2})c_a c_e[\partial\beta(\sigma, d)/\partial R]_0, \tag{3.18}$$

or, in the "pictorial" language,

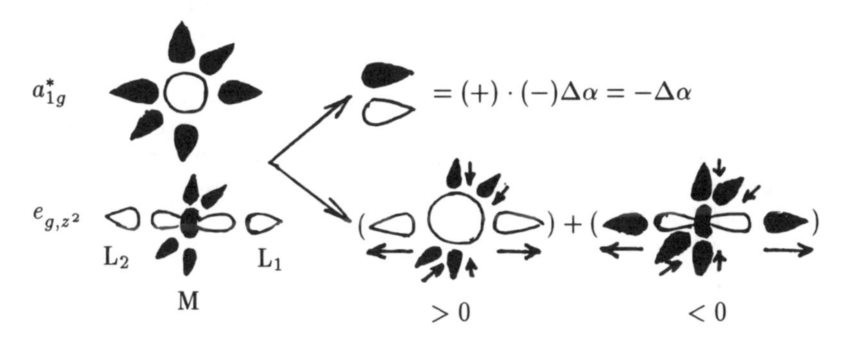

(Unlike the two preceding schemes, this one includes mode Q_3 rather than Q_{12}.) Equations (3.17) and (3.18) and their "pictorial" equivalent show that the sign of S is opposite to that of $\Delta\alpha$, but the sign of vibronic coupling constant A is not obvious *a priori*, because (3.18) contains two competing terms of opposite signs. To elucidate this situation, we must consider the character of chemical bonds in the complexes in more detail.

2. Recall that nontransition element ML_n complexes can be divided into two classes depending on the relative positions of ligand σ levels and the valence levels of central atom M (see Section 1.4). Systems of the first class include complexes formed by typical ligands and elements M situated in the left part of the Periodic Table of the Elements. In these complexes, the ns AOs of the central atom have energies approximately the same as or even higher than those of the σ AOs of typical ligands (Figs. 2 and 3). The second class includes complexes of elements M from the right part of the Periodic Table. The valence ns AOs of these elements have lower energies than the σ AOs of ligands, and central atom np AOs lie at approximately the same level as ligand σ AOs.

One more circumstance should be taken into account. The largest contribution to bonding with ligands is made by the *ns* and *np* nontransition element orbitals rather than its vacant *nd* AOs. In other words, the c_e coefficient in (3.16) is small compared with c_a, l_a, and l_e. This means that for complexes of the first class, the $c_a c_e$ product in (3.18) is small in comparison with $l_a l_e$. As a consequence, the first term dominates in (3.18), and $A > 0$. It then follows from (3.8) that the sign of the Q_3^f mode active in mutual ligand effects should be the same as that of S, that is, opposite to the sign of $\Delta\alpha$. Taking into account the form of the Q_3 mode, we arrive at the conclusion that when ligand L in ML_6 is replaced by a more electropositive ligand X, the octahedron shrinks along axis z and expands in the equatorial plane. Conversely, substituting a more electronegative ligand for L stretches the octahedron along z and contracts it in the equatorial directions.

It follows from additivity (3.12) of coordinates Q_ν^f with respect to substituents that the effect should be twice as large when two ligands L situated at the opposite octahedron vertices undergo replacement, $ML_6 \longrightarrow$ *trans*-ML_4X_2:

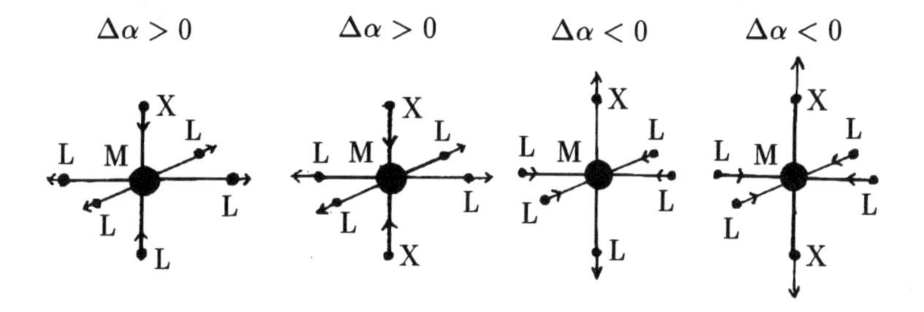

3. As mentioned, mutual influence of ligands of precisely this type, which is not very accurately termed *cis*-influence, is observed in compounds of nontransition elements situated in the left part of the Periodic Table of the Elements. The phenomenon was for the first time observed in the IR study of the *trans*-$SnL_4(CH_3)_2$ complexes, where L = F, Cl, Br, and NCS [65]. Later, numerous similar examples were discovered in X-ray structure studies of Sn(IV) complexes and complexes of some other elements such as Pb(IV) and Ga(III). Some of the bond length values are given in the schemes [85, 63]

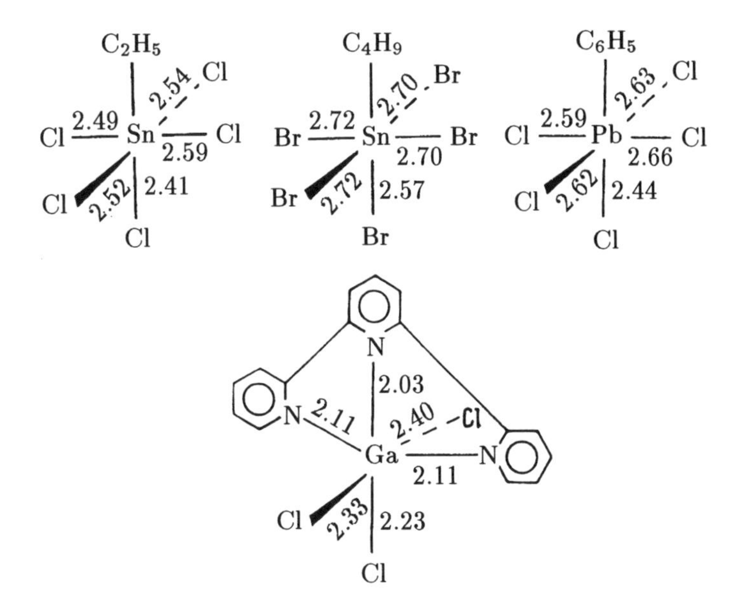

Depicted in the schemes are the $[EtSnCl_5]^{2-}$ and $[BuSnBr_5]^{2-}$ complexes in crystalline $(Me_4N)_2[EtSnCl_5]$ and $(Me_4N)_2[BuSnBr_5]$ and the $[PbPhCl_5]^{2-}$ complex in crystalline $Cs_2PhPbCl_5$. Note for comparison that the mean Sn—Cl and Sn—Br distances in $SnCl_6^{2-}$ and $SnBr_6^{2-}$ ($(Me_4N)_2SnHal_6$ crystals) are 2.45 and 2.62 Å, respectively, and the mean Pb—Cl distance in $PbCl_6^{2-}$ ($CsPbCl_6$ crystals) is 2.51 Å.

All these *cis*-influence examples are provided by Group III and IV element compounds. Although substantially weaker, *cis*-effects were also observed in Group V element derivatives [86]. Structural data are available for two monosubstituted phosphorus complexes [87, 88]

NH3 Py
F F
F—P—F F—P—F
F F
F F

In both compounds, the axial *trans* (with respect to the more electropositive substituent) P—F bond is shortened by 0.01–0.02 Å in comparison with the equatorial bonds. The axial P—F bond length in the complex with ammonia equals 1.581 Å, whereas the equatorial bond lengths are in the range 1.589–1.600 Å. In the phosphorus complex with pyridine, the bond length between P and fluorine positioned *trans* to pyridine is 1.594 Å, whereas P—F bond lengths in the equatorial plane are of 1.606–1.609 Å. Substitution-induced distortions of the same character are observed in the arsenic and antimony compounds [89, 90]

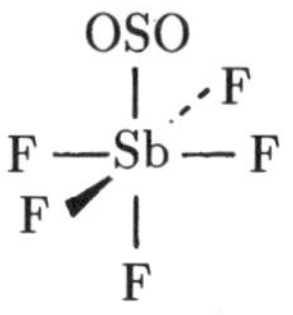

NMeOSF$_2$ / F—As—F with F, F, F (As complex); OSO / F—Sb—F with F, F, F (Sb complex)

Here, the *trans* M—F bonds are shorter by ~0.01 Å than the *cis* bonds. Note also *cis* effects in other similar complexes of Sb(V), the structure of which was determined in [91–96], such as $SbCl_5Ph^-$ and $SbBr_5Ph^-$ in some crystals, $SbCl_4Ph_2^-$, the $SbCl_5DPSO$ and $SbCl_3Ph_2DPSO$ diphenylsulfoxide complexes, and the complex

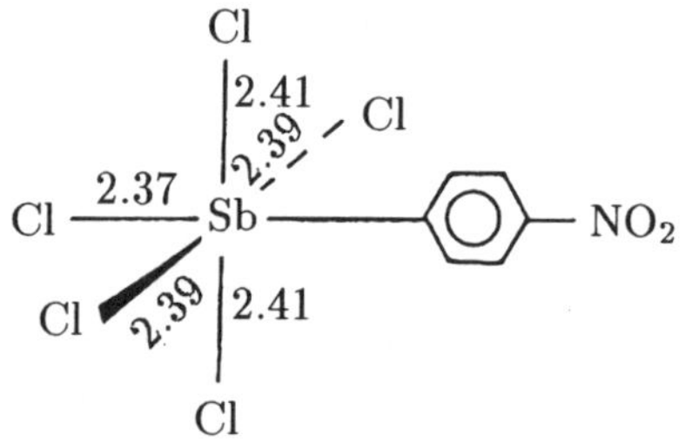

Cl (2.41), Cl (2.39), Cl (2.37)—Sb—C_6H_4—NO_2, Cl (2.39), Cl (2.41)

4. Up to this point, we have only considered 12-electron complexes of nontransition elements whose *ns* AOs have energies higher than or approximately the same as those of ligand σ AOs. Below, we discuss complexes with central atom *ns* AO energies lower that those of ligand σ AOs. Figure 3 shows that the formation of such complexes can be expected for the elements situated in the right halves of nontransition element series, that is, Group VI and VII elements and, under certain conditions, Group V elements. It is easy to understand that the change in the relative arrangement of the *ns*(M) and σ(L) levels can cause a change in the character of deformations of octahedral ML_6 complexes that accompany the $ML_6 \longrightarrow ML_5X$ substitution. It follows from (3.8), (3.16), and (3.17) that the type of mode determining the deformation and the sign of the S matrix element are independent of the position of the *ns*(M) level. However the sign of the vibronic coupling constant A (3.18) may change as a result of a decrease in the energy of central atom *ns* AOs, because the c_a coefficient in (3.16) and (3.18), that is, the contribution of the *ns* AO of atom M to the a_{1g} bonding MO, should

then increase, and, accordingly, the l_a coefficient should decrease.

Clearly, at a fairly low energy of the *ns* AO of atom M, the first term in (3.18) should become smaller than the second, although purely qualitative considerations do not allow us to specify complexes in which such an inversion might occur. The sign of constant A then changes, and the sign of mode Q_3^f coincides with that of $\Delta\alpha$. Taking into account the form of mode Q_3 (Fig. 4), we come to the conclusion that in such complexes, the $ML_6 \longrightarrow ML_5X$ substitution with replacement of ligand L_1 by a more electropositive one should cause $M-L_{trans}$ bond elongation rather than shortening in comparison with $M-L_{cis}$ bonds. Conversely, the introduction of a more electronegative ligand X should cause shortening of the *trans* bond in comparison with the *cis* ones. In the intermediate region between positive and negative constant A values, i. e., when this constant is zero, the *cis* and *trans* bond lengths in ML_5X should coincide more or less regardless of $\Delta\alpha$, that is, of substituent X.

5. In spite of their qualitative character, these conclusions are substantiated by experiment. For instance, in IOF_5, the *trans* bond is appreciably longer than the *cis* link,

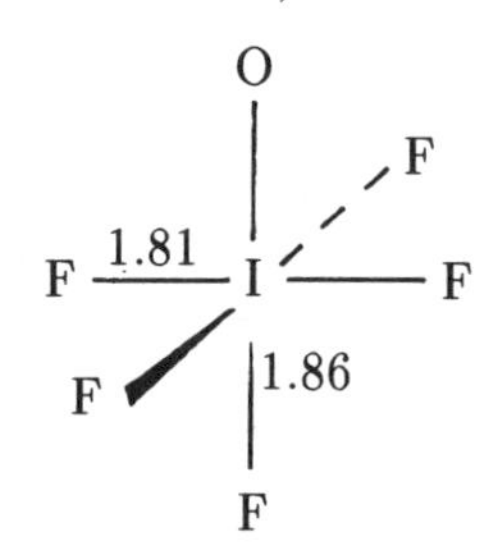

(some increase in I—O bond order is of little importance and, in the approximation of frontier orbitals, does not affect bond lengths at all; more details are given in Section 6.1). It should be stressed that this type of mutual influence of ligands corresponds with the position of the $5s$(I) valence level below the $2p$(F) one.

Figure 3 shows that the energy of the *ns* central atom valence AO increases from Group VII to Group VI elements. It should therefore be expected that steric changes in MF_6 molecules caused by the $MF_6 \longrightarrow MF_5X$ substitution, where M is a Group VI element, will be less manifest than in Group VII element compounds. Even the absence of any such manifestations cannot be ruled out. Indeed, the electron diffraction data [67, 97] show that in the SF_5Cl molecule, the lengths of the *cis* and *trans* S—F bonds coincide within 0.01 Å and equal the S—F bond length in SF_6,

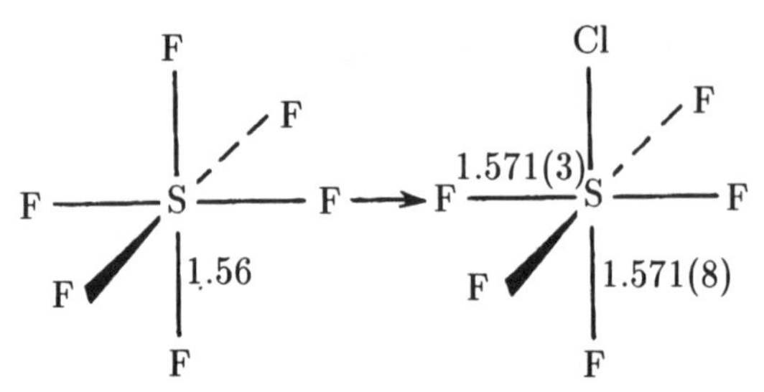

Similarly, the M — F bond lengths remain unchanged after the $MF_6 \longrightarrow MF_5X$ substitution in $SF_5(NCO)$, $SeF_5(NCO)$, $TeF_5(NCO)$ [98], and SF_5CF_3 [99] studied by the method of gas phase electron diffraction (the S — F, Se — F, and Te — F distances in SF_6, SeF_6, and TeF_6 are 1.56, 1.69, and 1.82 Å, respectively):

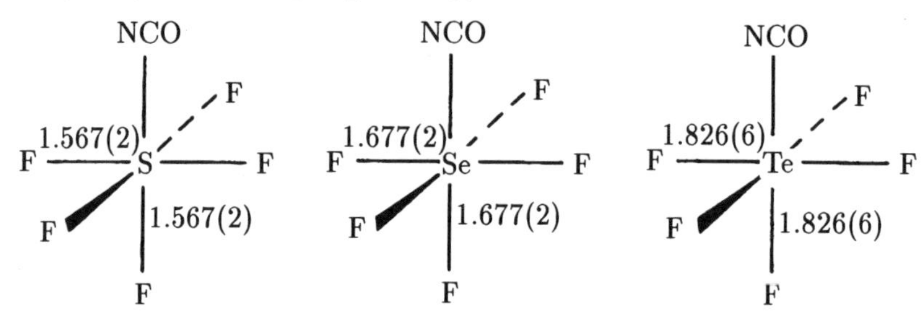

Only when the CF_3 group is introduced into SF_6, is a 0.01 Å difference of the *cis* and *trans* S — F bond lengths observed:

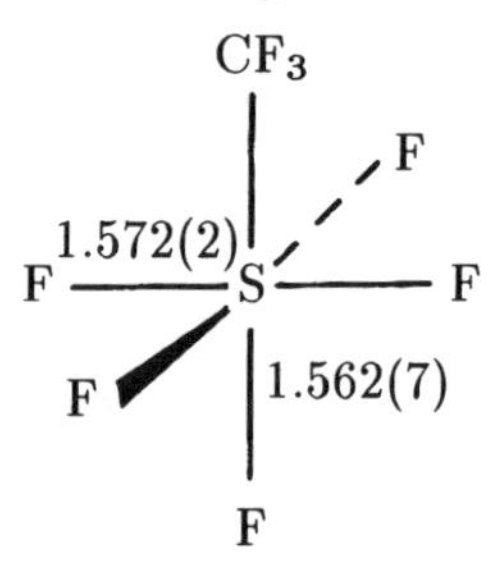

The considerations given above also explain why *cis*-influence is noticeably weakened in Group V element compounds. Because of the intermediate position of Group V between Groups IV and VI, mutual influence of ligands in P(V), As(V), and Sb(V) complexes, although retains the character of *cis*-effects, is close in magnitude to zero effects in S(VI), Se(VI), and Te(VI) complexes. However in the presence of some outer-sphere cations, we observe a *trans*-influence in Sb(V) complexes as simultaneous elongation of both *cis* and *trans* bonds with a relative increase in *trans* bond lengths by several hundredths of Å. Such effects were reported for $[SbPhHal_5]^-$ in $M^{(I)}(SbPhCl_5)$ (M(I) = Cs and NH_4), $Me_2NH_2(SbPhCl_5)$, and $M^{(I)}(SbPhBr_5)$ (M(I) = K and NH_4) [91]. These observations can easily be explained in terms of the theory taking into account all valence MOs (Section 3.6) rather than only frontier ones.

3.5. The Nature of Mutual Influence of Ligands

The simple examples considered in the preceding section can conveniently be used to elucidate the general problems related to mutual influence of ligands; some of them will be discussed in this section. The first question that arises is that of the origin of ligand effects. What are the forces that cause changes in the lengths of *trans* bonds in complexes of types a and b or the characteristic Q_3 deformation of coordination octahedra in complexes of type c (Table 5)? As mentioned, the minimum of the potential energy surface of an ML_n system shifts as a result of the $ML_n \longrightarrow ML_{n-k}XY...Z$ substitution because of the appearance of a component linear in Q_ν in expressions (2.47) and (2.55) for the lower sheet of the adiabatic potential,

$$E_2^{(1)}(Q) = -2\sum_{\nu}\sum_{j\neq 0}\frac{S_j A_j^\nu}{\omega_j}Q_\nu = -4\sum_{\nu}\sum_{n\neq m}\frac{S_{nm}A_{nm}^\nu}{\omega_{nm}}Q_\nu .$$

Because of this component, the Q_0 point, which is a minimum for the ML_n system, now lies on a slope of the potential energy surface. Along this slope, the system "rolls down" to a new minimum with coordinates (2.49) or, which is the same, (2.56). The position of the new minimum is determined by competition between the contributions to the total energy linear and quadratic in Q_ν. At small Q_ν, the linear contribution predominates; this contribution is negative and directed down the slope. At large Q_ν, the major contribution is made by the quadratic term. The new minimum is attained at the point where the rates of both contribution variations, that is, the derivatives of these contributions with respect to Q_ν, are equal in magnitude and have opposite signs, which causes the $\partial\mathcal{E}(Q)/\partial Q_\nu$ derivatives to vanish. It follows that the appearance of the $E_2^{(1)}$ linear component of the total energy can be considered the physical reason for mutual influence of ligands.

It is worth while stressing once more that precisely the linear dependence of energy $\mathcal{E}$ on Q_ν (at small Q_ν values) is responsible for the characteristic difference between mutual influence of ligands, which is ligand-substitution-induced deformation of molecular systems, and the strong pseudo-Jahn–Teller effect, which is spontaneous deformation. As mentioned earlier, the strong pseudo-Jahn–Teller effect occurs not always but only if certain inequalities are met, because the expansion of total energy $\mathcal{E}(Q)$ of an ML_n system at point Q_0 begins with terms quadratic in Q_ν. Conversely, for $ML_{n-k}XY...Z$ systems, this expansion begins with linear terms, which always results in deformation no matter what the width of energy gaps or force and vibronic coupling constant values. However according to (2.49) and (2.56), the A, ω, and K constant values affect the Q_ν^f distortion value similarly to the strong pseudo-Jahn–Teller effect.

2. The linear component of the adiabatic potential of a heteroligand system is of importance not only for explaining the phenomenon of mutual influence of ligands; it can also be used to study the character of this influence [19]. For instance, consider ML_5X complexes a, b, and c represented in Table 5. The linear component that we are considering may include six stretching modes (Q_1, Q_2, Q_3, Q_{10}, Q_{11}, and Q_{12}) with certain coefficients. Applying the approximation of frontier MOs and selection rules (2.43a) and (2.43b) reduces the number of modes active in mutual ligand effects to a single one, and each type of complexes is characterized by the mode of its own (see the last column in Table 5). As a result, the linear component takes the form $E_2^{(1)}(Q) = -4SA^\nu Q_\nu/\omega$. The only characteristic of the mode not defined yet is its sign, which is determined by the requirement that the deformation should decrease the energy of the complex, that is, that $E_2^{(1)}$ be negative. The sign of coefficient SA^ν of coordinate Q_ν is found from (3.10) and (3.11) for complexes of type a or similar formulas for complexes of types b and c. Following this line of reasoning, we arrive at the same conclusions on the character of mutual influence of ligands as those based on (3.8). This approach only differs from the one applied in Sections 3.2–3.4 in that the absolute Q_ν^f deformation value cannot be found from consideration of the $E_2^{(1)}$ linear contribution alone. Indeed, the terms quadratic in Q_ν should also be taken into account.

3. In more familiar terms, the physical meaning of the $E_2^{(1)}$ linear total energy component is as follows. As this component describes the potential energy of atomic nuclei in the neighborhood of point Q_0, its derivatives taken with the opposite sign, $-(\partial E_2^{(1)}/\partial Q_\nu)_0$, can be treated as forces acting on the system of atomic nuclei. Each such "force" acts on several nuclei, because normal coordinates describe collective displacements of atoms (in analytical mechanics, such forces are called generalized). In terms of forces, it can be said that the reason for geometric deformations of ML_n systems that accompany the $ML_n \longrightarrow ML_{n-k}XY...Z$ substitution is the appearance of forces acting on ligands.

In the approximation of frontier MOs, the generalized force is directed along Q_{12} in complexes of types a and b and along Q_3 in complexes of type c, and its sign is determined by the sign of the numerator in (3.8). From the forms of the normal coordinates (Table 2 and Fig. 4), it is easy to find formulas expressing the generalized forces in terms of usual forces acting on separate atoms (ligands) along Cartesian axes. The corresponding representation is actually given in Fig. 4, where each arrow corresponds to a force acting on a separate ligand. The figure shows what character of mutual influence of ligands corresponds to forces acting on ligands in complexes of types a, b, and c. Generalized and "usual" forces can also easily be found taking into account all valence MOs rather than only frontier ones. For the theory of mutual influence of ligands, of special interest are the F_{cis} and F_{trans} forces acting on the *cis* and *trans* ligands, because a comparison of these forces allows us to draw conclusions on the relative magnitudes and signs of *cis* and *trans* bond length changes. The approach based on the "method of forces" is described in detail in Section 3.6.

4. The physical nature of mutual influence of ligands is often treated in terms of electron density ρ redistributions. Intuitively, it appears likely that electron density changes $\Delta\rho$ accompanying ligand substitution are the reason for the appearance of forces that distort the geometry of homoligand systems. Indeed, according to the well-known Hellmann–Feynman theorem [7], a force acting on each atomic nucleus in a molecular system equals the sum of electrostatic forces of repulsion from the other atomic nuclei and attraction to the electronic cloud. It is, however, difficult to say which $\Delta\rho$ components are responsible for the forces acting on ligands. Computationally, the determination of these forces according to the laws of electrostatics poses no problem. But within the framework of qualitative considerations based on the LCAO approach, it is difficult to *a priori* find the source of these forces. It will be shown in Section 3.6 that mutual influence of ligands is related to changes in overlap populations, that is, electronic cloud regions in the space between atoms, rather than atomic charge changes.

5. Because of high symmetry of typical homoligand complexes ML_n, octahedral, tetrahedral, or square-planar, symmetry considerations are of importance in the theory of mutual influence of ligands. Of importance is not so much spatial symmetry (point symmetry group) as the symmetry properties of wave functions described by the irreducible representations of this group. For this reason, ML_5X complexes of types a or b and c exhibit essentially different mutual ligand effects, although the corresponding homoligand derivatives ML_6 have the same O_h point symmetry group. The reason for this difference is the difference in the symmetry properties of the frontier MOs. The HOMO and LUMO of

type a complexes transform under the T_{1u} and E_g representations, respectively, whereas those of type c complexes, under E_g and A_{1g}. In reality, the character of mutual influence of ligands is determined by an even subtler circumstance, *viz.*, the separation of the direct product of the irreducible representations corresponding to the frontier MOs into the sum of irreducible representations (see selection rules (2.25) and (2.43a)). For this reason, complexes of types a and b exhibit similar mutual ligand effects (*trans*-effects), although the symmetry types of the frontier orbitals in these complexes are different (T_{1u} and E_g in a and A_{1g} and T_{1u} in b).

Note one more property of mutual influence of ligands quite manifest in the examples considered above. Steric distortions of ML_n homoligand systems induced by the $ML_n \longrightarrow ML_{n-k}XY...Z$ substitution do not go too far, and the point symmetry group of the $ML_{n-k}XY...Z$ system remains as determined by the difference between ligands L, X, ..., Z. For instance, a one-ligand $ML_6 \longrightarrow ML_5X$ substitution in an octahedral complex of type a lowers symmetry to C_{4v}. The Q_{12} mode active in mutual ligand effects does not distort the ML_5X complex beyond this point symmetry group. A similar behavior is characteristic of complexes of type c. The Q_3 mode active for these complexes also causes a reduction in symmetry not further than to C_{4v}. On the whole, mutual influence of ligand satisfies the well-known Curie principle, according to which dissymmetry (symmetry descent) of consequences is contained in causes responsible for these consequences; for mutual ligand effects, symmetry of consequences equals symmetry of causes. The latter statement might seem trivial but for the observation that both the (static) Jahn–Teller effect and the strong pseudo-Jahn–Teller effect do not satisfy the Curie principle. Curiously, for mutual influence of ligands, this principle can not only be stated but in a sense "proved" by taking into account the symmetry properties of wave functions and normal modes. The above reasoning together with Table 5 in essence constitute such a proof (or, more exactly, specify the mechanism of action) of the Curie principle for quasi-octahedral complexes. Note that the Curie principle in turn facilitates an analysis of mutual influence of ligands for actual molecular systems. For instance, applying the Curie principle to ML_5X complexes with substituent X in position 1 allows us to conclude immediately (without recourse to Table 5) that the modes active in mutual ligand effects should be Q_1, Q_3, and Q_{12} but not Q_2 or Q_{11}.

6. In considering substituted complexes such as quasi-octahedral ML_5X complexes, a distinction is often drawn between directional mutual influence of ligands, when ligand X only affects ligands L in certain

positions with respect to it, and nondirectional influence involving all ligands. From the phenomenological standpoint, there can be no objections to such a terminology. It should, however, be clearly recognized that from the point of view of the mechanism of mutual influence of ligands, this terminology does not correspond with the essence of the problem, not only because ligand effects cannot be described either as direct or indirect (through the central atom) influence of one ligand on another. In reality, there is no difference in principle between directional and nondirectional effects, because the difference between these phenomena is determined by the type of the Q_ν mode dominating mutual influence of ligands rather than some difference in their natures, and the type of the predominant mode is in turn related to the character of frontier MOs. In quasi-octahedral ML_5X complexes, directional *trans*-influence results from the predominance of the contribution of the T_{1u} modes, whereas nondirectional influence is observed when the E_g and (as will be shown below) A_{1g} modes predominate.

3.6. Mutual Influence of Ligands in the Basis of All σ-MOs

1. In the preceding sections, the approximation of frontier MOs was used to give a clearer idea of the main points. Mutual influence of ligands can, however, be treated with the use of the basis set of all MOs built of valence (sometimes also vacant) central atom AOs and ligand σ orbitals [100]. For the first time, such an analysis was conducted in [46, 47] in terms of overlap populations on M—L bonds as a criterion of mutual influence of ligands; we will follow these works in our presentation.

This more complete treatment of mutual influence of ligands is not only of methodological interest. The results obtained in such a way complement and sometimes modify conclusions drawn in the approximation of frontier MOs. We stress that in principle, abandoning the approximation of frontier orbitals does not introduce fundamentally new features into the theory. The starting point is, as previously, equation (2.56), where we should take into account a larger number of terms. In addition, in each molecule, there will be several rather than one modes contributing to mutual influence of ligands. This immediately follows from the selection rules (2.25). Figures 1 and 2 show that occupied and unoccupied MOs of octahedral complexes can be mixed by modes of the A_{1g}, E_g, and T_{1u} types. Applying selection rules (2.43) for the matrix elements of the substitution operator decreases the number of such modes (as in the approximation of frontier MOs). Nevertheless, there remain several modes. For instance, for octahedral complexes singly substituted at position 1, the Q_1, Q_3, and

Q_{12} modes are active; these modes do not distort the resulting ML_5X heteroligand complex to lower than C_{4v} symmetry.

2. The equilibrium configuration of an $ML_{n-k}XY...Z$ system can in principle be calculated provided all values in (2.49) and (2.56) including force constants K_ν of normal modes are known. However qualitatively, all stretching force constants of ML_n may be considered equal. This assumption is not radically at variance with experiment and, as will be seen, leads to correct conclusions. (Another approach free of this assumption is described in Section 4.1.) In this approximation, the steric consequences of the $ML_n \longrightarrow ML_{n-k}XY...Z$ substitution can easily be estimated by analyzing the term linear in Q_ν in equation (2.55) for the adiabatic potential of the substituted complex.

We have already mentioned that the coefficient of Q_ν in (2.55) can be interpreted as a generalized force [101] acting on ligands along the Q_ν coordinate and responsible for ML_n system deformation accompanying the $ML_n \longrightarrow ML_{n-k}XY...Z$ substitution. In quasi-octahedral ML_5X complexes, *cis*- and *trans*-effects can conveniently be compared after recalculating these generalized forces into usual ones acting on the *cis* and *trans* ligands along Cartesian coordinates. More noticeable changes are to be expected for the bonds that experience the action of forces larger in magnitude. To perform the required calculations, let us use Table 1 and write explicit expressions for the MOs that can be formed from the s, p, and d AOs of atom M and ligand σ AOs. We have

$$\begin{aligned}
\psi_a &= c_a s + (l_a/\sqrt{6})(\sigma_1+\sigma_2+\sigma_3+\sigma_4+\sigma_5+\sigma_6),\\
\psi_a^* &= l_a s - (c_a/\sqrt{6})(\sigma_1+\sigma_2+\sigma_3+\sigma_4+\sigma_5+\sigma_6),\\
\psi_{e,z^2} &= c_e d_{z^2} + (l_e/2\sqrt{3})(2\sigma_1+2\sigma_2-\sigma_3-\sigma_4-\sigma_5-\sigma_6),\\
\psi_{e,z^2}^* &= l_e d_{z^2} - (c_e/2\sqrt{3})(2\sigma_1+2\sigma_2-\sigma_3-\sigma_4-\sigma_5-\sigma_6),\\
\psi_{t,z} &= c_t + (l_t/\sqrt{2})(\sigma_1-\sigma_2),\\
\psi_{t,z}^* &= l_t - (c_t/\sqrt{2})(\sigma_1-\sigma_2).
\end{aligned} \tag{3.19}$$

In (3.19), two (bonding and antibonding) MOs are given for each irreducible representation.

Let us consider ML_5X complexes with substituent X in position 1 (see Fig. 4). The sum of the linear terms in (2.55) is

$$E_2^{(1)} = 4\sum_{n,m} W_{nm}, \tag{3.20}$$

$$W_{nm} = -\sum_{\nu} \frac{S_{nm}A_{nm}^{\nu}}{\omega_{nm}} Q_\nu, \tag{3.21}$$

This sum virtually only includes the Q_1, Q_3, and Q_{12} modes,

$$\begin{aligned}
Q_1 &= (1/\sqrt{6})(Z_1 - Z_2 + X_3 - X_4 + Y_5 - Y_6) \\
&= (1/\sqrt{6})q_1, \\
Q_3 &= (1/2\sqrt{3})(2Z_1 - 2Z_2 - X_3 + X_4 - Y_5 + Y_6) \\
&= (1/2\sqrt{3})q_3, \\
Q_{12} &= (1/\sqrt{2})(Z_1 + Z_2) = (1/\sqrt{2})q_{12},
\end{aligned} \tag{3.22}$$

where q_1, q_3, and q_{12} are the non-normalized modes of the A_{1g}, $E_{g,\,z^2}$, and $T_{1u,\,z}$ types.

First consider 12-electron nontransition element complexes with 12 electrons on σ MOs. There are nine different W_{nm} values corresponding to different transitions from occupied to unoccupied MOs. Using (3.19) and formulas (2.39) and (2.27) to determine S_{nm} and A^{ν}_{nm} and taking into account (3.22), we obtain

$$\begin{aligned}
W_{aa^*} &= \frac{1}{6}\Big(\frac{l_a^2}{\omega_{aa^*}} - \frac{c_a^2}{\omega_{aa^*}}\Big)\mu_a \Delta\alpha q_1, \\
W_{ae^*} &= \frac{1}{6}\Big(-\frac{c_e^2}{\omega_{ae^*}}\mu_a + \frac{l_a^2}{\omega_{ae^*}}\mu_e\Big)\Delta\alpha q_3, \\
W_{at^*} &= \frac{1}{2}\Big(-\frac{c_t^2}{\omega_{at^*}}\mu_a + \frac{l_a^2}{\omega_{at^*}}\mu_t\Big)\Delta\alpha q_{12}, \\
W_{ea^*} &= \frac{1}{6}\Big(\frac{l_e^2}{\omega_{ea^*}}\mu_a - \frac{c_a^2}{\omega_{ea^*}}\mu_e\Big)\Delta\alpha q_3, \\
W_{ee^*} &= \frac{1}{3}\Big(\frac{l_e^2}{\omega_{ee^*}} - \frac{c_e^2}{\omega_{ee^*}}\Big)\mu_e \Delta\alpha q_1 + \frac{1}{6}\Big(\frac{l_e^2}{\omega_{ee^*}} - \frac{c_e^2}{\omega_{ee^*}}\Big)\mu_e \Delta\alpha q_3, \\
W_{et^*} &= \Big(-\frac{c_t^2}{\omega_{et^*}}\mu_e + \frac{l_e^2}{\omega_{et^*}}\mu_t\Big)\Delta\alpha q_{12}, \\
W_{ta^*} &= \frac{1}{2}\Big(\frac{l_t^2}{\omega_{ta^*}}\mu_a - \frac{c_a^2}{\omega_{ta^*}}\mu_t\Big)\Delta\alpha q_{12}, \\
W_{te^*} &= \Big(\frac{l_t^2}{\omega_{te^*}}\mu_e - \frac{c_e^2}{\omega_{te^*}}\mu_t\Big)\Delta\alpha q_{12}, \\
W_{tt^*} &= \frac{1}{2}\Big(\frac{l_t^2}{\omega_{tt^*}} - \frac{c_t^2}{\omega_{tt^*}}\Big)\mu_t \Delta\alpha q_1 + \frac{1}{2}\Big(\frac{l_t^2}{\omega_{tt^*}} - \frac{c_t^2}{\omega_{tt^*}}\Big)\mu_t \Delta\alpha q_3.
\end{aligned} \tag{3.23}$$

Here, the notation

$$\mu_a = \frac{1}{\sqrt{6}} c_a l_a \Big[\frac{\partial \beta(\sigma, s)}{\partial R} \Big]_0,$$
$$\mu_e = \frac{1}{2\sqrt{3}} c_e l_e \Big[\frac{\partial \beta(\sigma, d_\sigma)}{\partial R} \Big]_0, \tag{3.24}$$
$$\mu_t = \frac{1}{3\sqrt{2}} c_t l_t \Big[\frac{\partial \beta(\sigma, p_\sigma)}{\partial R} \Big]_0.$$

is used. Let us rewrite (3.20) in the form

$$\begin{aligned} E_2^{(1)} &= 2\big[(f_{1a}\mu_a + f_{1e}\mu_e + f_{1t}\mu_t) q_1/3 + (f_{3a}\mu_a + f_{3e}\mu_e + f_{3t}\mu_t) q_3/3 \\ &\quad + (f_{12a}\mu_a + f_{12e}\mu_e + f_{12t}\mu_t) q_{12} \big] \Delta\alpha \\ &= 2(1/3 f_1 q_1 + 1/3 f_3 q_3 + f_{12} q_{12}) \Delta\alpha, \end{aligned} \tag{3.25}$$

where the coefficients of q_i play the role of generalized forces acting along q_i, and the $f_{\nu i}$ values are

$$\begin{aligned} f_{1a} &= \Big[\frac{l_a^2}{\omega_{aa^*}} - \frac{c_a^2}{\omega_{aa^*}} \Big], \qquad f_{1e} = 2\Big[\frac{l_e^2}{\omega_{ee^*}} - \frac{c_e^2}{\omega_{ee^*}} \Big], \\ f_{1t} &= 3\Big[\frac{l_t^2}{\omega_{tt^*}} - \frac{c_t^2}{\omega_{tt^*}} \Big], \qquad f_{3a} = \Big[\frac{l_e^2}{\omega_{ea^*}} - \frac{c_e^2}{\omega_{ae^*}} \Big], \\ f_{3e} &= f_{3e}^{(1)} + f_{3e}^{(2)} = \Big[\frac{l_e^2}{\omega_{ee^*}} - \frac{c_e^2}{\omega_{ee^*}} \Big] + \Big[\frac{l_a^2}{\omega_{ae^*}} - \frac{c_a^2}{\omega_{ea^*}} \Big], \\ f_{3t} &= 3\Big[\frac{l_t^2}{\omega_{tt^*}} - \frac{c_t^2}{\omega_{tt^*}} \Big], \qquad f_{12a} = \Big[\frac{l_t^2}{\omega_{ta^*}} - \frac{c_t^2}{\omega_{at^*}} \Big], \\ f_{12e} &= 2\Big[\frac{l_t^2}{\omega_{te^*}} - \frac{c_t^2}{\omega_{et^*}} \Big], \\ f_{12t} &= f_{12t}^{(1)} + f_{12t}^{(2)} = 2\Big[\frac{l_e^2}{\omega_{et^*}} - \frac{c_e^2}{\omega_{te^*}} \Big] + \Big[\frac{l_a^2}{\omega_{at^*}} - \frac{c_a^2}{\omega_{ta^*}} \Big]. \end{aligned} \tag{3.26}$$

As mentioned, $E_2^{(1)}$ can also be expressed in terms of ligand displacements along Cartesian axes x, y, and z. The role of forces responsible for these displacements will be played by certain combinations of generalized forces present in (3.25). These combinations can easily be found through expressing q_i in terms of Cartesian coordinates X_i, Y_i, and Z_i. We then have

$$E_2^{(1)} = (2/3)\Delta\alpha(3f_{12} - 2f_3 - f_1) Z_2 + (2/3)\Delta\alpha(f_1 - f_3) X_3 + \dots$$

(this expression only contains terms corresponding to the Z displacement of the axial *trans* ligand and the X displacement of one of the equatorial *cis* ligands in the ML_5X complex). The coefficient of Z_2, which will be denoted as F_{trans}, determines the absolute elongation of the *trans* $M-L_2$ bond, and the coefficient of X_3 taken with the opposite sign, which will be denotes as F_{cis}, determines the absolute elongation of any of the *cis* bonds (Fig. 6). The $\Delta F = F_{trans} - F_{cis}$ difference determines the relative elongation of the *trans* bond with respect to the *cis* bonds,

$$F_{trans} = \frac{2}{3}\Delta\alpha(3f_{12} - 2f_3 - f_1),$$
$$F_{cis} = -\frac{2}{3}\Delta\alpha(f_1 - f_3), \qquad \Delta F = 2\Delta\alpha(f_{12} - f_3). \tag{3.27}$$

It is now easy to write the expressions for the F_{cis} and ΔF values, which are of key importance for analyzing mutual influence of ligands, in the explicit form

$$F_{cis} = \frac{2}{3}\Delta\alpha\Big\{\Big[\Big(\frac{l_a^2}{\omega_{ae^*}} - \frac{c_a^2}{\omega_{ea^*}}\Big) - \Big(\frac{l_e^2}{\omega_{ee^*}} - \frac{c_e^2}{\omega_{ee^*}}\Big)\Big]\mu_e$$
$$+ \Big[\Big(\frac{l_e^2}{\omega_{ea^*}} - \frac{c_e^2}{\omega_{ae^*}}\Big) - \Big(\frac{l_a^2}{\omega_{aa^*}} - \frac{c_a^2}{\omega_{aa^*}}\Big)\Big]\mu_a\Big\},$$
$$\Delta F = 2\Delta\alpha\Big\{\Big[\Big(\frac{l_t^2}{\omega_{ta^*}} - \frac{c_t^2}{\omega_{at^*}}\Big) - \Big(\frac{l_e^2}{\omega_{ea^*}} - \frac{c_e^2}{\omega_{ae^*}}\Big)\Big]\mu_a \tag{3.28}$$
$$+ \Big[2\Big(\frac{l_t^2}{\omega_{te^*}} - \frac{c_t^2}{\omega_{et^*}}\Big) - \Big(\frac{l_e^2}{\omega_{ee^*}} - \frac{c_e^2}{\omega_{ee^*}}\Big) - \Big(\frac{l_a^2}{\omega_{ae^*}} - \frac{c_a^2}{\omega_{ea^*}}\Big)\Big]\mu_e$$
$$+ \Big[2\Big(\frac{l_e^2}{\omega_{et^*}} - \frac{c_e^2}{\omega_{te^*}}\Big) + \Big(\frac{l_a^2}{\omega_{at^*}} - \frac{c_a^2}{\omega_{ta^*}}\Big) - 3\Big(\frac{l_t^2}{\omega_{tt^*}} - \frac{c_t^2}{\omega_{tt^*}}\Big)\Big]\mu_t\Big\}.$$

We do not give formulas for F_{trans}, which can be written through ΔF and F_{cis}.

3. Expressions similar to (3.28), which describe transition metal and nontransition element complexes with 12 electrons on σ MOs, can be obtained for 14-electron nontransition element complexes, where the a_{1g}^* orbital is no longer vacant. For this reason, the expression for $E_2^{(1)}$ contains eight rather than nine W_{nm} values corresponding to transitions from occupied MOs to unoccupied ones. Therefore we should exclude W_{aa^*}, W_{ea^*}, and W_{ta^*} from (3.23) but add $W_{a^*e^*}$ and $W_{a^*t^*}$ instead. The latter values have the form

$$W_{a^*t^*} = \frac{1}{2}\Big(\frac{c_t^2}{\omega_{a^*t^*}}\mu_a + \frac{c_a^2}{\omega_{a^*t^*}}\mu_t\Big)\Delta\alpha q_{12},$$
$$W_{a^*e^*} = \frac{1}{6}\Big(\frac{c_e^2}{\omega_{a^*e^*}}\mu_a + \frac{c_a^2}{\omega_{a^*e^*}}\mu_e\Big)\Delta\alpha q_3. \tag{3.29}$$

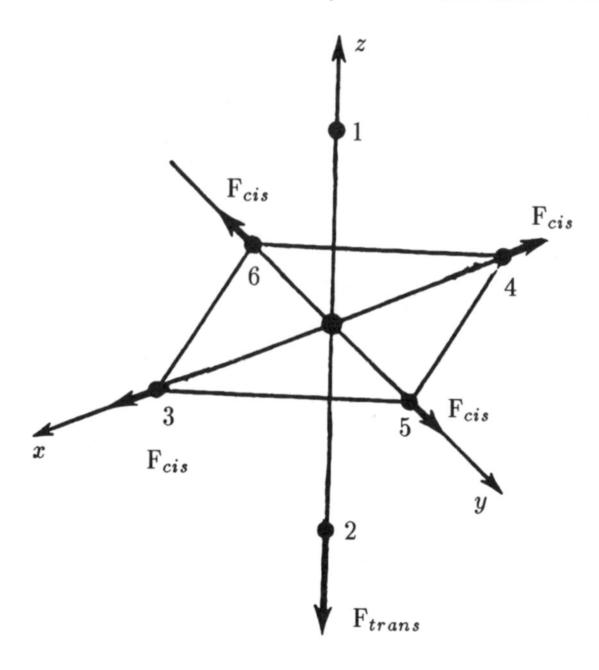

FIG. 6. FORCES ACTING ON *cis* AND *trans* LIGANDS IN OCTAHEDRAL COMPLEXES EXPERIENCING $ML_6 \longrightarrow ML_5X$ SUBSTITUTION.

After these changes, the expressions for F_{cis} and ΔF become

$$\begin{aligned} F_{cis} &= \frac{2}{3}\Delta\alpha\left\{ c_e^2\left(\frac{1}{\omega_{a^*e^*}} - \frac{1}{\omega_{ae^*}}\right)\mu_a \right. \\ &\quad \left. + \left[\left(\frac{l_a^2}{\omega_{ae^*}} + \frac{c_a^2}{\omega_{a^*e^*}}\right) - \left(\frac{l_e^2}{\omega_{ee^*}} - \frac{c_e^2}{\omega_{ee^*}}\right)\right]\mu_e \right\}, \\ \Delta F &= 2\Delta\alpha\left\{ \left[c_t^2\left(\frac{1}{\omega_{a^*t^*}} - \frac{1}{\omega_{at^*}}\right) - c_e^2\left(\frac{1}{\omega_{a^*e^*}} - \frac{1}{\omega_{ae^*}}\right)\right]\mu_a \right. \\ &\quad + \left[2\left(\frac{l_e^2}{\omega_{et^*}} - \frac{c_e^2}{\omega_{te^*}}\right) + \left(\frac{l_a^2}{\omega_{at^*}} + \frac{c_a^2}{\omega_{a^*t^*}}\right) - 3\left(\frac{l_t^2}{\omega_{tt^*}} - \frac{c_t^2}{\omega_{tt^*}}\right)\right]\mu_t \\ &\quad \left. + \left[2\left(\frac{l_t^2}{\omega_{te^*}} - \frac{c_t^2}{\omega_{et^*}}\right) - \left(\frac{l_a^2}{\omega_{ae^*}} + \frac{c_a^2}{\omega_{a^*e^*}}\right) - \left(\frac{l_e^2}{\omega_{ee^*}} - \frac{c_e^2}{\omega_{ee^*}}\right)\right]\mu_e \right\}. \end{aligned} \tag{3.30}$$

4. Let us apply equations (3.28) and (3.30) to analyze mutual influence of ligands. For transition metal complexes, we must use the

typical scheme of MOs. As is seen from Fig. 1, the valence levels of the central atom form the series $(n-1)d < ns < np$, whereas the valence σ AOs of typical ligands have lower energies than all central atom valence levels. We assume that the contributions of metal and ligand AOs to MOs (3.19) are approximately proportional to orbital electronegativities and, therefore,

$$c_e > c_a > c_t, \qquad l_e < l_a < l_t. \tag{3.31}$$

It is worth while mentioning that inequalities (3.31) obtained from qualitative considerations conform with computation results, which, in turn, give a more detailed idea of the ratio between c_i and l_i. According to [20], it would be more correct to write $c_e^2 \gtrsim c_a^2 > c_t^2$. This yields the estimates

$$\mu_e \gtrsim \mu_a > \mu_t, \tag{3.32}$$

because the $c_i l_i$ products decrease in the series

$$c_e l_e > c_a l_a > c_t l_t. \tag{3.33}$$

In addition, the ns and np transition metal AOs are more diffuse than the $(n-1)d$ AOs. It is therefore natural to expect that under atomic spacing variations, the $\beta(\sigma, s)$ and $\beta(\sigma, p_\sigma)$ resonance integrals should change more slowly than the $\beta(\sigma, d_\sigma)$ integral, that is

$$\partial\beta(\sigma, d_\sigma)/\partial R > \partial\beta(\sigma, s)/\partial R \approx \partial\beta(\sigma, p_\sigma)/\partial R. \tag{3.34}$$

Lastly, μ_t contains the smallest numeric multiplier, but the multiplier in μ_e is smaller than that in the expression for μ_a, which in part weakens the inequality $c_e l_e \beta'(\sigma, d_\sigma) > c_a l_a \beta'(\sigma, s)$. Let us use inequalities (3.31) and (3.32) and ignore the small μ_t and c_t^2 values in equation (3.28); in the basis set of metal $(n-1)d$ and ns AOs, these values are at all zero. The expression for ΔF determining the relative elongation of the *trans* bond then takes the form

$$\begin{aligned} \Delta F \approx\ & 2\Delta\alpha\left[\left(\frac{l_t^2}{\omega_{ta^*}} - \frac{l_e^2}{\omega_{ae^*}}\right) + \frac{c_e^2}{\omega_{ea^*}}\right]\mu_a \\ & + 2\Delta\alpha\left[\left(2\frac{l_t^2}{\omega_{te^*}} - \frac{l_e^2}{\omega_{ee^*}} - \frac{l_a^2}{\omega_{ae^*}}\right) + \frac{c_e^2}{\omega_{ee^*}} + \frac{c_e^2}{\omega_{ea^*}}\right]\mu_e. \end{aligned} \tag{3.35}$$

It follows that $\Delta F/\Delta\alpha > 0$, that is $\Delta F > 0$ if $\Delta\alpha > 0$ and $\Delta F < 0$ if $\Delta\alpha < 0$, for it can be inferred from Fig. 1 and inequalities (3.31) that all differences in parentheses are positive. (They remain positive even if the MO lowest in energy is a_{1g} rather then e_g.)

Next, let us show that the relative elongation (at $\Delta\alpha > 0$) or shortening (at $\Delta\alpha < 0$) of the *trans* bond in comparison with the *cis* bonds is accompanied by its absolute elongation or shortening. For this purpose, consider F_{cis} taking into account that in the complexes under discussion, the energy gap between the bonding a_{1g} and e_g levels is small in comparison with the gap between them and the a_{1g}^* and e_g^* antibonding orbitals. Using the notation $\omega_1 = (1/2)(\omega_{ee^*} + \omega_{ae^*})$ and $\omega_2 = (1/2)(\omega_{ea^*} + \omega_{aa^*})$ in (3.28) $(\omega_1 < \omega_2)$, we obtain

$$F_{cis} \approx \frac{2}{3}\Delta\alpha\left[(c_e^2 - c_a^2)\left(\frac{\mu_e}{\omega_1} + \frac{\mu_e}{\omega_2} - 2\frac{\mu_a}{\omega_2}\right) + c_e^2(\mu_e - \mu_a)\left(\frac{1}{\omega_1} - \frac{1}{\omega_2}\right)\right]. \quad (3.36)$$

It follows from (3.36) that $F_{cis} > 0$ if $\Delta\alpha > 0$ and $F_{cis} < 0$ if $\Delta\alpha < 0$.

To summarize, the $ML_6 \longrightarrow ML_5X$ substitution in octahedral σ-bonded transition metal complexes should cause elongation of both *cis* and *trans* bonds if X is a more electropositive ligand than L. Otherwise, shortening of bonds of both types should occur. Bond length variations should be more substantial for *trans* bonds than for *cis* ones. We see that the new results complement those obtained in the approximation of frontier MOs. As previously, the $ML_6 \longrightarrow ML_5X$ substitution causes changes in the length of *trans* bonds, but at the new level of theory, these changes are accompanied by similar although smaller changes in *cis* bond lengths.

5. As concerns mutual influence of ligands in 12-electron nontransition element complexes, recall that depending on the relative positions of the levels of valence central atom AOs and σ ligand AOs, these complexes can be divided into two classes (Section 3.5). Accordingly, we will consider the F_{trans}, F_{cis}, and ΔF values for two extremal cases: (1) when the energy of the *ns* AO of atom M in ML_6 is close to or somewhat higher than the energy of ligand σ AOs and (2) when the *ns* AO of atom M lies well below ligand σ AOs and the energy of *np*(M) AOs is close to that of ligand σ AOs.

Let us first analyze equation (3.28) for complexes of type (1). It follows from the relative arrangement of central atom and ligand AO levels that

$$c_a > c_t > c_e, \quad l_e > l_t > l_a, \quad c_a \approx l_a, \quad \mu_a > \mu_t \gg \mu_e. \quad (3.37)$$

These inequalities mean that in the complexes under consideration, *ns*(M) and ligand σ AOs make approximately equal contributions to bonding, whereas the contributions of *np*(M) and, especially, *nd*(M) AOs should be much smaller. This difference in the contributions of central atom AOs can be modeled by ignoring c_e (and, thereby, μ_e)

in the further estimates while retaining terms linear in c_t and first-, second-, and third-order in c_a. Equation (3.28) then becomes

$$\Delta F \approx 2\Delta\alpha\left[\left(\frac{1}{\omega_{ta^*}}-\frac{1}{\omega_{ea^*}}\right)\mu_a+\left(\frac{2}{\omega_{et^*}}-\frac{3}{\omega_{tt^*}}+\frac{l_a^2}{\omega_{at^*}}-\frac{c_a^2}{\omega_{ta^*}}\right)\mu_t\right]. \quad (3.38)$$

Taking into account the usual scheme of MOs (Fig. 2), we come to the conclusion that the coefficient of μ_a (in parentheses) is negative. If the contribution of np(M) AOs to bonding is not very large and $(\omega_{tt^*}/\omega_{et^*}) < 3/2$, the coefficient of μ_t is also negative. This means that in the complexes under discussion, $\Delta F < 0$ if $\Delta\alpha > 0$ and $\Delta F > 0$ if $\Delta\alpha < 0$. On the same assumptions, we simultaneously have

$$F_{cis} \simeq \frac{2}{3}\Delta\alpha\left[\left(\frac{1}{\omega_{ea^*}}-\frac{1}{\omega_{aa^*}}\right)+\frac{2c_a^2}{\omega_{aa^*}}\right]\mu_a, \quad (3.39)$$

that is, $F_{cis} > 0$ if $\Delta\alpha > 0$ and $F_{cis} < 0$ if $\Delta\alpha < 0$.

These inequalities are in agreement with the results of treatment in the approximation of frontier orbitals. They mean that when ligand L in ML_6 is replaced by a more electropositive ligand X, the *cis* bonds experience both relative and absolute elongation. An analysis of the expression for F_{trans} shows that this elongation may be accompanied by shortening of the *trans* bond. The substitution of a more electronegative ligand X for L leads to opposite steric consequences. However, the change in the *trans* bond length not necessarily equals twice the change in the *cis* one taken with the opposite sign, which distinguishes treatment given in this section from treatment in terms of the approximation of frontier MOs. This is easy to explain, because in the basis of all σ-MOs, several vibrational modes rather than a single Q_3 one contribute to mutual ligand effects. Note that the latter result is substantiated by structural data. In all complexes considered in Section 3.4.3, changes in the lengths of the *trans* bonds are smaller than twice the *cis* bond length changes.

6. In an analysis of mutual influence of ligands in nontransition element complexes formed by M atoms with low-energy *ns* AOs, it should be taken into account that when the energy of the element M *ns* AO decreases, energetic denominators ω_{ea^*} in expression (3.28) for ΔF also decrease. In addition, they decrease when the contribution of the element M *nd* AOs to bonding decreases. For this reason, a low contribution of *nd* AOs does not necessarily imply smallness of the $(c_a^2/\omega_{ea^*})\mu_e$ term. To estimate the sign of ΔF, let us represent ΔF in the form

$$\Delta F \approx 2\Delta\alpha\left\{\frac{c_a^2\mu_e - l_e^2\mu_a}{\omega_{ea^*}}+\left(2\frac{l_e^2}{\omega_{et^*}}-\frac{c_a^2}{\omega_{ta^*}}+\frac{c_t^2-l_t^2}{\omega_{tt^*}}\right)\mu_t+\Delta f'\right\}, \quad (3.40)$$

where $\Delta f'$ stands for the other terms from (3.28). The $\omega_{ea^*} = \varepsilon_a^* - \varepsilon_e$ energy will be written as

$$\omega_{ea^*} = (\varepsilon_a^* - \alpha_\sigma) + (\alpha_\sigma - \varepsilon_e), \tag{3.41}$$

$$(\alpha_\sigma - \varepsilon_e) = \frac{\alpha_d - \alpha_\sigma}{l_e^2 - c_e^2} c_e^2, \tag{3.42}$$

where α_d and α_σ are the Coulomb integrals of the nd(M) and σ(L) AOs. The validity of (3.42) follows from the solution to the two-orbital problem of finding energy ε_e of the e_g MO (Section 1.3).

If the contribution of nd AOs to bonding is small, and the ns level of atom M lies well below the σ level of ligands L, the c_e, l_a, μ_e, and μ_a values and, accordingly, the $\Delta f'$ term in (3.40) are small. However, in the first term in braces, the numerator behaves as $\sim B_1 c_e + \delta_1$, and the denominator, as $\sim B_2 c_e^2 + \delta_2$ (see (3.42)), where constants B_1 and B_2 are positive. (The $\delta_1 = -l_e^2 \mu_a$ and $\delta_2 = \varepsilon_a^* - \alpha_\sigma$ values tend to zero as the energy of the ns(M) AO decreases.) It follows that for sufficiently low-energy ns(M) AOs, the first term in (3.40), which has the same sign as $\Delta\alpha$, determines the sign of the whole right-hand side of the equation. Therefore in ML_5X with $\Delta\alpha > 0$ ($\Delta\alpha < 0$), we observe relative elongation (shortening) of the $M-L_{trans}$ bond in comparison with the $M-L_{cis}$ ones. There is one more reason for such a character of mutual ligand effects in the complexes under consideration. If the contribution of nd(M) AOs to bonding is insignificant and the energy of ns(M) AOs is fairly low, we have $\omega_{ta^*} \approx \alpha_\sigma - \varepsilon_t$ and $\omega_{et^*} \approx \varepsilon_t^* - \alpha_\sigma$. In addition, when np(M) levels are close in energy to σ(L) ones, $\varepsilon_t^* - \alpha_\sigma \approx \alpha_\sigma - \varepsilon_t$ and $l_t \approx c_t$. Then $\omega_{et^*} \approx \omega_{ta^*}$, and, therefore, the second term in braces in (3.40) is also positive.

To draw conclusions about absolute bond length changes, consider the F_{cis} and F_{trans} forces written as

$$F_{cis} \approx \frac{2}{3}\Delta\alpha\Big\{-\frac{c_a^2\mu_e - l_e^2\mu_a}{\omega_{ea^*}} + \frac{c_a^2 - l_a^2}{\omega_{aa^*}}\mu_a\Big\}, \tag{3.43}$$

$$F_{trans} \approx 2\Delta\alpha\Big\{\frac{2}{3}\frac{c_a^2\mu_e - l_e^2\mu_a}{\omega_{ea^*}} + \Big(2\frac{l_e^2}{\omega_{et^*}} - \frac{c_a^2}{\omega_{ta^*}} + \frac{c_t^2 - l_t^2}{\omega_{tt^*}}\Big)\mu_t + \frac{1}{3}\frac{c_a^2 - l_a^2}{\omega_{aa^*}}\mu_a + \Delta f'\Big\}. \tag{3.44}$$

It follows from (3.40) and (3.44) that apart from the coefficient of the first positive term, the expression for F_{trans} only differs from that for ΔF by the third term in braces, which is also positive. Therefore $F_{trans} > 0$ if $\Delta\alpha > 0$ and $F_{trans} < 0$ if $\Delta\alpha < 0$. At the same time, the expression in braces in (3.43) is negative if the energy of the ns(M) AO is fairly low and small and positive for more high-energy ns(M) levels.

To summarize, going from 12-electron nontransition element complexes of the first class to second-class complexes results in that in the limit, an "electropositive" $ML_6 \longrightarrow ML_5X$ substitution causes relative and absolute elongations of the *trans* bond not necessarily accompanied by shortening of the *cis* bonds, which may even become slightly longer. "Electronegative" $ML_6 \longrightarrow ML_5X$ substitution has exactly opposite effects. A comparison of these conclusions with those obtained in the approximation of frontier MOs (Section 3.4) shows that the analysis of 12-electron complexes conducted using the basis set of all σ-MOs substantiates and refines the results obtained in the approximation of frontier MOs. All examples of *cis*- and *trans*-influence in 12-electron nontransition element complexes considered in Section 3.5 are in agreement with this generalized treatment including weakening of *cis*-influence and its transformation into *trans*-influence as the position of atom M in the Periodic Table of the Elements changes.

7. Consider several examples of a similar character that refer to five-coordinate ML_5 complexes with tetragonal pyramidal structures, or ML_5E complexes in Gillespie's notation [102, 102a]. Here, E is the lone pair of electrons often treated as a pseudoligand with the limiting donor ability, such that the electrons of its σ AO are fully transferred to the MOs of the remaining ML_5 group. It should be stressed that certainly, lone pair E cannot be treated on equal terms with usual substituents X, because the influence of such a pseudoligand is outside the scope of perturbation theory. Nevertheless, even in this extreme situation, "ligand" E effects on ligands L follow trends characteristic of usual $ML_6 \longrightarrow ML_5$ substitution (see Table 6 [103]). Although *cis*-influence of ligand E does not transform into *trans*-influence, it distinctly weakens from SbF_5^{2-} to XeF_5^+ [102b, 103].

8. It remains to consider quasi-octahedral σ-bonded 14-electron complexes of nontransition elements. The discussion will be based on equations (3.30) for ΔF and F_{cis}. As central atom nd AOs make an insignificant contribution to bonding, we may write

$$\Delta F \approx 2\Delta\alpha\left\{ c_t^2\left(\frac{1}{\omega_{a^*t^*}} - \frac{1}{\omega_{at^*}}\right)\mu_a + \left[\frac{2}{\omega_{et^*}} + \left(\frac{l_a^2}{\omega_{at^*}} + \frac{c_a^2}{\omega_{a^*t^*}}\right) - 3\left(\frac{l_t^2}{\omega_{tt^*}} - \frac{c_t^2}{\omega_{tt^*}}\right)\right]\mu_t\right\}. \tag{3.45}$$

It follows from (3.45) that $\Delta F > 0$ if $\Delta\alpha > 0$ and $\Delta F < 0$ if $\Delta\alpha < 0$. The arguments for the validity of these inequalities are as follows [47].

Table 6. Bond lengths and $F_{ax}-M-F_{eq}$ bond angles in isoelectronic $MF_5^{n\pm}$ fluorides averaged on the assumption of C_{4v} symmetry [103]

Bond (Å), angle (deg)	SbF_5^{2-}	TeF_5^-	IF_5	XeF_5^+
$M-F_{ax}$	1.92	1.86	1.82	1.81
$M-F_{eq}$	2.07	1.95	1.87	1.84
$F_{ax}-M-F_{eq}$	79.4	78.8	80.9	79.2

The scheme of levels shown in Fig. 2 implies that the $(1/\omega_{a^*t^*})$ term in the coefficient of μ_a is always (and substantially at that) larger than the second term, $(1/\omega_{at^*})$. As concerns the coefficient of μ_t, it is only negative if the unrealistic condition

$$c_t^2 < \frac{1}{3}\left(l_t^2 - \frac{\omega_{tt^*}}{\omega_{at^*}} l_a^2 - \frac{\omega_{tt^*}}{\omega_{a^*t^*}} c_a^2\right) \tag{3.46}$$

is satisfied. It is easy to see that simultaneously, the approximate equality $F_{cis} \approx 0$ holds. Therefore, the 14-electron ML_5X complexes that we are discussing should exhibit *trans*-influence, that is, relative and absolute elongation of the *trans* bond if $\Delta\alpha > 0$ and relative and absolute shortening of this bond if $\Delta\alpha < 0$. The experimental data on such complexes were considered in Section 3.3.

9. In conclusion, let us return to the question of the origin of forces acting on ligands when the $ML_n \longrightarrow ML_{n-k}XY...Z$ substitution occurs. The problem will be treated from the point of view of the Hellmann–Feynman theorem. Compare equations (3.28) or (3.30) for F_{cis}, F_{trans}, and ΔF with similar equations obtained in [46, 47], where overlap populations of AOs participating in bonding were used as a criterion of bond strength. It is easy to see that the expressions for $F_{cis}/\Delta\alpha$, $F_{trans}/\Delta\alpha$, and $\Delta F/\Delta\alpha$ coincide in form with the expressions for overlap population variations

$$\Delta N(M-L_{cis})/\Delta\alpha, \qquad \Delta N(M-L_{trans})/\Delta\alpha,$$
$$[\Delta N(M-L_{trans}) - \Delta N(M-L_{cis})]/\Delta\alpha,$$

if μ_a, μ_e, and μ_t are identified (to within a common proportionality factor) with η_a, η_e, and η_t [46, 47]; also see Section 3.2. To understand the meaning of this correspondence between two approaches, recall that η_i are overlap populations between central atom AOs and ligand group orbitals of appropriate symmetry types. On the other hand, it follows from (3.24) that the μ_i values can be written in the form

$$\mu_i = C\partial\varepsilon_i/\partial R, \tag{3.47}$$

where ε_i is the level of the ith occupied MO, and C is the coefficient independent of i (actually, $C = 1/2$).

In other words, μ_i are proportional to the derivatives of the energies of the a_{1g}, e_g, and t_{1u} occupied MOs of the ML_6 complex with respect to the length of the M—L bond taken along the totally symmetric Q_1 mode. Note that for the e_g and t_{1u} MOs, the derivatives of the energies of individual MOs rather than all MOs of a given symmetry type are implied. Accordingly, the proportionality of μ_i to η_i ($\mu_i \sim \eta_i$, $i = a$, e, t) means that changes in MO energies under totally symmetric deformation are proportional to changes in central atom–ligands overlap populations. The $\mu_i \sim \eta_i$ proportionality can in turn easily be given a clear interpretation [100] in terms of the Hellmann–Feynman theorem [7]. According to (3.47), the $(-\mu_i)$ value is proportional to the partial generalized force that acts on the ligands and counterbalances mutual repulsion of atomic cores. This force is determined by attraction between atomic cores and the part of electronic density ρ localized between ligands and the central atom and approximately proportional to overlap population η_i. The total generalized force acting on the ligands, $-\partial \sum_i 2\varepsilon_i/\partial R$, is determined by the sum of the populations over all occupied MOs. Accordingly, additional forces, not counterbalanced by mutual repulsion of atomic cores and arising as a result of ligand substitution in ML_n, are determined by electron density changes in the regions of M—L bonds.

3.7. Bond Angles in Heteroligand Complexes

1. In the preceding sections, we considered changes in interatomic distances caused by the replacement of one or several ligands in a molecule or a complex. Meanwhile, ligand substitution in a homoligand molecular system is also accompanied by bond angle deformations. Curiously, this aspect of mutual influence of ligands is usually circumvented in theoretical works, probably, because of the absence of simple quantum-chemical values characterizing bond angles in the same way as bond orders, overlap populations, or Wiberg indices characterize bond lengths. Adiabatic potential $\mathcal{E}(Q)$ of a molecular system, however, is also a function of bond angles, and the vibronic theory of mutual influence

of ligands that we use can also be applied to study angular deformations [105]. General formulas (2.49) and (2.56) for coordinates Q^f_ν remain unchanged, but the number of normal modes to be analyzed should be extended by inclusion of bending modes involving angle variations (see Fig. 7).

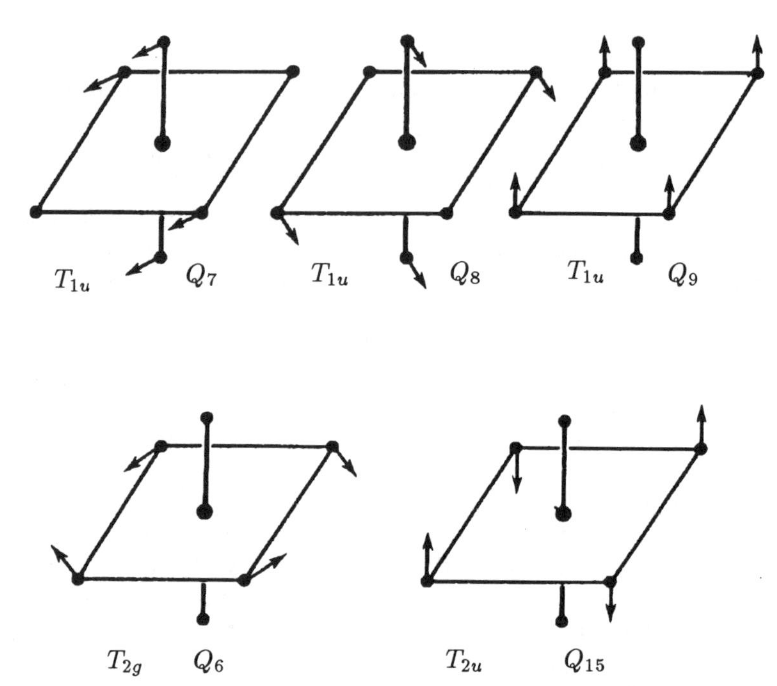

FIG. 7. NORMAL (SYMMETRIZED) BENDING MODES FOR OCTAHEDRAL ML_6 COMPLEXES (ALL $Q_\nu > 0$).

2. First, consider angular distortions in 12-electronic quasi-octahedral complexes of nontransition elements with central atoms M in the higher oxidation states. The approximation of frontier MOs will be used. As usual, our point of departure will be the typical scheme of MOs shown in Fig. 2 and the form of vibrational modes characteristic of octahedral ML_6 complexes (Fig. 7 and Table 2). We assume that ligand substitution occurs at position 1 (Fig. 4). Even without applying selection rules (2.25), it is clear that the only bending mode contributing to mutual ligand effects is mode Q_9 of symmetry T_{1u} (Fig. 7). The sign and the magnitude of angular deformations are determined by (3.8) with $Q^f_\nu = Q^f_9$. In (3.8), the matrix element of the H_S operator and vibronic coupling constant A are calculated for frontier MOs of the

ML_6 complex, but not for the e_g HOMO and the a^*_{1g} LUMO, because these orbitals are not mixed by bending vibration Q_9. We should use either the e_g and t^*_{1u} or the t_{1u} and a^*_{1g} pair or orbitals as frontier MOs, depending on which of the energy gaps, $e_g - t^*_{1u}$ or $t_{1u} - a^*_{1g}$, is narrower. As we cannot *a priori* decide in favor of one of these variants, both should be analyzed, especially as the two gaps may be of approximately the same width.

Let us first derive an expression for the off-diagonal orbital vibronic coupling constant $A = \langle\psi_g|(\partial H/\partial Q_9)_0|\psi_z\rangle$, where ψ_z is an odd MO involving the p_z AO of M, and ψ_g is an even MO of type $e_{g,\,z^2}$ or a_{1g}. These MOs can be written in the general form

$$\begin{aligned}\psi_g &= c_g\chi_g + l_{g,eq}(\sigma_3 + \sigma_4 + \sigma_5 + \sigma_6) + l_{g,ax}(\sigma_1 + \sigma_2),\\ \psi_z &= c_u p_z + l_u(\sigma_1 - \sigma_2)\end{aligned} \tag{3.48}$$

(*eq* stands for equatorial, and *ax*, for axial). As mode Q_9 only involves motions of equatorial ligands, we have

$$A = \langle\psi_g|(\partial H/\partial Q_9)_0|\psi_z\rangle = 4c_u l_{g,eq}\langle p_z|(\partial H/\partial Q_9)_0|\sigma_3\rangle. \tag{3.49}$$

Let, for the time being, $\beta(z_3)$ denote the resonance integral between the p_z AO of central atom M and the AO of ligand L_3 shifted by z_3 as a result of deformation Q_9. Then

$$\begin{aligned}\langle p_z|(\partial H/\partial Q_9)_0|\sigma_3\rangle &= (1/2)(\partial\beta(z_3)/\partial z_3)_{z_3=0}\\ &= (1/2)[\beta(\sigma, p_\sigma)/R]_0.\end{aligned} \tag{3.50}$$

It follows from (3.50) that the matrix element in the right-hand side of (3.49) is negative, because $\beta(\sigma, p_\sigma) < 0$. Note however that the inequality $\langle p_z|(\partial H/\partial Q_9)_0|\sigma_3\rangle < 0$ is fairly obvious, because when z_3 changes from negative to positive values, the overlap integral between the p_z and σ_3 AOs changes likewise. Simultaneously, the resonance integral changes from positive to negative, and its derivative with respect to z_3 is negative:

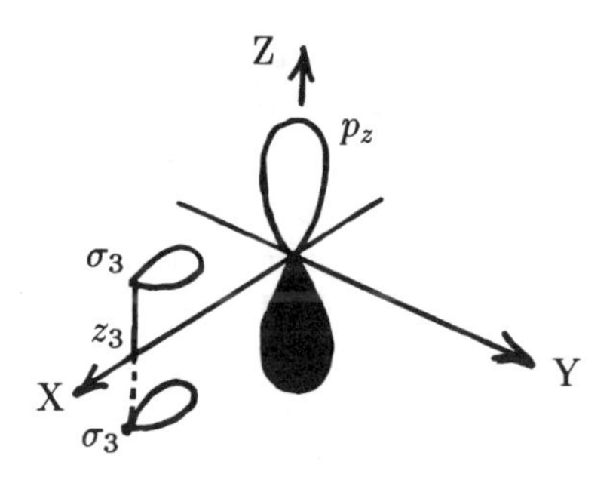

Let us write the frontier MOs for the (e_g, t_{1u}^*) and (t_{1u}, a_{1g}^*) variants,

$$\begin{cases} \text{HOMO: } e_g, \ \psi_{z^2} = c_e d_{z^2} + (l_e/\sqrt{12})(2\sigma_1 + 2\sigma_2 \\ \qquad\qquad - \sigma_3 - \sigma_4 - \sigma_5 - \sigma_6), \\ \text{LUMO: } t_{1u}^*, \ \psi_z^* = l_t p_z - (c_t/\sqrt{2})(\sigma_1 - \sigma_2), \end{cases}$$

$$\begin{cases} \text{HOMO: } t_{1u}, \ \psi_z = c_t p_z + (l_t/\sqrt{2})(\sigma_1 - \sigma_2), \\ \text{LUMO: } a_{1g}^*, \ \psi_a^* = l_a s - (c_a/\sqrt{6})(\sigma_1 + \sigma_2 + \sigma_3 \\ \qquad\qquad + \sigma_4 + \sigma_5 + \sigma_6), \end{cases} \tag{3.51}$$

and calculate matrix elements S for both,

$$\text{HOMO: } e_g, \quad \text{LUMO: } t_{1u}^*, \quad S = -c_t l_e \Delta\alpha/\sqrt{6}, \tag{3.52}$$

$$\text{HOMO: } t_{1u}, \quad \text{LUMO: } a_{1g}^*, \quad S = -c_a l_t \Delta\alpha/\sqrt{12}. \tag{3.53}$$

It follows from (3.8) and (3.49)–(3.53) that $Q_\nu^f < 0$ if $\Delta\alpha > 0$ and $Q_\nu^f > 0$ if $\Delta\alpha < 0$.

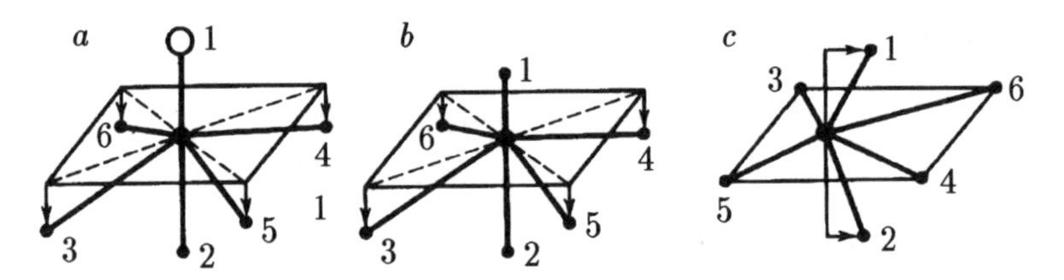

FIG. 8. ANGULAR DEFORMATIONS IN 12-ELECTRON QUASI-OCTAHEDRAL NONTRANSITION ELEMENT ML_5X COMPLEXES: (*a*) THE INFLUENCE OF A MORE ELECTROPOSITIVE SUBSTITUENT X, (*b*) THE INFLUENCE OF A SHORTENED M—X BOND, AND (*c*) "TETRAHEDRAL DISTORTION."

3. We arrive at the conclusion that the substitution of a more electropositive axial ligand X for L in a σ-bonded 12-electron quasi-octahedral complex ML_6 with nontransition element M in its higher oxidation state results in that equatorial $M—L_{eq}$ bonds slant off from the substituent (Fig. 8a). The introduction of a more electronegative substituent has the opposite steric effect. Table 7 shows that these conclusions agree with experiment.

The relevant experimental data can be augmented by including 12-electron ML_5 tetragonal pyramidal molecules and complexes such as

Table 7. Bond angle distortions in ML_5X and ML_5E nontransition element complexes

Complex	$L_{ax}-M-L_{eq}$, degrees	Refs.
SF_5Cl	89	[67]
TeF_5Cl	88	[106, 107]
IOF_5	72	[68]
SbF_5^{2-}	82	[102]
$SbCl_5^{2-}$	85	[102]
TeF_5^-	79	[102]
BrF_5	84	[102]
XeF_5^+	79	[102]

SbF_5^{2-}, BrF_5, XeF_5^+, etc., that is, ML_5E systems in Gillespie's notation [102]. Such molecular systems were considered in Section 3.6, where it was noted that their treatment is beyond the scope of perturbation theory. Nevertheless, equation (3.8) correctly predicts the character of angular distortions for these systems too (see Table 7). In all ML_5E compounds, the equatorial $M-L_{eq}$ bonds slant off from lone pair E toward axial ligand L_{ax} (Fig. 9). This is in agreement with the well-known and fairly effective Gillespie–Nyholm stereochemical model [102, 102a], which postulates that electron pairs of $M-L$ bonds are more strongly repelled from lone pair E than from each other. It is noteworthy that this result can be obtained without additional postulates on the strength of mutual repulsion of bonding and lone electron pairs within the framework of a simple MO theory not including electron correlation. Parallels between the Gillespie model and the theory of molecular orbitals are discussed in [108, 109].

4. Substitution-induced deformation of bond angles is related to another phenomenon that we come across in structural chemistry. It is known that angular distortions are characteristic of complexes with bonds of unequal lengths, for instance, when some of the bonds are terminal and the others form bridges. A complete statement of the problem requires bond lengths and bond angles to be analyzed simultaneously. A more restricted approach is, however, also of interest [105]. For instance, angle distortions can be analyzed on certain assumptions about changes in the lengths of some bonds (in the same way as we treat substitution as a change in the donor ability of the ligand in-

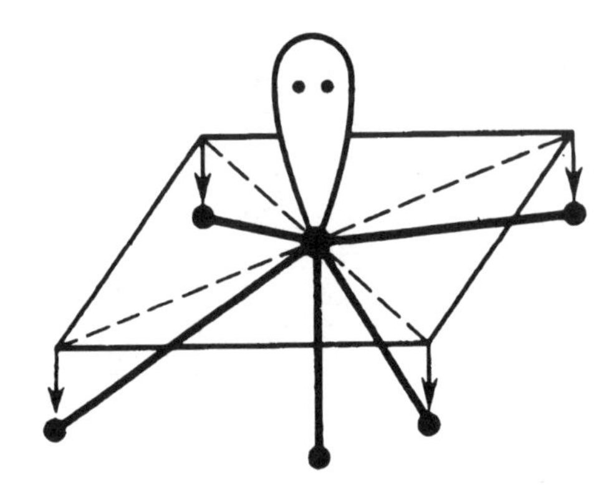

FIG. 9. GEOMETRY OF ML_5E COMPLEXES.

volved). The problem can then easily be solved by applying the general approach developed in Sections 2.2–2.4. Elongation or shortening of some M—L bonds is modeled by using a nondiagonal matrix of substitution operator H_S, which describes changes in the resonance integrals corresponding to the bonds whose lengths are varied (Section 2.3).

Consider such an artificial change of bond lengths in 12-electron σ-bonded quasi-octahedral complexes of nontransition elements. It will first be assumed that only the $M-L_1$ bond length is changed. This change is described by changes in the $\beta_{s1} = \beta(s, \sigma_1)$, $\beta_{p1} = \beta(p, \sigma_1)$, and $\beta_{d1} = \beta(d, \sigma_1)$ resonance integrals, where σ_1 stands for the σ AO of ligand L_1. It is easy to see that in the approximation of frontier MOs used to obtain formulas (3.52) and (3.53), the angular deformation of the complex is determined by (3.8) with vibronic coupling constant A given by (3.49) and (3.50). Matrix element S of the substitution operator can then be written as

$$S = S_s + S_p + S_d, \tag{3.54}$$

where the form of S_s, S_p, and S_d depends on the MOs chosen as frontier MOs. For the e_g, t_{1u}^* frontier MOs, we have

$$S_s = 0, \quad S_p = l_e l_t \Delta\beta_{p1}/\sqrt{3}\,, \quad S_d = c_e c_t \Delta\beta_{d1}/\sqrt{2}\,. \tag{3.55}$$

Similarly, for the t_{1u}, a_{1g}^* frontier MOs,

$$S_s = l_a l_t \Delta\beta_{s1}/\sqrt{2}\,, \quad S_p = -c_a c_t \Delta\beta_{p1}/\sqrt{6}\,, \quad S_d = 0. \tag{3.56}$$

This eventually gives

$$Q_9^f = -\frac{4}{3} l_e l_t \frac{l_e l_t \Delta\beta_{p1} - \sqrt{3/2}\, c_e c_t \Delta\beta_{d1}}{K\omega_{et^*}} \left[\frac{\beta(\sigma, p_\sigma)}{R}\right]_0 \tag{3.57}$$

for the e_g, t_{1u}^* pair and

$$Q_9^f = -\frac{4\sqrt{3}}{3} c_a c_t \frac{l_a l_t \Delta\beta_{s1} - (1/\sqrt{3}) c_a c_t \Delta\beta_{p1}}{K\omega_{ta^*}} \left[\frac{\beta(\sigma, p_\sigma)}{R}\right]_0. \quad (3.58)$$

for t_{1u}, a_{1g}^*. Consider formula (3.57). In nontransition element ML_n complexes, the nd AO of atom M makes an insignificant contribution to bonding. We can therefore assume that the c_e coefficient is small in comparison with the other coefficients in expression (3.51). This yields

$$Q_9^f = -\frac{4 l_e^2 l_t^2}{3K\omega_{et^*}} \left[\frac{\beta(\sigma, p_\sigma)}{R}\right]_0 \Delta\beta_{p1}. \quad (3.59)$$

Equation (3.59) implies that when the $M-L_1$ bond becomes shorter, that is, when $\Delta\beta_{p1} < 0$, the equatorial bonds slant off toward the *trans* ligand, as in the $ML_6 \longrightarrow ML_5X$ substitution of a more electropositive substituent for L_1. Elongation of the $M-L_1$ bond has the opposite effect.

The same conclusion follows from an analysis of formula (3.58). First we will show that the coefficient of $\Delta\beta_{s1}$ in (3.58) is larger than the coefficient of $\Delta\beta_{p1}$. Indeed, it has already been mentioned that for the σ levels of ligands L and the valence ns and np levels of the central atom, we have either $\alpha_\sigma < \alpha_s < \alpha_p$ or $\alpha_s < \alpha_\sigma < \alpha_p$. The first inequalities are characteristic of, e. g., Group IV element compounds and Group V element and Te fluorides; they give $c_a \approx l_a$ and $c_t < l_t$, whence $c_a c_t < l_a l_t$ and, the more so, $(\sqrt{3}/3) c_a c_t < l_a l_t$. The second inequalities, which are met by Group VI (except Te) and Group VII element fluorides, yield $\alpha_\sigma \lesssim (1/2)(\alpha_s + \alpha_p)$. We assume that the contribution of a given AO of atom M to $M-L$ bonding is approximately proportional to the difference of the electronegativities of the AOs involved; therefore $c_a c_t \lesssim l_a l_t$, which again gives $(\sqrt{3}/3) c_a c_t \lesssim l_a l_t$. Next, because of the more contracted character of the low-lying ns AO of M in comparison with its higher np AOs, the $\Delta\beta_{s1}$ value should more rapidly change with interatomic distance variations than the $\Delta\beta_{p1}$ one, that is, $|\Delta\beta_{p1}| < |\Delta\beta_{s1}|$ (although it would suffice that the weaker requirement $|\Delta\beta_{p1}| < \sqrt{3}(l_a l_t / c_a c_t)|\Delta\beta_{s1}|$ be satisfied). The first term in the numerator in (3.58) then prevails over the second, and, therefore, when the length of the axial $M-L_1$ bond changes, the equatorial $M-L_{equ}$ bonds show the same behavior as previously when the e_g and t_{1u}^* MOs were considered frontier.

It follows that in 12-electron nontransition element complexes, the effect of artificially changing the length of an $M-L$ bond (Fig. 8b) is the same no matter what the character of frontier MOs; the effect also remains unchanged when terms corresponding to both energy denominators, ω_{et^*} and ω_{ta^*}, are simultaneously included in the formula for Q^f_ν.

4. The analysis of the effects of artificially changing the length of one of the M—L bonds is easily extended to changes of the lengths of several bonds. As for the $ML_n \longrightarrow ML_{n-k}XY...Z$ ligand substitution, it follows from the right-hand sides of (2.49) and (2.56) that the effect of artificially changing the lengths of several bonds is a superposition of the effects of changes in the lengths of separate bonds. In view of the available experimental data, of interest is the effect of artificially shortening or stretching equatorial bonds in octahedral 12-electron σ-bonded nontransition element complexes. For instance, let the $M-L_3$ and $M-L_5$ bonds be shortened and (or) the $M-L_4$ and $M-L_6$ bonds elongated (Fig. 8c). Consider the $M-L_1$ and $M-L_2$ axial bonds. It follows from the foregoing that the complex should experience angular deformations induced by the Q_7 and Q_8 modes (Table 2 and Fig. 7). As a result, the axial bonds should slant off from the shortened $M-L_3$ and $M-L_5$ bonds toward the longer $M-L_4$ and $M-L_6$ ones. If the effect is sufficiently strong, the octahedron acquires features of a distorted $ML_1L_2L_3L_5$ tetrahedron with two additional $M-L_4$ and $M-L_6$ bonds (the phenomenon was described in [66] and review [63], see Fig. 8c). That the effect does exist was convincingly demonstrated for the example of six-coordinate Sn(IV) complexes [63, 66, 110] (Table 8). Similar distortions were observed in complexes of other elements. For instance, in the $[Sb(CH_3)_2Cl_3]_2$ pentavalent antimony binuclear complex,the axial $Sb-CH_3$ bonds slant off from the terminal equatorial *cis* Sb—Cl bonds toward elongated bridge equatorial *cis* Sb—Cl bonds [111]. Another example is the SbF_6^- complex with two bridge fluorine atoms in $BrF_2^+SbF_6^-$ crystals [112]:

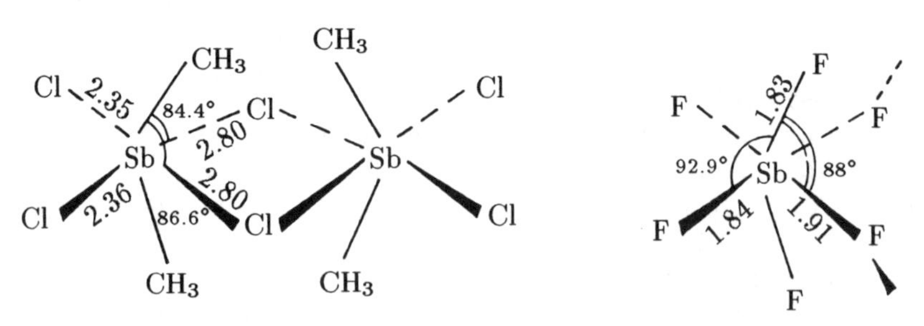

5. At first sight, tetrahedral distortions of octahedra occur for steric rather than electronic reasons. Indeed, a shortening of $M-L_{eq}$ bonds should be accompanied by deflection of axial bonds because of repul-

Table 8. "Tetrahedral" distortions of octahedra in $Sn^{(IV)}L_2L'_2X_2$ complexes [66]

Compound	X_{ax}	X − Sn − X	Δ(Sn − L)	L	L − Sn − L	Δ(Sn − L′)	L′	L′ − Sn − L′	Δ(Sn − X)
$SnPh_2Cl_2$	Ph	127	−0.07	Cl	97.8	+1.35	Cl	—	0.00
$SnMe_2Cl_2$	Me	123.5	−0.02	Cl	93.8	+1.12	Cl	—	0.00
$Sn(CH_2Cl)_2(\mu\text{-}Cl)_2$	C	135.0	−0.05	Cl	93.0	+0.29	Cl	—	−0.02
$SnMe_2(S_2CNMe)_2$	Me	136.0	0.07(av)	S	82.7	+0.53(av)	S	149.1	−0.05(*av*)
$SnMe_2(NO_3)_2$	Me	143.6	0.07(av)	O	74.7	+0.47(av)	O	176.0	−0.08
$SnMe_2(ONMeCOMe)_2$	Me	145.8	−0.03(av)	O	73.2	+0.29(av)	O	144.1	−0.09(*av*)
$SnMe_2(cis\text{-}Cl_2)(LNI)$	Me	161.0	0.06(av)	Cl	97.1	+0.39(av)	O	61.3	−0.08
$SnMe_2(cis\text{-}Cl_2)(DMEA)_2$	Me	165.0	0.05	Cl	94.3	0.30	O	85.9	−0.07
$SnMe_2(cis\text{-}Cl_2)(DMSO)_2$	Me	170.0	0.08(av)	Cl	96	0.26	O	84	−0.12
$SnPh_2(cis\text{-}Cl_2)(bipy)_2$	Ph	173.5	0.09(av)	Cl	103.5	0.23	N	69.0	−0.08(*av*)

Note: Δ(Sn − L) deviations of experimental distances from standard Sn − L bond lengths; Δ(Sn − L) are in Å, angles in degrees; LNI stands for N,N′-ethylene-*bis*(salicylideneimine), DMSO is dimethylsulfoxide, and bipy is bipyridine.

sion between axial and equatorial ligands. The matter is, however, not so simple. Consider a curious example of angular deformations in 14-electron nontransition element quasi-octahedral complexes. In the approximation of frontier MOs, the roles of the HOMO and LUMO are played by the a_{1g}^* and t_{1u}^* MOs given by (3.51). First consider the replacement of ligand L_1 or an artificial change in the $M-L_1$ bond length. The angular distortion Q_9^f value is then given by (3.8), where vibronic coupling A is determined by (3.49) and (3.50) and the form of the frontier MOs. It is easy to see that the A value is positive. The matrix elements of the H_S operator describing ligand L_1 replacement or $M-L_1$ bond elongation are

$$S = (1/2\sqrt{3})c_a c_t \Delta\alpha, \tag{3.60}$$

$$S = S_s + S_p + S_d, \quad S_s = -(1/\sqrt{2})c_t l_a \Delta\beta_{s1},$$
$$S_p = -(1/\sqrt{6})c_a l_t \Delta\beta_{p1}, \quad S_d = 0. \tag{3.61}$$

It follows from these formulas that in a 14-electron ML_6 complex, the replacement of the L_1 axial ligand by a more electropositive ligand X or a shortening of the $M-L_1$ bond causes displacements of equatorial ligands toward the substituent or the shortened axial bond, which is opposite to the effects characteristic of 12-electron complexes. The replacement of ligand L_1 by a more electronegative substituent or an elongation of the $M-L_1$ bond change the sign of the effect (Fig. 10). It is hardly necessary to analyze the effects of shortening or elongation of equatorial bonds in 14-electron complexes. The axial bonds then, in spite of steric hindrances, slant off from the longer equatorial bonds toward shorter ones. Such "antitetrahedral" distortions occur in, e. g., $[TeCl_5^-]_\infty$ polymeric anions formed by six-coordinate Te(IV) atoms [113]:

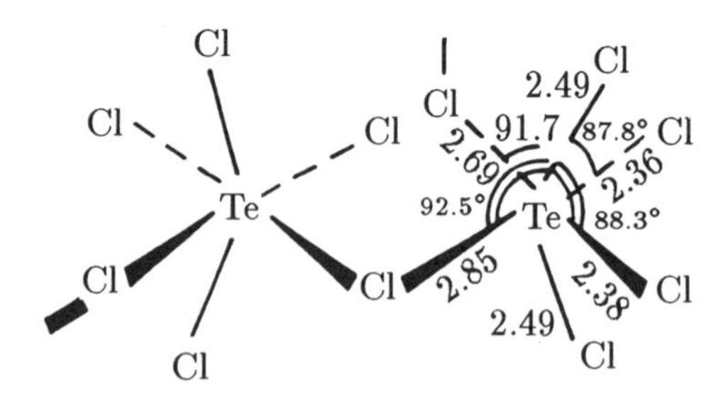

6. A detailed analysis of angular distortions in transition metal ML_5X complexes with 12 electrons on σ MOs is hardly necessary. The situation is similar to that with 12-electron nontransition element compounds. The only difference is that in transition metal complexes, the

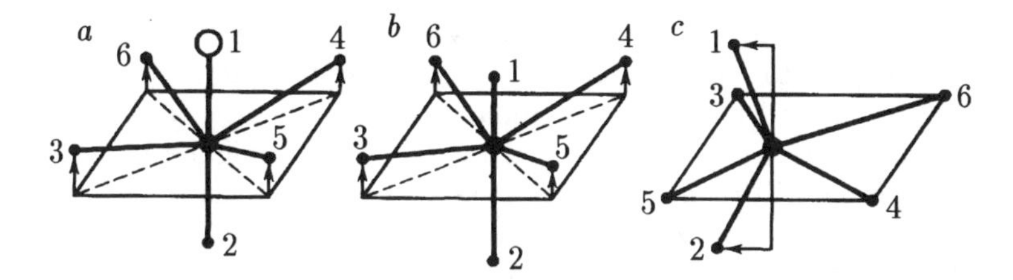

FIG. 10. ANGULAR DEFORMATIONS IN 14-ELECTRON QUASI-OCTAHEDRAL NONTRANSITION ELEMENT ML_5X COMPLEXES: (*a*) THE INFLUENCE OF A MORE ELECTROPOSITIVE SUBSTITUENT X, (*b*) THE INFLUENCE OF A SHORTENED BOND, AND (*c*) "ANTITETRAHEDRAL DISTORTION."

frontier MOs are t_{1u} and e_g^* (Fig. 1), whereas in the derivatives of nontransition elements, these are e_g and t_{1u}^* or t_{1u} and a_{1g}^*. Clearly, the replacement of the e_g and t_{1u}^* frontier MOs in (3.51) and (3.52) by the t_{1u} and e_g^* ones has no effect on the signs of S and A and, consequently, the type of angular distortions. In ML_5X transition metal complexes, the equatorial $M{-}L_{eq}$ bonds slant off from the axial substituent X when X is more electropositive than L and shift toward X when X is more electronegative than L. This conclusion is well illustrated, e.g., by the structural data on OsF_5Cl^{2-} [83a], where all Cl—Os—F angles are larger than 90° and vary from 91.1° for the $Cl{-}Os{-}F_2$ angle to 91.79° for the $Cl{-}Os{-}F_3$ one:

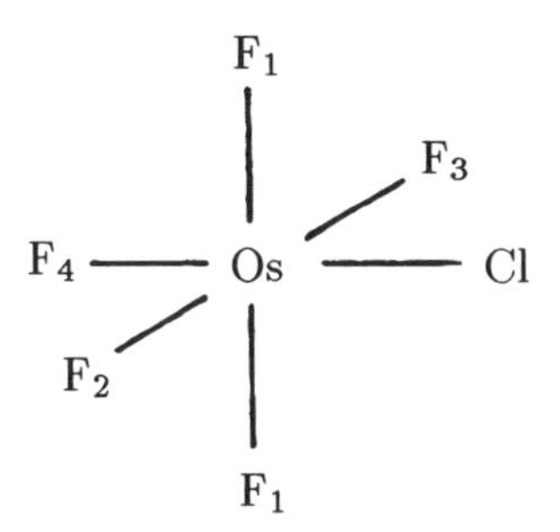

3.8. Isomers of Transition Metal d^0 and d^6 Complexes

1. The use of high-resolution NMR spectroscopy in studies of the structure of transition metal complexes [114] led to the accumulation of data on the stereochemistry and the relative stabilities of the *cis* and *trans* isomers of transition metal d^0 and d^6 complexes. We will analyze the relative stabilities of σ-bonded d^0 and d^6 complexes by comparing the

total energies of their isomeric forms, which will allow us to interpret experimental data. In this analysis, the general approach described in Section 2.4 and based on a comparison of the depths of adiabatic potential minima for different isomeric forms of heteroligand molecules and complexes will be used. First, the dependence of the total energy of d^0 $ML_{6-k}X_k$ complexes on the number and mutual arrangement of substituents X will be considered. Apart from the problem of the relative stabilities of *cis* and *trans* isomers, the character of deviations from the additivity rule [115] for the total energy as a function of $k = 0$–6 will be discussed. (According to this rule, the energy should be a linear function of k from ML_6 to MX_6.)

2. Recall that in (2.50) and (2.57), the S_0 term, which appears in first-order perturbation theory, depends on the number of substituents and is independent of their mutual arrangement. The other terms in (2.50) and (2.57) are structurally sensitive second-order corrections. Therefore S_0 is an additive value, and deviations from additivity are due to the presence of structurally sensitive terms which determine the differences in stabilities of various molecular system isomers. As in analyzing mutual influence of ligands, we will start with applying the approximation of frontier MOs to obtain results representable in a simple and physically clear form.

According to Fig. 1, ML_6 homoligand predecessors of σ-bonded transition metal heteroligand complexes have the t_{1u} HOMO and the e_g^* LUMO, which transform under the T_{1u} and E_g irreducible representations of the O_h point symmetry group of ML_6, as the frontier MOs. Consider the action of substitution operator H_S describing the transition from the ML_6 complex to the $ML_{6-k}X_k$ heteroligand system on these MOs. It is assumed (up to Chapter 5, inclusive) that the H_S operator is given by (2.38) and (2.39) and is representable in form (2.42). Let us apply selection rules (2.43a). The direct product of the T_{1u} and E_g irreducible representations is $T_{1u} \times E_g = T_{1u} + T_{2u}$. The decomposition of the H_S operator into symmetrized components, however, contains T_{1u} and no T_{2u} component no matter what the relative arrangement of substituents X (see Table 3; data on many-electron operators $\mathcal{H}_S$ are also valid for one-electron H_S ones).

It follows that the frontier orbitals can only be mixed by the T_{1u} component of H_S. For ML_6, *trans*-ML_4X_2, and *trans*-ML_2X_4, H_S does not contain T_{1u}-type contributions, and for such complexes, the structurally sensitive terms in (2.57) vanish. However, for ML_5X, *cis*-ML_2X_4, *cis*- and *trans*-ML_3X_3, *cis*-ML_2X_4, and MLX_5, the H_S operator contains a T_{1u} component (Table 3). This H_S operator mixes the t_{1u} and e_g^* orbitals, which stabilizes the complexes in comparison with predictions based on the additivity rule; the *cis* form of the ML_3X_3 complex is more stable than the *trans* one. For the H_S substitution operator that we are considering, the t_{2g} MOs are inactive, and the described dependence of the energy of σ-bonded complexes on their composition and structure is therefore also valid for d^3 and low-spin d^6 complexes.

This reasoning can be translated into the simpler pictorial language as

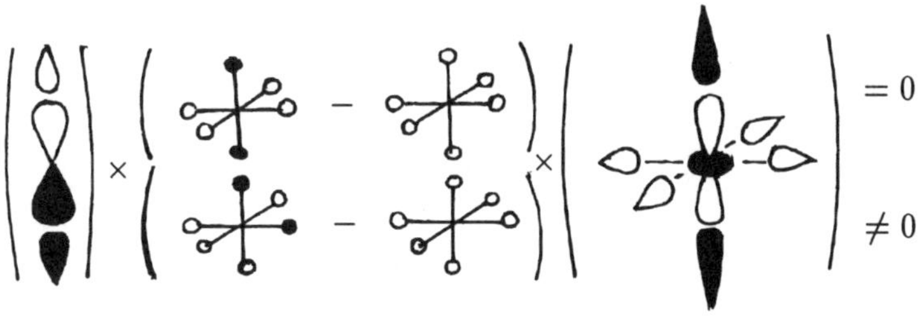

$= 0$

$\neq 0$

$t_{1u,z}$ e^*_{g,z^2}

This scheme sheds light on the formation of the integrand in the matrix element $\langle t_{1u,\,z}|H_S|e^*_{g,\,z^2}\rangle = \int \psi_z H_S \psi^*_{z^2}\, dv$ for the *cis* and *trans* isomers of ML_4X_2. For the *trans* isomer, the integrand is the product of two even and one odd functions, and the integral therefore vanishes. It follows that the $t_{1u,\,z}$ and $e^*_{g,\,z^2}$ MOs are not mixed by the H_S(*trans*-ML_4X_2) operator, although they can be mixed by the H_S(*cis*-ML_4X_2) one.

It is worth while mentioning that this mechanism of stabilization of *cis* isomers has a bearing on static *trans*-influence in transition metal complexes, which (see Section 3.3) is due to geometric relaxation of complexes accompanying mixing of odd (t_{1u}) and even (e^*_g) MOs under the action of a suitable substitution operator (for instance, in monosubstituted ML_5X complexes). The stabilization of *cis* isomers is an energetic manifestation of the same geometric (and electronic) relaxation of complexes after the $ML_6 \longrightarrow ML_{6-k}X_k$ substitution.

We considered the problem of the relative stabilities of the isomers of d^0 $ML_{6-k}X_k$ complexes in terms of the electronic structure of ML_6 complexes. Treatment in terms of the electronic structure of heteroligand complexes themselves seems to be even simpler. In a heteroligand complex, the HOMOs are the MOs derivative from the t_{1u} MOs of its homoligand predecessor. In *cis*-$ML_{6-k}X_k$, these MOs on average have lower energies than in its *trans* counterpart because of a more effective mixing of the t_{1u} and e^*_g MOs under the action of H_S describing the transition from a homoligand complex to $ML_{6-k}X_k$. The total energy of the *cis* complex at point Q_0 is therefore lower than that of the *trans* one. Geometric relaxation of complexes results in further *cis* isomer stabilization.

3. The results of the analysis of deviations of the total energy of heteroligand d^0 complexes from additivity are shown in Fig. 11. According to this figure, the total energy coincides with that calculated by the additivity rule for *trans*-ML_4X_2 and *trans*-ML_2X_4 and is lower than additive values for the other complexes. It follows that the additivity rule only gives the upper estimates of the energies of d^0 complexes.

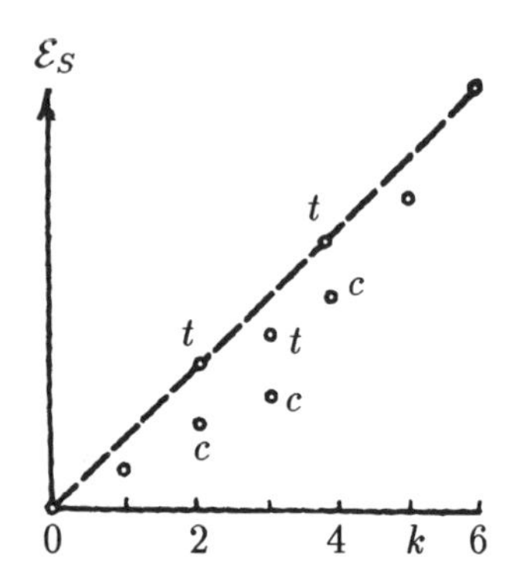

FIG. 11. THEORETICAL DEPENDENCE OF THE TOTAL ENERGY $\mathcal{E}_0$ OF A d^0 $ML_{6-k}X_k$ COMPLEX ON THE NUMBER AND ARRANGEMENT OF SUBSTITUENTS IN THE APPROXIMATION OF FRONTIER MOs (t STANDS FOR *trans*, AND c, FOR *cis*).

A comparison of Figs. 11 and 12 (the latter one contains the thermochemical data from [117]) is indicative of satisfactory agreement between theory and experiment. The total energy of $WF_{6-k}(OMe)_k$ complexes shows noticeable deviations from additivity of the same character as those predicted theoretically [115]. On the whole, theoretical results should be considered predictive, because so far as we know, for none of the series of $ML_{6-k}X_k$ $(0 \leq k \leq 6)$ complexes, are the energies of all possible 10 isomeric forms available.

4. The discussion of stability of isomers given above was based on the approximation of frontier MOs. This approximation may be abandoned to analyze the relative stabilities of various isomers using the basis set of all valence σ MOs of ML_6 complexes [116]. As in Section 3.6, we assume that ligand X differs from L by the Coulomb integral value for the σ AOs $(\alpha(X) - \alpha(L) = \Delta\alpha)$. Let us calculate the energy of the transformation of *trans*-ML_4X_2 into the *cis* form. Using (2.58), we

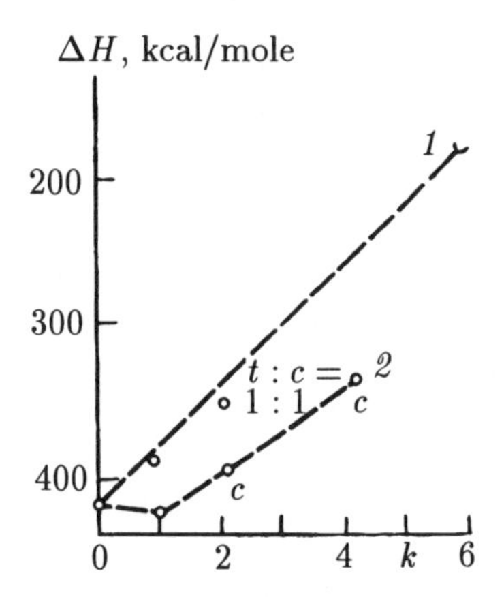

FIG. 12. EXPERIMENTAL DEPENDENCE OF THE ENTHALPY OF FORMATION OF $WF_{6-k}F_k$ COMPLEXES ON THEIR COMPOSITION [117]; *1*, Cl AND *2*, OMe.

obtain

$$\Delta\mathcal{E}_{t\to c} = 2\Delta\alpha^2\left\{\frac{c_e^2}{6}\left(\frac{2}{\omega_{te^*}} - \frac{l_a^2}{\omega_{ae^*}} - \frac{l_e^2}{\omega_{ee^*}}\right) + \frac{c_a^2}{6}\left(\frac{1}{\omega_{ta^*}} - \frac{l_e^2}{\omega_{ea^*}}\right)\right\}$$
$$+ \frac{24\Delta\alpha^2}{K_T}\left\{\frac{2\mu_e}{\omega_{te^*}} + \frac{\mu_a}{\omega_{ta^*}}\right\}^2 \tag{3.62}$$
$$- \frac{24\Delta\alpha^2}{K_E}\left\{\mu_e\left(\frac{c_a^2}{\omega_{ea^*}} - \frac{l_a^2}{\omega_{ae^*}} + \frac{c_e^2 - l_e^2}{\omega_{ee^*}}\right) + \mu_a\left(\frac{c_e^2}{\omega_{ae^*}} - \frac{l_e^2}{\omega_{ea^*}}\right)\right\}^2.$$

Here, the same notation as in Section 3.6 is used, and K_T and K_E are the force constants for the T_{1u} and E_g modes.

To determine the sign of (3.62), consider the scheme of MOs of ML_6 (Fig. 1). We must also use qualitative estimates of the relative values of typical force constants. These can be obtained based on Pearson's concept of the relative lability coordinates [32], according to which $K_T \leq K_E$. The first expression in braces in (3.62) is then positive, because $\omega_{te^*} < \omega_{ae^*} < \omega_{ee^*}$ and $\omega_{ta^*} < \omega_{ea^*}$. Next, the second term in (3.62) is larger than the third, because for the scheme of levels shown in Fig. 1, we have

$$(1/\omega_{te^*}) > (c_a^2/\omega_{ea^*}) - (l_a^2/\omega_{ae^*}),$$
$$(1/\omega_{te^*}) > (c_e^2 - l_e^2)/\omega_{ee^*}, \tag{3.63}$$
$$(1/\omega_{ta^*}) > c_e^2/\omega_{ae^*} - l_e^2/\omega_{ea^*},$$

and $K_T \leq K_E$. This leads us to conclude that the $\Delta\mathcal{E}_{t\to c}$ value is positive for d^0 complexes, and *cis* isomers should therefore be more stable, in agreement with the results of applying the approximation of frontier MOs. Note that according to (3.62), stabilization of *cis* isomers of the complexes under consideration rather heavily (quadratically) depends on the difference of the orbital ionization potentials of ligands X and L. Therefore noticeable stabilization of *cis* isomers should be observed for ML_4X_2 complexes with L and X that are removed from each other as far as possible on the scale of electronegativities or in the static *trans*-influence series. The stabilization of the *cis* isomer with respect to its *trans* counterpart should rapidly decrease with decreasing $|\Delta\alpha|$, that is, the "distance" between L and X in the static *trans*-influence series.

A similar approach can be applied to ML_3X_3 isomer stabilities [116]. It is easy to see that for these complexes, the isomer with all three substituents positioned *cis* with respect to each other should possess the highest stability (the *cis*, or facial, isomer). The expression for the energy of isomerization of ML_3X_3 complexes is the same as for ML_4X_2 and, as previously, is given by (3.62).

5. Compare our theoretical conclusions with experimental data. Data on the heats of isomerization, such as used in Chapter 4 in discussing quasi-square-planar platinum(II) complexes, would be most suitable for this purpose. However, systematic data on the heats of isomerization of transition metal complexes under consideration are lacking. For this reason, we will largely rely on structural studies and NMR data on complexes in solutions. We will use the results of works where the formation of only the most stable isomer was observed or, if two isomers were formed, the ratio between their concentrations was determined and isomerization constants $K_i = [cis\text{-}ML_{6-k}X_k]/[trans\text{-}ML_{6-k}X_k]$ used to characterize the relative stabilities of the isomers were calculated [118–135]. (It is assumed for ML_4X_2 complexes from statistical considerations that the $K_i > 4$ value is indicative of predominant stability of the *cis* isomer, and $K_i < 4$, of predominant stability of the *trans* one. For ML_3X_3 complexes, the *cis* isomer has the higher stability if $K_i > 2/3$.)

As follows from the analysis given above, maximal stabilization of *cis* isomers should be observed when the coordination sphere of a complex simultaneously contains ligands with strong (such as alkyl, aryl, and triorganosilyl groups or hydride ions) and weak (halogen and pseudohalogen ions or NH_3, OH, and H_2O groups) static *trans*-influence. In agreement with the theory, the overwhelming majority of dihydride d^6 complexes MH_2L_4 (M = Fe(II), Ru(II), and Os(II) and L = PF_3, $P(OR)_2$, and PR_3) have *cis* configurations. Only solutions of Fe(II) and Ru(II) complexes with L = $P(OEt)_2Ph$ or $P(OMe)_2Ph$ simultaneously contain both geometric isomers [136]. According to review [136], most of $Ir^{III}H_3L_3$ ($L = PR_3$ and AsR_3) trihydride complexes also have the *cis* structure, and only some of them occur in both isomeric forms. The *cis* structure is also characteristic of thoroughly studied Pt(IV) derivatives of the compositions PtR_2L_4, PtR_3L_3, and PtR_4L_2, where L is a phosphine group or a halogen, and R is an alkyl group [81].

One more fairly numerous group of complexes includes substituted fluoro complexes of Ti(IV), Ta(V), and W(VI) with the d^0 metal configuration. According to the ^{19}F NMR data, these compounds are also characterized by a higher stability of the *cis* form in comparison with *trans* (Table 9). The $[PtF_4Br_2]^{2-}$ d^6 complex [132] and the tetrafluoro complex of tantalum with picolinic acid (see below) [86] also have the *cis* configuration.

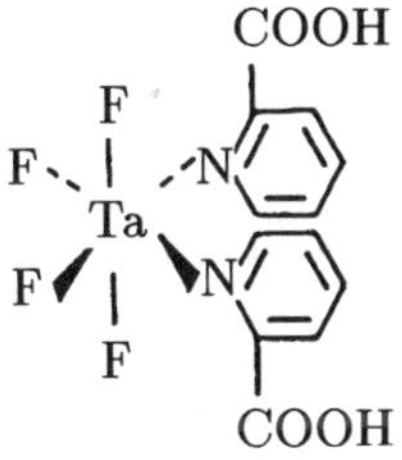

COOH F F Ta N N F F COOH

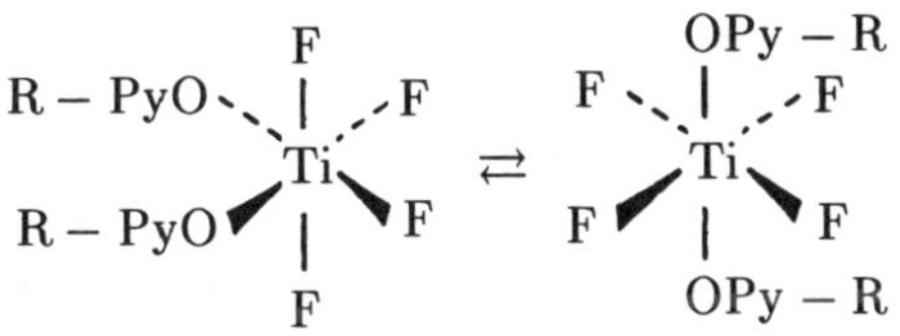

R – PyO F F Ti R – PyO F F F ⇄ OPy – R F F Ti F F OPy – R

Lastly, complexes of the composition MF_3X_3 with M = Ti(IV) and X = NCO and NCS and also with M = W(VI) and X = OPh and OMe [124, 86] occur in solutions as mixtures of geometric isomers. The isomerization constant K_i values exceed the statistical value of 2/3, which, as mentioned, is evidence of predominant stabilization of *cis* isomers. Similar enhanced stabilities of *cis* isomers were observed for complexes $TiF_4(OPyR)_2$, where, depending on the nature of substituent R, the concentration of the *cis* isomers varied in the range 100–85 mol %, whereas the concentration of the *trans* form did not exceed 15 mol %.

6. As the distance between ligands L and X in the static *trans*-influence series (or on the scale of electronegativities) decreases, the relative stability of the *cis* isomers of the $ML_{6-k}X_k$ composition should also decrease. Factors ignored by the model (direct ligand–ligand interactions or solvent effects) can even cause stabilization of the *trans* isomers of such complexes. Accordingly, only *trans* isomers were observed for the $[TiF_4(2,6\text{-}Me_2C_5H_3NO)_2]$ and $TaF_4X_2^+$ complexes, where X are molecular N-donors of the type of azoles [123, 133]. An analysis of structure models performed in [123] showed that the *cis* configuration of the titanium complex with four fluorine atoms and two 2,6-dimethylpyridineoxide molecules could not exist because of strong

Table 9. Predominant isomers and isomerization constants for quasi-octahedral MF_4X_2 transition metal fluoro complexes

Metal	X	Isomer type, K_i	Refs.
Ti(IV)	H_2O	*cis*-isomer	[118]
	MeOH	*cis*-isomer	[119]
	EtOH	*cis*-isomer	[119]
	PrOH	*cis*-isomer	[119]
	THF	*cis*-isomer	[120]
	TBP	*cis*-isomer	[121]
	TBPO	*cis*-isomer	[121]
	TPPO	*cis*-isomer	[121]
	HMPA	*cis*-isomer	[121]
	2-ClC_5H_4N	*cis*-isomer	[122]
	2-BrC_5H_4N	*cis*-isomer	[122]
	C_5H_5NO	*cis*-isomer	[123]
	2-MeC_5H_4NO	13.3	[123]
	2-EtC_5H_4NO	6.1	[123]
	4-MeC_5H_4NO	*cis*-isomer	[123]
	$2,4\text{-Me}_2C_5H_3NO$	5.7	[123]
	$3,5\text{-Me}_2C_5H_3NO$	*cis*-isomer	[123]
	NCO	10	[124]
	NCS	~100	[124]
Ta(V)	OEt	*cis*-isomer	[125]
	OPh	*cis*-isomer	[126]
	TBPO	4	[127]
	$OPPh_2SEt$	8	[128]
	OMe	*cis*-isomer	[128]
	OEt	*cis*-isomer	[130]
	OPh	*cis*-isomer	[126, 127, 130]
	$ONC(Me)_2$	*cis*-isomer	[86]
	$OOCCF_3$	*cis*-isomer	[86]
W(VI)	OH	*cis*-isomer	[86]
	OMe	*cis*-isomer	[86]
	OEt	*cis*-isomer	[86]
	OPh	*cis*-isomer	[86]
Mo(VI)	OH	*cis*-isomer	[86]
	OCH_2CF_3	*cis*-isomer	[86]
Nb(V)	$OOCCF_3$	*cis*-isomer	[86]

Note: THF is tetrahydrofuran, TBP is tributyl phosphate, TBPO is tributylphosphine oxide, TPPO is triphenylphosphine oxide, and HMPA is hexamethylphosphorotriamide.

steric hindrances between methyl groups and fluorine atoms.

A study of $ScL_{6-k}X_k$ d^0 complexes with trialkylphosphine ligands L and X = Cl performed in [134, 135] was concerned with solvent effects on the relative stabilities of isomers. First it was found that in acetonitrile, *cis* and *trans* isomers were formed in concentrations somewhat deviating from the statistical values in favor of the *trans* isomers [134] which was evidence of preferable stabilization of the latter in comparison with *cis* isomers. However later [135], a decrease in solvent molecule polarity was observed to cause a rapid increase in the concentration of *cis* isomers, see Table 10. It is reasonable to assume that a decrease in solvent molecule polarity is accompanied by weakening of solvent effects on the properties of solutes. The obtained data therefore provide evidence for the predominant stability of *cis* isomers of trivalent scandium complexes.

Table 10. Dependence of the $K_i = [cis]/[trans]$ ratio for the $ScCl_3[P(CH_3)_3]_3$ complex on dipole moment μ of solvent molecules [134, 135]

Solvent	μ, D	K_i
Diethyl ether	1.25	∞
Tetrahydrofuran	1.70	13 : 1
Ethyl acetate	1.85	10 : 1
Acetone	2.86	1 : 2.8
Nitromethane	3.57	1 : 2
Acetonitrile	3.96	1 : 3

Note also that in $MBr_{6-k}Cl_k$ complexes of Pt(IV) and Nb(V), the ratio K_i between isomer concentrations coincides with the statistical value [137, 138]. The closeness of the energies of the *cis* and *trans* isomers of these compounds is explained by the closeness of the Br and Cl ligands in the static *trans*-influence series. A similar situation is characteristic of mixed fluorohalo Nb(V), Ta(V), and W(VI) d^0 complexes of the type

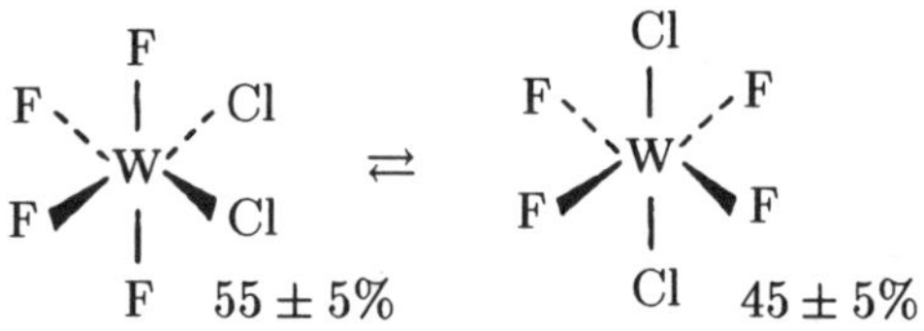

Solutions of these compounds contain both their *cis* and *trans* isomers. The concentration ratio between the isomers is close to unity, which even appears to indicate some stabilization of the *trans* form [86].

7. To conclude this section, consider quasi-octahedral $ML_{6-k}XY...Z$ transition metal complexes with ligands of more than two types. Applying the approximation of frontier MOs makes it clear [115] that for σ-bonded d^0 and d^6 complexes, the more stable forms are those in which ligands with the largest difference of electronegativities (most distant from each other in the static *trans*-influence series) are positioned *trans* with respect to each other. Indeed, relaxation stabilization of the complexes under consideration involves mixing of the t_{1u} and e_g^* orbitals under the action of the antisymmetric component of operator $\hat{H}_S$, which is maximum for the *trans* arrangement of such ligands.

The mutual arrangement of strong- and weak-*trans*-effect ligands predicted theoretically was observed experimentally for complexes with alkyl and hydride substituents [81]. For instance, for $PtHal_2(PR_3)_2Me_2$ complexes, geometric considerations allow the existence of five isomers I–V:

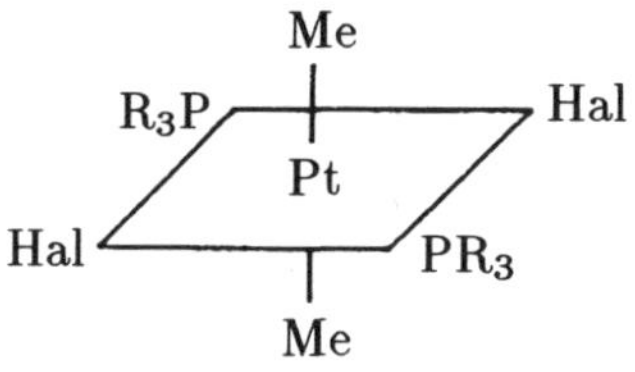

I

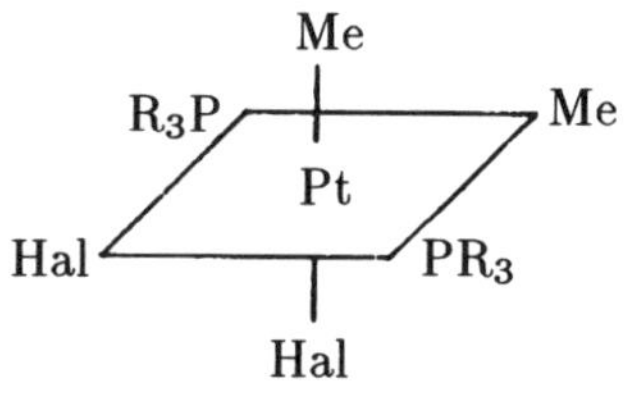

II

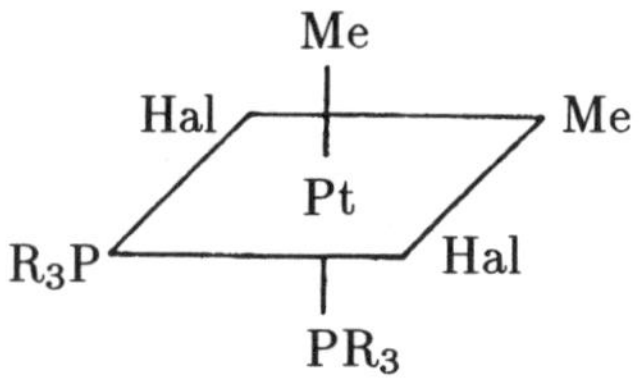

III

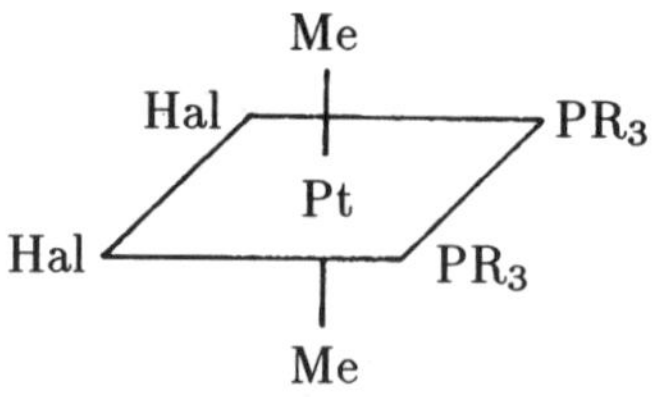

IV

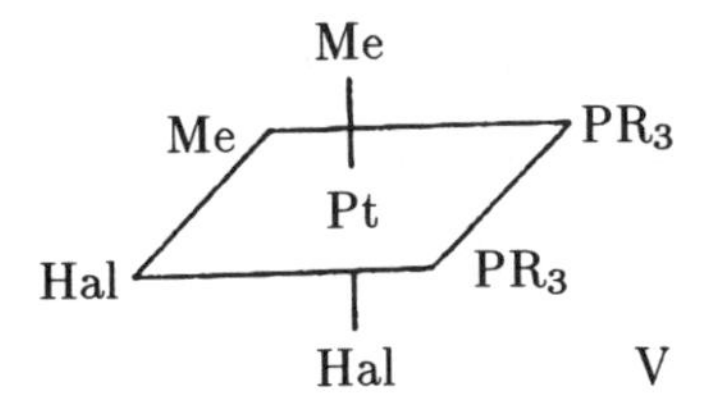

As the t_{1u} and e_g^* frontier MOs of σ-bonded d^6 complexes possess different parities, and the transition from ML_6 to complex I is described by an even substitution operator, the sum of the energies of electronic and geometric relaxations for complex I should be zero. For the other complexes, this sum is nonzero and approximately proportional to the squares of the following differences of the Coulomb orbitals of ligand σ AOs:

$$\text{isomer II}: \quad 2\{\alpha(\text{Me}) - \alpha(\text{Hal})\}^2 = 2\Delta\alpha^2_{\text{Me,Hal}},$$

$$\text{isomer III}: \quad 2\{\alpha(\text{Me}) - \alpha(\text{PR}_3)\}^2 = 2\Delta\alpha^2_{\text{Me,PR}_3},$$

$$\text{isomer IV}: \quad 2\{\alpha(\text{PR}_3) - \alpha(\text{Hal})\}^2 = 2\Delta\alpha^2_{\text{PR}_3\text{,Hal}},$$

$$\text{isomer V}: \quad \{\alpha(\text{Me}) - \alpha(\text{Hal})\}^2 + \{\alpha(\text{Me}) - \alpha(\text{PR}_3)\}^2 + \{\alpha(\text{PR}_3) - \alpha(\text{Hal})\}^2 = 2\big(\Delta\alpha^2_{\text{Me,Hal}} - |\Delta\alpha^2_{\text{Me,PR}_3}\Delta\alpha^2_{\text{PR}_3\text{,Hal}}|\big).$$

Judging from the strength of the *trans*-effects produced by the ligands (see Sections 4.1 and 4.4), the $|\Delta\alpha_{\text{Me, Hal}}| > |\Delta\alpha_{\text{Me, PR}_3}|$ and $|\Delta\alpha_{\text{Me, Hal}}| > |\Delta\alpha_{\text{PR}_3\text{, Hal}}|$ inequalities should hold. Therefore, the energy of geometric and electronic relaxation should be maximum for II. It follows that this complex should possess the highest stability, which fully agrees with the experimental data [81, 131].

To summarize, it follows from the data on the structure and stability of d^0 and d^6 σ-bonded complexes that on the whole, experiment substantiates theory within the scope of theory applicability. When ligands of different kinds have noticeably different orbital electronegativities (are situated far apart in the *trans*-influence series), *cis* isomers of $ML_{6-k}X_k$ have the highest stabilities. The opposite is true when against the background of an insignificant energy of geometric and electronic relaxation, factors that were not taken into account in deriving (2.58) come to the fore. Such factors may include direct interligand interactions, solvent effects, etc. In complexes containing ligands of three kinds, the *trans* arrangement of the ligands characterized by the largest difference of their electronegativities is stabilized.

3.9. Isomers of Nontransition Element Complexes

1. Consider isomers of quasi-octahedral nontransition element complexes and molecules [139, 145]; first, it should be noted that the relative stability of the *cis* and *trans* forms of these species has received much attention from researchers. From the point of view of steric interligand interactions, the problem was discussed in [140] for the example of SnL_4X_2 complexes. It was shown that if monoatomic ligands L and X had substantially different steric characteristics, the *cis* isomers were stabilized.

Another approach based on the Gillespie–Nyholm model of repulsion between electron pairs [102, 102a, 102b] was developed in [141, 142], where it was shown that *trans*-ML_4X_2 complexes were stabilized when the electronegativities of L and X were substantially different. The authors of [143] explained a decrease in stability of *cis* isomers of $SnCl_4X_2$ complexes with dimethylchalcogenide ligands X in the series $Me_2O > Me_2S > Me_2Se$ by destabilization of Sn(IV) $3d$ orbitals as a result of a decrease in ligand electronegativity.

The scheme of MOs in nontransition element complexes was used to discuss the relative stabilities of *cis* and *trans* isomers in terms of the model of three-center bonds. It was noted in [144] that in ML_4X_2 complexes, stability of *trans* isomers increased from less to more electropositive M atoms provided that the donor properties of L and X were fairly different.

2. After this short digression into the history of the problem, let us turn to an analysis of the relative stabilities of the isomers of quasi-octahedral 12-electron σ-bonded complexes of nontransition elements based on the theory described in Section 2.4. As in the preceding section, we will again use the approximation of frontier MOs and equation (2.57). As was mentioned in Section 3.5, the HOMOs in these complexes are the even e_g MOs. The a_{1g}^* LUMO is also even. The even e_g and a_{1g}^* MOs can only be mixed by a perturbation of E_g symmetry. Therefore the formation of the more stable $ML_{6-k}X_k$ complex from its homoligand predecessor ML_6 should be described by the H_S substitution operator with the largest E_g component.

According to Table 3, the E_g component of operator H_S is zero for *cis*-ML_3X_3 and nonzero for *trans*-ML_3X_3. It follows that electronic and geometric relaxations stabilize the *trans* form of ML_3X_3 complexes under consideration. This conclusion is also valid for ML_4X_2 complexes, which, however, require a more detailed analysis. According to Table 3, the H_S operator mixes the a_{1g}^* MO only with the e_{g,z^2} one and does not mix it with the e_{g,x^2-y^2} MO, because the decomposition of H_S does not contain the $H_{E_g}^2$ component for either *cis*- or *trans*-ML_4X_2.

According to (3.16), the corresponding matrix elements are

$$S_{z^2,a^*}(cis\text{-}\mathrm{ML_4X_2}) = (c_e l_e/3\sqrt{2})\Delta\alpha,$$
$$S_{z^2,a^*}(trans\text{-}\mathrm{ML_4X_2}) = (2c_e l_e/3\sqrt{2})\Delta\alpha.$$

Applying (2.57) shows that electronic and geometric relaxations accompanying the $ML_6 \longrightarrow ML_{6-k}X_k$ substitution occur no matter whether the *cis* or the *trans* isomer is formed, but the energy of relaxation-induced stabilization is four times larger for the *trans* isomer.

With certain reservations (see Section 3.6), the conclusion of predominant stabilization of *trans* isomers can be applied to 12-electron ML_4E_2 complexes with lone pairs E. For these complexes, ligand vacancies E (lone pairs) should be positioned *trans* with respect to each other in octahedra, which, as in the Gillespie–Nyholm model [102], leads to square-planar configurations of ML_4E_2 complexes.

3. The relevant experimental material includes structural data, the enthalpies of some *cis-trans* isomerization reactions, and isomerization constants. Clearly, for complexes containing ligands with maximally different orbital electronegativities (for instance, when L = halogen and X = hydride ion or alkyl radical), we should expect maximal stabilization of *trans* isomers. The complexes $SnMe_2Hal_4^{2-}$ with Hal = F, Cl, Br, and I; $PH_2F_4^-$ [65, 146, 147]; $PR_2F_4^-$ with R = Me and Ph [148, 149]; and SR_2F_4 with R = Pr and Ph [150] have the expected *trans* configurations. Square-planar structures of 12-electron ML_4E_2 complexes were observed in structure studies of the XeF_4 molecule and Te(II) complexes [102, 102a, 151] and the ClF_4^-, BrF_4^-, IF_4^- [102a, 152], and ICl_4^- [102a] interhalide anions.

When ligand L and X electronegativities in 12-electron nontransition element complexes approach each other, we should observe weakening of the stabilization of *trans* complexes in comparison with *cis*. Experimental data show that in conformity with this conclusion, such complexes occur as mixtures of both isomers. The relative isomer stabilities can be inferred from the enthalpies of isomerization. The enthalpies of isomerization were measured for $SnHal_4X_2$ (Hal = Cl and Br and X = Me_2Se and hexamethylphosphorotriamide) in [143], where it was found that the *trans* form was more stable than the *cis* by several tenths of kcal/mol. Stabilization of the *trans* form was also reported in [153, 154]; according to these works, temperature lowering shifted the *cis–trans* equilibria in $SnF_4(EtOH)_2$ and $SnF_4(PyO)_2$ (PyO stands for substituted pyridine N-oxides) in favor of the *trans* form.

In [143, 153, 155–157], the relative stabilities of the isomers of the complexes under consideration were characterized by isomerization constant values. Unlike similar constants for transition metal complexes, those discussed in this section were calculated as the ratio of the concentration of *trans* isomers to that of their *cis* counterparts, $K_i = [trans\text{-}ML_4X_2]/[cis\text{-}ML_4X_2]$; the higher stability of *trans* isomers therefore corresponded to $K_i > (1/4)$, and the higher stability of *cis* ones, to $K_i < (1/4)$.

Table 11. Isomerization constants and predominant isomer types for quasi-octahedral σ-bonded nontransition element complexes

Complex	K_i, isomer type	T, K	Refs.
$SnCl_4(Me_2O)_2$	0.64	175	[143]
$SnCl_4(Me_2S)_2$	0.79	190	[143]
$SnCl_4(Me_2Se)_2$	1.37	190	[143]
$SnBr_4(Me_2S)_2$	1.38	170	[143]
$SnBr_4(Me_2Se)_2$	2.33	180	[143]
$SnF_4(EtOH)_2$	1.0	—	[153]
$SnF_4(NSF_2)_2$	3.3	—	[155]
$SnF_4(CF_3)_2^{2-}$	1.5	—	[156]
$TeF_4(OR)_2$	< 0.25	—	[157]
$PF_4(CF_3)_2^-$	*trans*-isomer	—	[145]
$SF_4(CF_3)_2$	1.5	—	[145]
$GeF_4(CF_3)_2^{2-}$	9	—	[145]
$AsF_4(CF_3)_2^-$	*cis*-isomer	—	[145]
$SiF_4(ROH)_2$	*trans*-isomer	—	[145]
$GeF_4(ROH)_2$	0.18–0.33	—	[145]
$SnF_4(ROH)_2$	0.25	—	[145]

The isomerization constants of neutral complexes, for which medium effects are less significant than for charged complexes, are given in Table 11; the statistical values being taken into account, these data evidence stabilization of *trans* isomers. For MF_4X_2 complexes with weak-*trans*-effect ligands such as $X = H_2O$ [158], MeOH [159], EtOH, and ethyl acetate [160], the isomerization constants were close to 1/4 at temperatures of 183 to 238 K, that is, the isomers virtually had the same stability, in agreement with theoretical predictions. Lastly, note that for ionic complexes with ligands L and X more or less close to each other in the *trans*-influence series and in electronegativities (for instance, for $SnF_4I_2^-$), data on K_i are indicative of stabilization of their *cis* isomers [161]. This discrepancy between theory and experiment can be due to either medium or interligand interaction effects. As noted in [143], an increase in solvent polarity shifts *cis–trans* equilibria toward *cis* isomers even in solutions of neutral Sn(IV) complexes. We conclude that on the whole, theoretically predicted stabilization of *trans* isomers in σ-bonded 12-electron nontransition element complexes is in conformity with experimental data.

4. For nontransition element complexes, data for the dependence of isomerization constants K_i on the position of the element in the Periodic Table of the Elements are available. A comparison of isomerization constants measured in solutions for complexes of different elements M having the same ligand composition (Table 11) shows that along periods, relative contents of *trans* isomers decrease. For instance, $K_i[PF_4(CF_3)_2]^- \gg K_i[SF_4(CF_3)_2]$ and $K_i[GeF_4(CF_3)_2]^{2-} \gg K_i[AsF_4(CF_3)_2]^-$, although for the first three complexes, the *trans* isomers are more stable than the *cis* isomers. The data on K_i are also indicative of a decrease in the relative stabilities of *trans* isomers along Groups of elements, e.g., $K_i[SiF_4(ROH)_2] \gg K_i[GeF_4(ROH)_2] \gg K_i[SnF_4(ROH)_2]$ and $K_i[PF_4(CF_3)_2]^- \gg K_i[AsF_4(CF_3)_2]^-$. For this reason, the *cis* isomers of the Te fluoro complexes of the compositions $TeF_4(OR)_2$ and $TeF_4(NMe_2)_2$ possess higher stabilities than their *trans* counterparts [145]. Because tellurium is situated in the bottom right corner of the Periodic Table of the Elements, both tendencies manifest themselves to a maximum degree.

This dependence of the relative stabilities of isomers in nontransition element complexes can be understood if the relation between isomer stabilities and mutual ligand effects is taken into consideration. As mentioned in Sections 3.5 and 3.6, the ML_5X complexes of elements M situated in the right part of the Periodic Table of the Elements are characterized by weakening of steric substitution effects even to their disappearance for Group VI elements. This decrease and then disappearance of geometric relaxation of heteroligand $ML_{6-k}X_k$ complexes tends to make the relative stabilities of different geometric isomers equal for the elements of the second half of the Periodic Table, in agreement with experimental observations.

It is more difficult to explain a decrease in the stability of *trans* isomers along Groups of elements and the reversal of the relative stabilities of the *cis* and *trans* isomers for tellurium (if these observations characterize the complexes rather than solvent effects). The phenomenon can be related to an increase in the contribution of the central atom d orbitals to the bonding e_g MO. It follows from (2.58) that the stabilization energy of the *trans* isomer is $\mathcal{E}_t(ML_{6-k}X_k) \sim c_a^2 l_e \Delta\alpha/\omega_{ea^*}$, where the coefficient $l_e = \sqrt{1 - c_e^2}$ of the group ligand orbital in the e_g MO decreases with increasing contribution c_e of vacant nd(M) AOs.

Lastly, we wish to stress once more that the treatment given above is valid for nontransition element complexes with σ-bonded ligands and does not apply to complexes with multiply bonded ligands X such as $SnHal_4X_2$ containing π-acceptor ligands (X = Me_2CO, MeCN, and Me_3PO) [143]. The experimental data [143] show that for these complexes, *cis* isomers are more stable. An analysis of such systems, however, requires using a scheme of MOs different from that described above.

CHAPTER 4

MUTUAL INFLUENCE OF LIGANDS AND RELATIVE ISOMER STABILITY IN σ-BONDED QUASI-SQUARE-PLANAR AND QUASI-TETRAHEDRAL COMPLEXES

In the preceding chapter, we considered quasi-octahedral complexes and molecules. This one will be concerned with quasi-square-planar and quasi-tetrahedral species. An analysis of mutual influence of ligands in quasi-square-planar molecules and complexes will be followed by a consideration of the relative stabilities of their *trans* and *cis* isomers. Mutual influence of ligands in quasi-tetrahedral complexes, which do not form *trans* and *cis* isomers, will be discussed in the final section. Apart from *cis–trans* isomerization in quasi-square-planar complexes, thermochemical data on isomerization of chain platinum complexes of the Magnus salt type will be considered.

4.1. Planar Transition Metal Complexes: Bond Lengths

1. So far as mutual influence of ligands in quasi-square-planar complexes is concerned [56, 162], note that, as with quasi-octahedral complexes, the characteristic feature of the approach that will be used is the possibility of treating the qualitative picture of steric distortions due to the replacement of a part of the ligands in a complex without recourse to detailed computational data on the energies of molecular levels, the coefficients of AOs, etc. For the most part, we will only need a qualitative scheme of one-electron levels and some general data on the AO composition of MOs.

A typical scheme of σ MOs for homoligand square-planar transition metal complexes ML_4 of the $Pt^{II}L_4$ type is shown in Fig. 13. Using the same notation as in Chapter 3 and the numbering of ligands given in

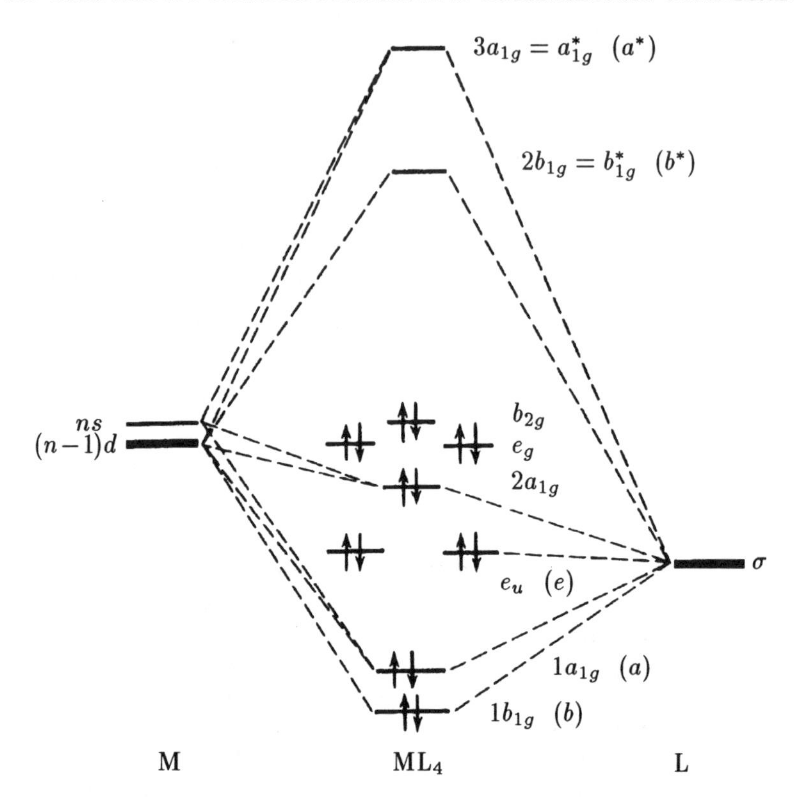

FIG. 13. QUALITATIVE ENERGY LEVEL DIAGRAM OF σ MOs FOR SQUARE-PLANAR $Pt^{II}L_4$-TYPE COMPLEXES; THE b_{2g}, e_g, AND $2a_{1g}$ ORBITALS EITHER DO NOT INCLUDE OR INCLUDE SMALL CONTRIBUTIONS FROM LIGAND σ AOs.

Fig. 14, we can write these MOs as

$$
\begin{aligned}
1a_{1g} &: \ \psi_a = c_a\chi_g + (l_a/2)(\sigma_1+\sigma_2+\sigma_3+\sigma_4), \\
2a_{1g} &: \ \psi_{2a} \approx \chi_g', \\
3a_{1g} &: \ \psi_a^* = l_a\chi_g - (c_a/2)(\sigma_1+\sigma_2+\sigma_3+\sigma_4), \\
1b_{1g} &: \ \psi_b = c_b d_{x^2-y^2} + (l_b/2)(\sigma_1+\sigma_2-\sigma_3-\sigma_4), \\
2b_{1g} &: \ \psi_b^* = l_b d_{x^2-y^2} - (c_b/2)(\sigma_1+\sigma_2-\sigma_3-\sigma_4), \\
e_u &: \ \psi_{e,x} = (1/\sqrt{2})(\sigma_1-\sigma_2), \quad \psi_{e,y} = (1/\sqrt{2})(\sigma_3-\sigma_4),
\end{aligned}
\tag{4.1}
$$

where χ_g and χ_g' are two different mutually orthogonal central atom AOs of A_{1g} symmetry formed from its valence $(n-1)d$ and ns AOs. (Unlike octahedral ML_6 complexes, square-planar ML_4 ones have three

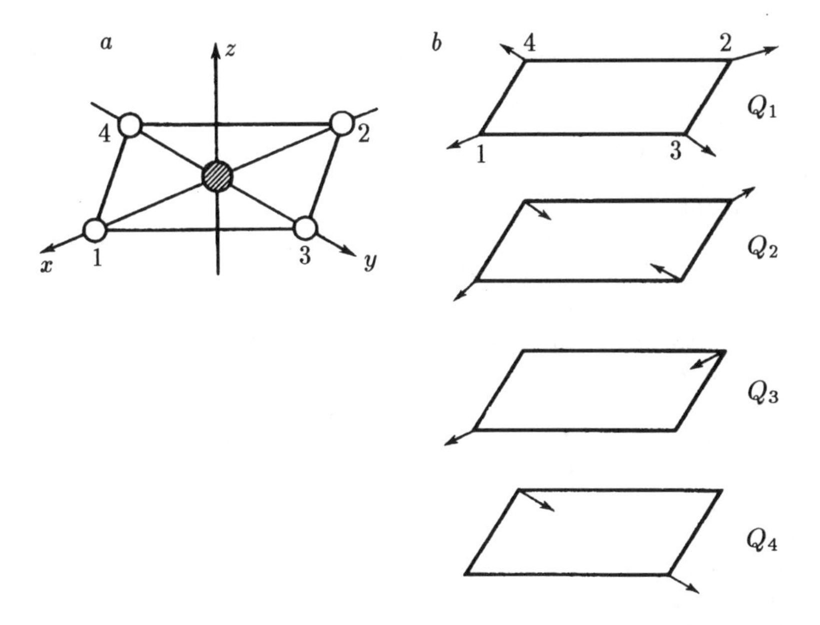

FIG. 14. (*a*) FRAME OF REFERENCE AND NUMBERING OF LIGANDS FOR SQUARE-PLANAR ML_4 COMPLEXES AND (*b*) NORMAL STRETCHING MODES FOR SQUARE COMPLEXES (ALL $Q_\nu > 0$).

MOs corresponding to the A_{1g} irreducible representation, and generally, each of the three a_{1g} MOs can contain $(n-1)d_{z^2}$ and ns central atom AOs and symmetrized combinations of ligand σ AOs that transform under the A_{1g} representation, see Section 1.4.) Specific expressions for χ_g and χ_g' in terms of d_{z^2} and s AOs will not be needed.

Formulas (4.1) take into account the photoelectron and X-ray spectroscopy data [163, 164] and the results of quantum-chemical calculations [20, 163, 165, 166], according to which vacant metal np AOs make an insignificant contribution to bonding, which allows us to use basis sets only including the $(n-1)d$ and ns metal AOs. In addition, the expressions for the $1a_{1g}$, $2a_{1g}$, and $3a_{1g}$ MOs follow from the calculations [163], the results of which are in agreement with the photoelectron and X-ray spectroscopy data. The same expressions are obtained by applying the standard procedure of composing MOs from $(n-1)d$ and ns central atom AOs and ligand σ AOs on the assumption that $|\alpha_s - \alpha_d| \ll |\alpha_s - \alpha_\sigma| \approx |\alpha_d - \alpha_\sigma|$; this inequality holds for the energies of metal $(n-1)d$ and ns and ligand σ orbitals.

Apart from the scheme of MOs, analyzing mutual influence of ligands requires knowledge of deformation types possible for square-planar ML_4

complexes; M—L stretching vibration modes and symmetry types are represented in Fig. 14 and Table 2.

It has repeatedly been noted that mutual influences of ligands can most conveniently be treated in the approximation of frontier MOs. As in Chapter 3, it will be assumed that substituents X differ from L by Coulomb integral values for σ AOs. This leads us to select the e_u HOMO and the b_{1g}^* LUMO as the frontier MOs in σ-bonded complexes of the $Pt^{II}L_4$ type. Indeed, the e_g, b_{2g}, and $2a_{1g}$ MOs have higher energies than the e_u MO, but they do not contain contributions from ligand σ AOs and cannot therefore play the role of the HOMO.

Let substituent X occupy position 1 in ML_3X (Fig. 14). Then it follows from (4.1) and selection rules (2.25) and (2.43b) that in the approximation of frontier MOs, there remains a single term in sum (2.56); this term describes mixing of the $e_{u,x}$ and b_{1g}^* orbitals (See Table 5). Accordingly, the adiabatic potential minimum is characterized by one nonzero coordinate $Q_\nu^f = Q_3^f$. To describe the distortion qualitatively, it suffices to determine the sign of Q_3^f (compare Section 3.3), which reduces to finding the sign of the numerator in (3.8). The necessary matrix element of the substitution operator and the vibronic coupling constant can easily be found by substituting the expression for the frontier MOs into (2.39) and (2.27). This yields

$$SA = S_{xb^*} A_{xb^*}^3 = -(\sqrt{2}/4) c_b l_b [\partial\beta(\sigma, d)/\partial R]_0 \Delta\alpha, \qquad (4.2)$$

where the sign of $\Delta\alpha$ depends on the orbital electronegativities of ligand L and substituent X, $\Delta\alpha = \alpha(X) - \alpha(L)$. In the right-hand side of (4.2), the c_b and l_b coefficients are positive by definition, and the derivative of the resonance integral with respect to interatomic distance R is also positive, because the integral is negative and decreases in magnitude as R increases, $[\partial\beta(\sigma, d)/\partial R]_0 > 0$. Consequently, $S_{xb^*} A_{xb^*}^3 < 0$.

We arrive at the conclusion that the $ML_4 \longrightarrow ML_3X$ replacement of ligand L by a more electropositive ligand X ($\Delta\alpha > 0$) at position 1 should deform the system along the Q_3 coordinate, and the sign of Q_3^f should be negative. Conversely, if ligand X is more electronegative than L, then the deformation along the same Q_3 coordinate should be positive. Figure 14 shows that the inequality $Q_3^f < 0$ corresponds to elongation of the $M-L_2$ bond positioned *trans* with respect to the substituent, whereas if $Q_3^f > 0$, this bond becomes shorter. Deformations along the Q_3 coordinate leave the $M-L_3$ and $M-L_4$ *cis* bonds intact. These considerations show that the well-documented *trans*-influence of X in square-planar complexes of the $Pt^{II}L_3X$ type can be qualitatively explained even in the approximation of frontier MOs.

2. We mentioned in Section 3.6 that sometimes it is worthwhile to go beyond the approximation of frontier MOs in order to check or complement the conclusions drawn within the framework of this approximation by taking into account transitions between all MOs possible in the selected basis set of AOs. The simplest way to do this is through using the "method of forces", which implies an analysis of the $E_2^{(1)}$ term, linear in Q_ν, in equation (2.55) for the adiabatic potential of the substituted complex (see Section 3.6).

Consider a d^8 ML_3X complex taking into account all σ MOs. The expression for $E_2^{(1)}$ then contains six terms (the b_{2g}, e_g, and $2a_{1g}$ MOs do not contribute to $E_2^{(1)}$):

$$\begin{aligned}
W_{aa^*} &= (1/4)(l_a^2\omega_{aa^*}^{-1} - c_a^2\omega_{aa^*}^{-1})\mu_a q_1 \Delta\alpha,\\
W_{ab^*} &= (1/4)(l_a^2\omega_{ab^*}^{-1}\mu_b - c_b^2\omega_{ab^*}^{-1}\mu_a) q_2 \Delta\alpha,\\
W_{ba^*} &= (1/4)(l_b^2\omega_{ba^*}^{-1}\mu_a - c_a^2\omega_{ba^*}^{-1}\mu_b) q_2 \Delta\alpha,\\
W_{bb^*} &= (1/4)(l_b^2\omega_{bb^*}^{-1} - c_b^2\omega_{bb^*}^{-1})\mu_b q_1 \Delta\alpha,\\
W_{ea^*} &= (1/2)\omega_{ea^*}^{-1}\mu_a q_3 \Delta\alpha,\\
W_{eb^*} &= (1/2)\omega_{eb^*}^{-1}\mu_b q_3 \Delta\alpha.
\end{aligned} \tag{4.3}$$

Here, as in Section 3.6, q_i are the nonnormalized normal coordinates, that is, $Q_1 = (1/2)q_1$ etc., and the notation

$$\begin{aligned}
\mu_a &= (1/2)c_a l_a \left[\partial\beta(\sigma, \chi_g)/\partial R\right]_0,\\
\mu_b &= (1/2)c_b l_b \left[\partial\beta(\sigma, d_{x^2-y^2})/\partial R\right]_0.
\end{aligned} \tag{4.4}$$

is used. Combining the coefficients of identical q_i terms in the expression for $E_2^{(1)}$ and performing the transformation to Cartesian coordinates, we obtain expressions for forces F_{trans} and F_{cis} acting on *trans* and *cis* ligands. (Recall that $F_{trans} > 0$ and $F_{cis} > 0$ imply elongation of *trans* and *cis* bonds, respectively.) Let us also introduce the $\Delta F = F_{trans} - F_{cis}$ value; if $\Delta F > 0$, the *trans* bond is elongated to a greater extent than the *cis* ones. We then have

$$\begin{aligned}
\frac{F_{trans}}{\Delta\alpha} &= \left(\frac{2}{\omega_{ea^*}} - \frac{l_a^2}{\omega_{aa^*}} - \frac{l_b^2}{\omega_{ba^*}} + \frac{c_a^2}{\omega_{aa^*}} + \frac{c_b^2}{\omega_{ab^*}}\right)\mu_a\\
&\quad + \left(\frac{2}{\omega_{eb^*}} - \frac{l_a^2}{\omega_{ab^*}} - \frac{l_b^2}{\omega_{bb^*}} + \frac{c_b^2}{\omega_{bb^*}} + \frac{c_a^2}{\omega_{ba^*}}\right)\mu_b, \qquad (4.5)\\
\frac{F_{cis}}{\Delta\alpha} &= \left(\frac{l_b^2}{\omega_{ba^*}} - \frac{l_a^2}{\omega_{aa^*}} + \frac{c_a^2}{\omega_{aa^*}} - \frac{c_b^2}{\omega_{ab^*}}\right)\mu_a
\end{aligned}$$

$$-\left(\frac{l_b^2}{\omega_{bb^*}} - \frac{l_a^2}{\omega_{ab^*}} + \frac{c_a^2}{\omega_{ba^*}} - \frac{c_b^2}{\omega_{bb^*}}\right)\mu_b, \qquad (4.6)$$

$$\frac{\Delta F}{\Delta\alpha} = \left(\frac{2}{\omega_{ea^*}} - \frac{2l_b^2}{\omega_{ba^*}} - \frac{2c_b^2}{\omega_{ab^*}}\right)\mu_a$$

$$+\left(\frac{2}{\omega_{eb^*}} - \frac{2l_a^2}{\omega_{ab^*}} + \frac{2c_a^2}{\omega_{ba^*}}\right)\mu_b. \qquad (4.7)$$

Generally, an analysis of these expressions is a mere repetition of what was said in Section 3.6. Taking into account the relative arrangement of energy levels reproduced in Fig. 13 and the trivial inequalities $c_i^2 < l_i^2 < 1$, we arrive at the conclusion that the coefficients of μ_a and μ_b in (4.5) and (4.7) are positive. The μ_a and μ_b values are also positive, and therefore, $(F_{trans}/\Delta\alpha) > 0$ and $(\Delta F/\Delta\alpha) > 0$. That is, the $ML_4 \longrightarrow ML_3X$ replacement of ligand L by a more electropositive ($\Delta\alpha > 0$) ligand X should cause elongation of the *trans* bond, both absolute and relative to the *cis* bonds. The extent of elongation is proportional to $\Delta\alpha$, that is, to the donor ability of X in comparison with L. When ligand L is replaced by a more electronegative ligand X ($\Delta\alpha < 0$), absolute and relative (in comparison with the *cis* bonds) shortening of the *trans* bond occurs.

The inclusion of all MOs gives a more detailed description of mutual influence of ligands in the complexes that we consider than the approximation of frontier MOs. Because of involvement of the Q_1 and Q_2 modes in the "complete" variant of the theory, the *cis* bonds do not remain unchanged. Their possible elongation is, however, smaller in comparison with the *trans* bond. Moreover, it follows from (4.6) that the expression for F_{cis} is obtained as the difference of two comparable values and is, therefore, sensitive to the special features of the electronic structure of ML_4. We cannot,therefore, *a priori* rule out the possibility that, under certain conditions, the $F_{cis}/\Delta\alpha$ value will turn negative, and the substitution with $\Delta\alpha > 0$ will cause *cis* bond shortening rather than elongation.

These conclusions can be drawn directly from (2.56) taking into account all σ MOs and the relative force constant values for vibrational modes of different symmetry types. It follows from the expressions for the vibrational modes of a square-planar complex that a change in the length of the bond positioned *trans* with respect to substituent 1 (Fig. 14) can be written as

$$\Delta R_{trans} = Q_1^f/2 + Q_2^f/2 - Q_3^f/\sqrt{2}\,. \qquad (4.8)$$

Similarly, the *cis* bond length will change by

$$\Delta R_{cis} - Q_1^f/2 - Q_2^f/2 + Q_4^f/\sqrt{2}\,. \qquad (4.9)$$

(Positive ΔR_{trans} and ΔR_{cis} values correspond to bond elongation, and negative ones, to bond shortening.) The $\Delta R_{trans} - \Delta R_{cis}$ difference determines the relative elongation of the *trans* bond in comparison with the *cis* ones. We have

$$\begin{aligned}\frac{\Delta R_{trans}}{\Delta\alpha} &= \left\{\frac{2}{K_E\omega_{ea^*}} - \frac{l_a^2}{K_A\omega_{aa^*}} - \frac{l_b^2}{K_B\omega_{ba^*}} + \frac{c_a^2}{K_A\omega_{aa^*}} + \frac{c_b^2}{K_B\omega_{ab^*}}\right\}\mu_a + \left\{\frac{2}{K_E\omega_{eb^*}} - \frac{l_b^2}{K_A\omega_{bb^*}}\right.\\ &\quad \left. - \frac{l_a^2}{K_B\omega_{ab^*}} + \frac{c_a^2}{K_B\omega_{ba^*}} + \frac{c_b^2}{K_A\omega_{bb^*}}\right\}\mu_b, \qquad (4.10)\end{aligned}$$

$$\begin{aligned}\frac{\Delta R_{cis}}{\Delta\alpha} &= \left(\frac{l_b^2}{K_B\omega_{ba^*}} - \frac{l_a^2}{K_A\omega_{aa^*}} + \frac{c_a^2}{K_A\omega_{aa^*}} - \frac{c_b^2}{K_B\omega_{ab^*}}\right)\mu_a\\ &\quad - \left(\frac{l_b^2}{K_A\omega_{bb^*}} - \frac{l_a^2}{K_B\omega_{ab^*}} + \frac{c_a^2}{K_B\omega_{ba^*}} - \frac{c_b^2}{K_A\omega_{bb^*}}\right)\mu_b, \qquad (4.11)\end{aligned}$$

$$\begin{aligned}\frac{\Delta R}{\Delta\alpha} &= \left(\frac{2}{K_E\omega_{ea^*}} - \frac{2l_b^2}{K_B\omega_{ba^*}} + \frac{2c_b^2}{K_B\omega_{ab^*}}\right)\mu_a\\ &\quad + \left(\frac{2}{K_E\omega_{eb^*}} - \frac{2l_a^2}{K_B\omega_{ab^*}} + \frac{2c_a^2}{K_B\omega_{ba^*}}\right)\mu_b. \qquad (4.12)\end{aligned}$$

Here, K_E, K_A, and K_B are the force constants for the E_u, A_{1g}, and B_{1g} modes, respectively. For the scheme of levels shown in Fig. 13, applying the Pearson concept of relative coordinate labilities [32] yields $K_E < K_B < K_A$. This estimate makes it easy to show that like $F_{trans}/\Delta\alpha$ and $\Delta F/\Delta\alpha$, $\Delta R_{trans}/\Delta\alpha$ and $\Delta R/\Delta\alpha$ are positive. For instance, a comparison of (4.5) and (4.10) shows that taking into account force constants (in (4.10)) results in a relative increase of the positive (first) term in braces in comparison with the negative (second or second and third) ones. Like the sign of F_{cis}, the sign of ΔR_{cis} and is determined by the difference of two comparable values.

We see that for $Pt^{II}L_3X$ complexes, the method of forces and direct use of (2.56) taking into account all σ-MOs lead to qualitatively similar conclusions and predict relative and absolute elongation (shortening) of the *trans* bond as a result of an electropositive (electronegative) $ML_4 \longrightarrow ML_3X$ substitution. Recall also that the type of geometric deformations resulting from multiple substitution in any heteroligand derivatives of closed-shell complexes is determined by superposition of deformations characteristic of monosubstituted species (see Section 3.3). As regards the transition metal complexes under consideration, this means that if $\Delta\alpha > 0$, the $ML_4 \longrightarrow$ *cis*-ML_2X_2 substitution should be accompanied by elongation of both M—L bonds. At the

Table 12. Metal–ligand bond lengths in quasi-square-planar Pt(II), Pd(II), and Te(II) complexes

Complex	Distance, Å			Refs.
	M—L_{cis}	M—L_{trans}	M—X	
MCl_4^{2-} (M = Pt, Pd)	2.30	2.30	—	[167a]
$PtCl_3(SMe_2)^-$	2.291–2.303 (Cl)	2.316 (Cl)	2.246 (S)	[167b]
cis-$MCl_2(PR_3)_2$ (M = Pt, Pd)	—	2.36–2.39 (Cl)	2.22–2.26 (P)	[168a]
cis-$Pd(P_2C_{26}H_{48})Cl_2$	—	2.376 (Cl)	2.233 (P)	[168b]
trans-$MCl_2(PR_3)_2$ (M = Pt, Pd)	2.31 (Cl)	—	2.30–2.38 (P)	[167a]
$[PtCl_3PEt_3]^-$	2.30 (Cl)	2.38 (Cl)	2.21 (P)	[167a]
trans-$PtHCl(PR_3)_2$	2.27 (P)	2.42 (Cl)	—	[169]
trans-$Pt(CH_2R)Cl(PR_3)_2$	2.29 (P)	2.42 (Cl)	—	[170a]
trans-$Pt(NH_2R)_2$(1,3-dimethyluracil-C_5)	2.02–2.05 (N)	2.40 (Cl)	1.90 (C)	[170b]
trans-$Pt(SiR_3)Cl(PR_3)_2$	2.30 (P)	2.46 (Cl)	2.32 (Si)	[171]
trans-$Pt(PEt_3)_2Br_2$	2.31 (P)	2.43 (Br)	—	[63]
trans-$PtHBr(PEt_3)_2$	2.26 (P)	2.56 (Br)	—	[63]
cis-$Pt(SnCl_3)_2Cl_2$	—	2.35 (Cl)	—	[172]
$PdCl_3SnCl_3$	2.37 (Cl)	2.38 (Cl)	—	[173]
$Te(tu)_4^{2+}$	2.69 (S)	2.69 (S)	—	[64]
cis-$Te(tu)_2Cl_2$	—	2.48 (S)	2.92 (Cl)	[64]
cis-$Te(tu)_2Br_2$	—	2.47 (S)	3.05 (Br)	[64]
trans-$Te(tu)_2ClPh$	2.61–2.74 (S)	3.61 (Cl)	2.11 (C)	[64]

same time, in *trans*-ML_2X_2 complexes, the M—L bonds will change less substantially and can even be shortened. These conclusions are in agreement with the *trans* influence patterns in quasi-square-planar Pt^{II} and Pd^{II} complexes reliably established by diffraction methods (Table 12) and the data on Rh^I complex with the bis(phosphine)-

3,5-bis(diphenylphosphinomethyl)pyridine chelating ligand [173b] and Au^{III} complex of the composition $[Au^{III}(dmtc)(damp)]$, where dmtc stands for Me_2NCS_2 and damp, for o-$C_6H_4CH_2NMe_2$ [173c]:

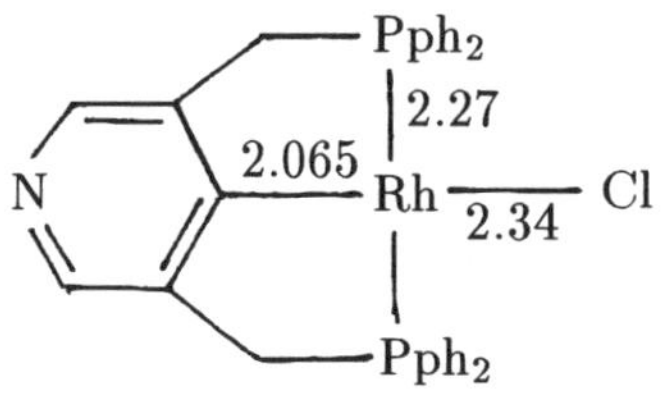

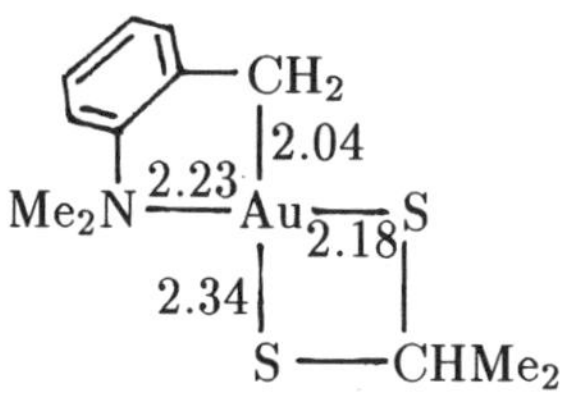

4.2. Planar Complexes of Nontransition Elements: Bond Lengths

1. The mutual influence of ligands in quasi-square-planar nontransition element complexes of the type of Te^{II} compounds is treated in the same way as for transition metal derivatives. A typical scheme of σ MOs for such ML_4 complexes is given in Fig. 15; these MOs can be written in the general form (the e_u and e_u^* MOs are represented by one orbital each)

$$\begin{aligned}
a_{1g} &: \ \psi_a = c_a s + (l_a/2)(\sigma_1 + \sigma_2 + \sigma_3 + \sigma_4), \\
a_{1g}^* &: \ \psi_a^* = l_a s - (c_a/2)(\sigma_1 + \sigma_2 + \sigma_3 + \sigma_4), \\
b_{1g} &: \ \psi_b = (1/2)(\sigma_1 + \sigma_2 - \sigma_3 - \sigma_4), \\
e_u &: \ \psi_{e,x} = c_e p_x + (l_e/\sqrt{2})(\sigma_1 - \sigma_2), \\
e_u^* &: \ \psi_{e,x}^* = l_e p_x - (c_e/\sqrt{2})(\sigma_1 - \sigma_2).
\end{aligned} \tag{4.13}$$

Equation (4.13) does not include vacant nd AOs of atom M; in nontransition element compounds, these AOs make smaller contributions to bonding than the valence ns and np AOs. In any event, the presence or absence of nd AOs in the basis set is of little consequence for the complexes under consideration, because their inclusion does not introduce qualitative changes in the character of frontier MOs (see below), which make the predominant contribution to the right-hand side of (2.56)–(2.58).

First, consider the problem in the approximation of frontier MOs, that is, the a_{1g}^* HOMO and the e_u^* LUMO. Let substituent X in ML_3X occupy position 1 (Fig. 14). We will again use formula (2.56) and apply selection rules (2.25) and (2.43b). This leaves only one term mixing the a_{1g}^* and $e_{u,x}^*$ MOs in (2.56). The minimum of the adiabatic potential of ML_3X is again characterized by a single coordinate, $Q_\nu^f = Q_3^f \neq 0$

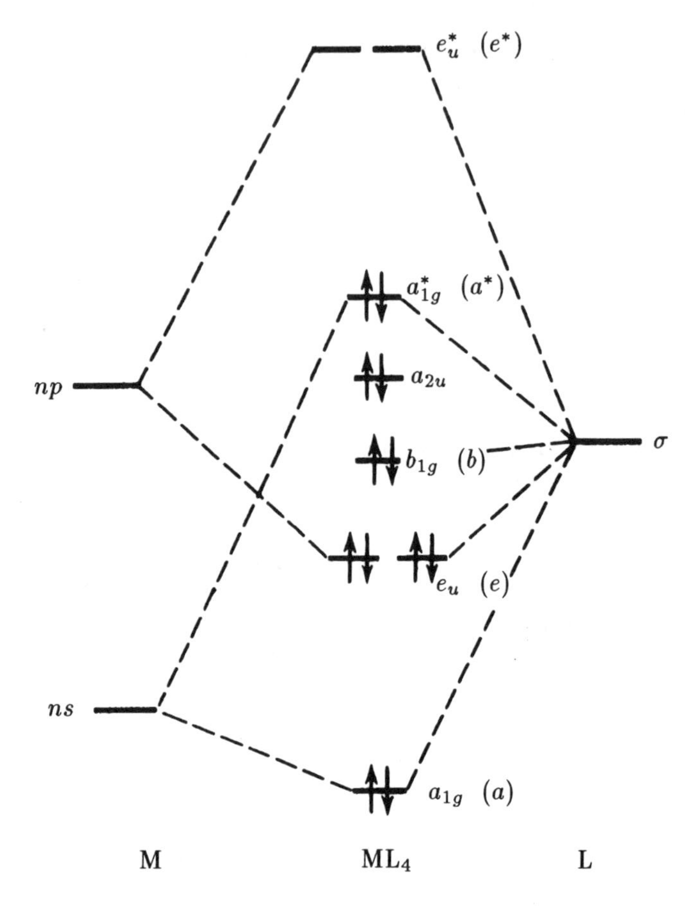

FIG. 15. QUALITATIVE SCHEME OF σ MOs FOR $Te^{II}L_4$-TYPE COMPLEXES.

(see Table 5); to determine its sign, it suffices to determine the sign of the product

$$SA = S_{a^* e^*_{u,x}} A^3_{a^* e^*_{u,x}} \tag{4.14}$$
$$= -\frac{\sqrt{2}}{4}\left\{ c_a l_a c_e^2 \left[\frac{\partial\beta(\sigma,s)}{\partial R}\right]_0 + \frac{c_a^2 c_e l_e}{\sqrt{2}} \left[\frac{\partial\beta(\sigma,p)}{\partial R}\right]_0 \right\}\Delta\alpha.$$

All values in (4.14) are positive (including the derivatives of resonance integrals), which means that the sign of SA is opposite to that of $\Delta\alpha$. This leads us to conclude that in nontransition element ML_3X complexes of the $Te^{II}L_3X$ type, *trans*-influence should be the same as in $Pt^{II}L_3X$-type transition metal complexes (see Section 4.1).

2. The same conclusion is reached when all σ MOs are included. Let us apply the method of forces to analyze mutual influence of ligands in such complexes. As follows from Fig. 15, the corresponding expression for $E_2^{(1)}$ contains four terms

$$\begin{aligned} W_{ae^*} &= (1/4)(l_a^2\omega_{ae^*}^{-1}\mu_e - c_e^2\omega_{ae^*}^{-1}\mu_a)q_3\Delta\alpha, \\ W_{a^*e^*} &= (1/4)(c_e^2\omega_{a^*e^*}^{-1}\mu_a + c_a^2\omega_{a^*e^*}^{-1}\mu_e)q_3\Delta\alpha, \\ W_{ee^*} &= (1/4)(l_e^2\omega_{ee^*}^{-1} - c_e^2\omega_{ee^*}^{-1})\mu_e(q_1+q_2)\Delta\alpha, \\ W_{be^*} &= (1/4)\omega_{be^*}^{-1}\mu_e q_3\Delta\alpha. \end{aligned} \tag{4.15}$$

Here, μ_a and μ_e have the form

$$\begin{aligned} \mu_a &= c_a l_a\big[\partial\beta(\sigma,s)/\partial R\big]_0, \\ \mu_e &= (c_e l_e/\sqrt{2})\big[\partial\beta(\sigma,p)/\partial R\big]_0. \end{aligned} \tag{4.16}$$

This gives

$$\begin{aligned} \frac{F_{trans}}{\Delta\alpha} &= \Big(\frac{c_e^2}{\omega_{a^*e^*}} - \frac{c_e^2}{\omega_{ae^*}}\Big)\mu_a \\ &+ \Big(\frac{1}{\omega_{be^*}} + \frac{c_a^2}{\omega_{a^*e^*}} - \frac{2(l_e^2 - c_e^2)}{\omega_{ee^*}} + \frac{l_a^2}{\omega_{ae^*}}\Big)\mu_e. \end{aligned} \tag{4.17}$$

According to Fig. 15, the coefficient of μ_a in (4.17) is positive. The coefficient of μ_e can also be shown to be positive. Indeed, the σ levels of typical ligands lie between the *ns* and *np* levels of nontransition elements such as Te (Fig. 3), which implies that the predominant contribution to the a_{1g} MO is made by the *ns* AO of M, whereas the e_u MOs largely consist of ligand AOs, that is, $c_a^2 \approx l_e^2$. It follows that in the second term in parentheses, even the sum of the first three terms is larger than zero, and, therefore, $(F_{trans}/\Delta\alpha) > 0$.

It follows that in agreement with the conclusions drawn in the approximation of frontier MOs, ligand X, more electropositive than L, should cause elongation of the *trans* bond in ML_3X square-planar complexes of nontransition elements in low oxidation states. Conversely, ligand X, more electronegative than L, should shorten the *trans* bond. As with transition metal complexes, this conclusion extends to multiply substituted species. For instance, the $ML_4 \longrightarrow$ *cis*-ML_2X_2 replacement of two ligands L by more electropositive (electronegative) ligands X should be accompanied by elongation (shortening) of two M$-$L bonds. The validity of these predictions and the results obtained in the approximation of frontier MOs is illustrated by Table 12.

4.3. Isomers of Planar Transition Metal ML_2X_2 Complexes

1. It is long since the suggestion was made that the relative stabilities of *cis* and *trans* isomers of quasi-square-planar transition metal complexes (experimental data on nontransition element derivatives is scanty) are related to the mutual influence of ligands. Not accidentally, the first attempt at obtaining thermochemical data on the relative stabilities of the *cis* and *trans* isomers of Pt^{II} complexes was made as early as in 1954 (Chernyaev *et al.*, [174]; also see [175]).

In this and the following sections, let us turn to a detailed consideration of the problem of the relative stabilities of *cis* and *trans* isomers of quasi-square-planar d^8 complexes of the ML_2X_2 and ML_2XY compositions. We will discuss complexes with M = Pt(II), Pd(II), and Au(III). In addition, we will consider $Cu^{II}L_2X_2$ d^9 complexes and thermal transformations of $[ML_4][MX_4]$ (M = Pt(II) or Pd(II)) salts.

As usual, the simplest and most comprehensible approach to the problem is through the use of the approximation of frontier MOs. We will proceed from the scheme of MOs of square-planar σ-bonded complexes of the $PtCl_4^{2-}$ type shown in Fig. 13; the basis set used to construct this scheme includes the $(n-1)d$ and ns AOs of the central atom and the σ AOs of ligands. Applying selection rules (2.43b), we find that in the approximation of frontier MOs, only the terms corresponding to the transition between the e_u and b_{1g}^* MOs should be retained in (2.57) and (2.58). The simple situation that we are considering can more conveniently be analyzed using (2.57) to estimate the energy of electronic and geometric relaxation for the *cis* and *trans* forms of ML_2X_2 rather than via isomerization energy calculations by (2.58). Let us use (4.1) for the e_u HOMO and the b_{1g}^* LUMO bearing in mind that the ψ_x and ψ_y MOs are mixed with the ψ_b^* one by the Q_3 and Q_4 modes of E_u symmetry, respectively (Fig. 14). This yields the orbital vibronic coupling constants and the matrix elements of the substitution operator in the form

$$A_{xb^*}^3 = A_{yb^*}^4 \neq 0, \quad S_{xb^*}^c = S_{yb^*}^c \neq 0, \quad S_{xb^*}^t = S_{yb^*}^t = 0.$$

It follows from (2.57) that in the approximation of frontier MOs, the change in the total energy of an ML_4 complex resulting from the $ML_4 \longrightarrow$ *trans*-ML_2X_2 substitution is given by the first, structurally independent, component S_0. At the same time, the formation of the *cis* isomer is characterized by an additional energy lowering described by the second and third terms on the right-hand side of (2.57). This leads us to conclude that the energy of *cis*-ML_2X_2 should be lower than that of *trans*-ML_2X_2 [55, 56]. We again stress that this result has a simple physical meaning and in essence, is a consequence of the symmetry properties of MOs. Because S_0 is independent of the structure, stabilization of one or another isomer is only possible when the second and third terms in (2.57) are nonzero, that is, when the HOMO and LUMO of ML_4 experience mixing as a result of ligand substitution. However, *trans* substitution symmetrical with respect to inversion (see Table 4) cannot mix the odd e_u MOs with the even b_{1g}^* one, whereas asymmetric *cis* substitution can. Stabilization of the *cis* isomer can also be explained in terms of the electronic structure of the heteroligand complex, *viz.*, at point Q_0, the energy of the HOMO is lower for *cis*-ML_2X_2 than for its *trans* counterpart.

Lastly, we again (at this point, at the level of the approximation of frontier orbitals) call reader's attention to interrelation between the stabilization of the *cis*-ML_2X_2 isomer and *trans*-influence of X [56]. As mentioned, static *trans*-influence involved in the $ML_4 \longrightarrow ML_3X$ substitution is explained by relaxation of bond lengths in the complex. The relaxation is due to forces acting on ligands in an unrelaxed molecule. In the approximation of frontier MOs, these forces act along the $X-M-L_{trans}$ coordinate and result in a E_u-type deformation (see Fig. 14). The replacement of two *trans* ligands in ML_4 causes balancing of the forces, and no relaxation occurs. However, when two *cis* ligands are replaced, the complex experiences deformation along two coordinates with a gain in geometric relaxation energy. According to (2.57), the energy effect of geometric and electronic relaxation increases approximately as $\Delta\alpha^2$. It can be inferred that *cis* isomer stabilization should heavily (quadratically) depend on $\Delta\alpha$, which characterizes *trans*-influence of substituent X.

2. Next, consider the relative isomer stabilities taking into account all σ MOs [56]. We will only be interested in stretching modes in (2.58), as usual, leaving bending modes (valence angle deformations) out of consideration. The σ MOs of the complexes will again be represented by (4.1). According to (2.43b), $S_{mn}^+ = S_{mn}^- = 0$ for the MOs that do not contain ligand AOs, which allows us to exclude the terms containing the b_{2g}, e_g, and $2a_{1g}$ MOs from (2.57). Further simplifications follow from selection rules (2.25) for vibronic coupling constants A_{mn}^{ν} and similar selection rules (2.43a) for S_{mn}. Lastly, the S_{mn}^- values can be expressed through S_{mn}^+. We eventually obtain

$$\Delta\mathcal{E}_{t\to c} = 2\Delta\alpha^2\left\{2\frac{(S^+_{xa})^2}{\omega_{ea^*}} + 2\frac{(S^+_{xb^*})^2}{\omega_{eb^*}} - \frac{(S^+_{ab^*})^2}{\omega_{ab^*}} - \frac{(S^+_{ba^*})^2}{\omega_{ba^*}}\right\}$$
$$+ \frac{16\Delta\alpha^2}{K_E}\left\{\frac{A^3_{xa^*}}{\omega_{ea^*}}S^+_{xa^*} + \frac{A^3_{xb^*}}{\omega_{eb^*}}S^+_{xb^*}\right\}^2$$
$$- \frac{8\Delta\alpha^2}{K_B}\left\{\frac{A^2_{ba^*}}{\omega_{ba^*}}S^+_{ba^*} + \frac{A^2_{ab^*}}{\omega_{ab^*}}S^+_{ab^*}\right\}^2. \quad (4.18)$$

The sign of $\Delta\mathcal{E}_{t\to c}$ is determined by substituting the S^+_{mn} values and the expressions for A^ν_{mn} into (4.18). With

$$A^2_{ba^*} = l_a l_b\left[\beta'(\sigma,\chi_g)\right]_0 - (\sqrt{3}/2)\left[\beta'(\sigma,d)\right]_0,$$
$$A^2_{ab^*} = (\sqrt{3}/2)l_a l_b\left[\beta'(\sigma,d)\right]_0 - c_a c_b\left[\beta'(\sigma,\chi_g)\right]_0,$$
$$A^3_{xb^*} = \sqrt{3}\,l_b\left[\beta'(\sigma,d)\right]_0,$$
$$A^3_{xa^*} = l_a\left[\beta'(\sigma,\chi_g)\right]_0,$$

where $\beta' = \partial\beta/\partial R$, (4.18) becomes

$$\Delta\mathcal{E}_{t\to c} = 2\Delta\alpha^2\left\{\frac{c_b^2}{4}\left(\frac{1}{\omega_{eb^*}} - \frac{l_b^2}{\omega_{eb^*}}\right) + \frac{c_a^2}{4}\left(\frac{1}{\omega_{ea^*}} - \frac{l_b^2}{\omega_{ba^*}}\right)\right\}$$
$$+ \frac{16\Delta\alpha^2}{K_E}\left\{\frac{1}{\omega_{ea^*}}\mu_a + \frac{1}{\omega_{eb^*}}\mu_b\right\}^2 \quad (4.19)$$
$$- \frac{8\Delta\alpha^2}{K_B}\left\{\left(\frac{l_b^2}{\omega_{ba^*}} - \frac{c_b^2}{\omega_{ab^*}}\right)\mu_a + \left(\frac{l_a^2}{\omega_{ab^*}} - \frac{c_a^2}{\omega_{ba^*}}\right)\mu_b\right\}^2,$$

where

$$\mu_a = (\sqrt{3}/4)l_a c_a\left[\beta'(\sigma,d)\right]_0,$$
$$\mu_b = (1/2)l_b c_b\left[\beta'(\sigma,\chi_g)\right]_0.$$

To analyze (4.19), consider the scheme of MOs. According to Fig. 13, $\omega_{ea^*} < \omega_{ba^*}$ and $\omega_{eb^*} < \omega_{ab^*}$. As, in addition, $l_a^2 < 1$ and $l_b^2 < 1$, we have

$$\frac{1}{\omega_{ea^*}} > \frac{l_b^2}{\omega_{ba^*}} - \frac{c_b^2}{\omega_{ab^*}}, \qquad \frac{1}{\omega_{eb^*}} > \frac{l_b^2}{\omega_{ab^*}} - \frac{c_a^2}{\omega_{ba^*}},$$

that is, the first expression in braces in (4.19) is positive, and the second expression in braces is larger than the third one. (In the limit of a purely covalent distribution of charges, $l_i^2 \approx c_i^2 \approx 1/2$, $\omega_{ba^*} \approx \omega_{ab^*}$, and the third expression in braces is close to zero.) The K_E and K_B force constants in the coefficients of the braced values usually have the same order of magnitude. Moreover, for the scheme of levels shown in Fig. 13, $K_E < K_B$ according to Pearson's concept of relative coordinate labilities [32].

We arrive at the conclusion that for the complexes that we are considering, $\Delta\mathcal{E}_{t\to c} > 0$, and the energy of the *cis* isomer is, as in the approximation of frontier MOs, lower than that of the *trans* one. It is, however, difficult to give an easily comprehensible interpretation of the result obtained because of the large number of MOs involved. Nevertheless, the conclusion that the stabilization of the *cis* isomer and *trans*-influence of ligand X are interrelated remains valid, because as previously, isomerization energy $\Delta\mathcal{E}_{t\to c}$ increases with the force acting on the *trans* ligand in ML_3X (or the deformation value).

3. The discussion given above shows that from the point of view of the theory that we use, the *cis* isomers of $Pt^{II}L_2X_2$-type d^8 complexes should be the more stable. This conclusion may, however, turn false under certain conditions [56], because our model ignores, e. g., electrostatic interligand repulsion, which favors stabilization of the *trans* form, where the repulsion is minimum. For this reason, *trans* isomers are sometimes more stable thermodynamically than *cis* ones.

We can draw conclusions on probable stabilization of either the *cis* or the *trans* form of ML_2X_2 from considerations based on approximate parallelism between static *trans*-influence of ligand X in ML_3X and *cis*-ML_2X_2 stabilization. According to (2.56), *trans* bond elongation Δl linearly depends on $\Delta\alpha$; therefore $\Delta\mathcal{E}_{t\to c} \sim (\Delta l)^2$. It follows that isomerization energy $\Delta\mathcal{E}_{t\to c}$ chainges fairly substantially (following a quadratic law) with the position of X in the *trans*-influence series. We may therefore expect that strong *trans*-influence of ligands X should stabilize the *cis* form, whereas additional factors such as electrostatic interligand repulsion may stabilize *trans*-ML_2X_2 isomers containing ligands X that have weak *trans*-influence on L.

5. The conclusions drawn above are valid for isolated complexes and should only with caution be extended to complexes in solutions and crystals. Nevertheless, a comparison with data on condensed phases shows that the theoretical results are not at variance with experiment. According to the structure data on $ML_{4-k}X_k$ d^8 complexes (L is Cl, and X is alkyl (R), PR_3, SR_2, or NH_3), alkyl groups cause maximum elongation of *trans* bonds (Table 13), which, in view of the considerations given above, means that *bis*-alkyl ML_2X_2 derivatives should be characterized by maximum stabilization of their *cis* forms. Indeed, Pt(II) and Au(III) d^8 *bis*-alkyl complexes have *cis* structures [81,

176]. Elongation of the Pt—Cl bond under the action of *trans*-PR_3 is about one third smaller than that caused by alkyl substituents, and the *trans* → *cis* isomerization energy for *bis*-phosphine complexes should be approximately two times lower than for *bis*-alkyl derivatives. The noticeable decrease in the isomerization energy corresponds with the existence of both isomers of $Pt(PR_3)_2Cl_2$ complexes, although the *cis* form is as previously more stable [74]; according to thermochemical data, the *trans* → *cis* isomerization reaction is exothermic for these complexes. The same holds true of $PtHal_2(AsR_3)_2$ and $PtHal_2(SbR_3)_2$ complexes [74, 177].

Table 13. Elongation (Δl) of Pt—Cl_{trans} bond in $Pt^{II}Cl_3X$ as a result of introduction of X into $PtCl_4^{2-}$ (R is alkyl)

X	Δl, Å
H, SiR_3, CR_3	0.11–0.12
SbR_3, AsR_3, PR_3	0.06–0.07
SR_2, SeR_2	0.02
NH_3	0

There exist data on isomerization of $Pd(PR_3)_2L_2$ complexes [178–180]. Here, L is Cl, N_3, 5-methyltetrazole, 5-trifluoromethyltetrazole, or NCO, and PR_3 is $(AC_6H_4)_nPMe_{3-n}$, where A = H, CH_3, CH_3O, or Cl. The heats of isomerization were measured accurate to 0.1 kcal/mol in a large number (11) of polar and nonpolar solvents. An analysis of combined data on isomerization of $Pt(PR_3)_2L_2$ and $Pd(PR_3)_2L_2$ in solutions led the authors to conclude that the *cis* isomers of the compounds were the more stable, although a considerable contribution to the observed heats of isomerization was made by interactions with solvents [178].

Thermochemical data on isomerization of Pt(II) and Pd(II) *bis*-phosphine complexes in the solid state also provide evidence of predominant stabilization of *cis* isomers. For instance, endothermic *cis* → *trans* isomerization was observed in heating $M(PR_3)_2X_2$ (X = Hal or $SnCl_3$, M = Pt or Pd) [181–184]. The only exception was exothermic transformation of *cis*-$Pt(PEt_3)_2)I_2$ into the *trans* form. Note, however, that this isomerization reaction occurs in melt, and one exothermic peak of the DTA curve corresponds to two processes, crystallization of the *trans* isomer and isomerization. A similar conclusion was drawn in [185], where isomerization of $Pt(SR_2)_2Hal_2$ in solutions was studied. As to *bis*-phosphine complexes, the transition of the *cis* into the *trans* form was found to proceed with heat absorption.

Lastly, consider works [186–188], where thermal transformations of *bis*-thioether complexes were studied in the solid state. The *trans* → *cis* isomerization of the pentamethylenesulfide and thioxane complexes was accompanied by heat evolution. However for the dimethyl- and diethylsulfide complexes, heat was released in the *cis* → *trans* process in the solid phase, although in solution, the *cis* isomer of the dimethylsulfide complex was more stable. This behavior was explained [187] by the difference in crystal lattice energies and its effect on isomerization in the solid state. The approach adopted in this work is not at variance with such an explanation. The SR_2 group exerts weaker influence on *trans* ligands than the PR_3 one, and *cis* isomers of sulfide complexes are stabilized to a less extent. For this reason, interactions of sulfide complexes with the medium may have a strong effect on the heat of isomerization.

The experimental data considered above substantiate the conclusion that $Pt^{II}Hal_2X_2$-type complexes with substituents X characterized by strong *trans*-influence are more stable in the *cis* form. The situation is different when ligands X have an insignificant, if any effect, on the *trans* bonds, as, e. g., when X is NH_3. This group virtually does not change the length of $Pt-Cl$ bonds in $[PtCl_3NH_3]^-$ in comparison with $PtCl_4^{2-}$ (Table 13). As mentioned, the relative stability of isomers may then be determined by factors disregarded by the model that we use. Indeed, according to the experimental data [189, 190], the *trans* isomers of the *bis*-ammino and *bis*-pyridine complexes, $PtCl_2(NH_3)_2$ and $PdCl_2(Py)_2$, are the more stable, which can be explained by electrostatic interligand repulsion not taken into account in our model.

To conclude the discussion of $Pt^{II}L_2X_2$ and $Pd^{II}L_2X_2$ complexes, let us estimate the order of magnitude of the corresponding isomerization energies [56]. In the approximation of frontier MOs at $c_b^2 \approx l_b^2 \approx 1/2$, equation (2.58) yields

$$\Delta\mathcal{E}_{t\to c} \approx \frac{1}{4}\left(\frac{\Delta\alpha}{\omega_{eb^*}}\right)\Delta\alpha + \frac{1}{2K_E}(\beta')^2\left(\frac{\Delta\alpha}{\omega_{eb^*}}\right)^2. \tag{4.20}$$

The substitution into (4.20) of typical parameter values ($\Delta\alpha \approx 2$ eV, $K_E \approx 2$ mdyn/Å, $\Delta\alpha/\omega \approx 1/5$, and $\beta' =$ several eV/Å) yields the heat of isomerization (several kcal/mol) which is in satisfactory agreement with experimental values. This result can be considered evidence in favor of predominantly intramolecular origin of measured heats of isomerization of square-planar complexes.

4.4. Isomers of Planar Transition Metal ML_2XY Complexes

1. In this section, we analyze stability of isomers of $Pt^{II}L_2XY$-type complexes with two different substituents,

```
      L                    X
      |                    |
  X — M — Y            Y — M — L
      |                    |
      L                    L
```

trans-isomer   *cis*-isomer

This analysis will not require the introduction of new principles, because equations (2.57) and (2.58) were obtained for heteroligand molecular systems $ML_{n-k}XY...Z$ of the most general form. For simplicity, we will restrict ourselves to the approximation of frontier MOs. The presence of ligands of three types, however, complicates the problem, which ceases to be solvable based on symmetry considerations alone. According to (2.57), the energy of the substituted complex under consideration can be written in terms of the frontier MOs as

$$\begin{aligned}\mathcal{E}(ML_2XY) = S_0 - 2\frac{S_{xb^\bullet}^2 + S_{yb^\bullet}^2}{\omega_{eb^\bullet}} \\ - 8\frac{1}{K_E\omega_{eb^\bullet}^2}\{S_{xb^\bullet}A_{xb^\bullet}^3 + S_{yb^\bullet}A_{yb^\bullet}^4\}^2.\end{aligned} \tag{4.21}$$

Substituting the $S_{xb^\bullet}$ and $S_{yb^\bullet}$ matrix elements calculated by (2.39) into (4.21) yields [191]

$$\begin{aligned}\mathcal{E}(trans\text{-}ML_2XY) = S_0 - \frac{c_b^2}{4\omega_{eb^\bullet}}(\Delta\alpha_X - \Delta\alpha_Y)^2 \\ - \frac{c_b^2}{K_E}\left(\frac{A_{eb^\bullet}^E}{\omega_{eb^\bullet}}\right)^2(\Delta\alpha_X - \Delta\alpha_Y)^2, \\ \mathcal{E}(cis\text{-}ML_2XY) = S_0 - \frac{c_b^2}{4\omega_{eb^\bullet}}(\Delta\alpha_X + \Delta\alpha_Y)^2 \\ - \frac{c_b^2}{K_E}\left(\frac{A_{eb^\bullet}^E}{\omega_{eb^\bullet}}\right)^2(\Delta\alpha_X + \Delta\alpha_Y)^2,\end{aligned} \tag{4.22}$$

where $A_{eb^\bullet}^E = A_{eb^\bullet}^{(3)} = -A_{eb^\bullet}^{(4)}$. Applying (4.22) to find the $\Delta\mathcal{E}_{t\to c} =$

$\mathcal{E}(trans) - \mathcal{E}(cis)$ isomerization energy for the *trans*-$ML_2XY \longrightarrow$ *cis*-ML_2XY transformation yields

$$\Delta\mathcal{E}_{t\rightarrow c} = \frac{c_b^2}{\omega_{eb^*}}\Delta\alpha_X\Delta\alpha_Y + 4\frac{c_b^2}{K_E}\left(\frac{A_{eb^*}^E}{\omega_{eb^*}}\right)^2\Delta\alpha_X\Delta\alpha_Y. \qquad (4.23)$$

It follows that the sign of the heat effect is determined by the signs of $\Delta\alpha_X$ and $\Delta\alpha_Y$. If both X and Y are more electropositive than L ($\Delta\alpha_X > 0$ and $\Delta\alpha_Y > 0$) or more electronegative than L ($\Delta\alpha_X < 0$ and $\Delta\alpha_Y < 0$), then $\mathcal{E}_{t\rightarrow c} > 0$, and the *cis* isomer is the more stable. (When X = Y, this result coincides with that obtained for ML_2X_2 complexes.) However if one of the substituents is more electropositive than L, whereas the other, more electronegative ($\Delta\alpha_X$ and $\Delta\alpha_Y$ have different signs), then the *trans* isomer is the more stable. To gain better understanding of these conclusions, consider two examples. If *trans* substituents have close electronegativities (*trans*-ML_2XY with $\Delta\alpha_X \approx \Delta\alpha_Y$), then perturbation H_S will be close to even and incapable of noticeably mixing the e_u and b_{1g} orbitals, which have different parities. Such a mixing can only be caused by *cis* substitution, and the *cis* isomer will therefore be the more stable, as when X = Y. If substituent electronegativities are noticeably different, the H_S operator of the $ML_4 \longrightarrow$ *trans*-ML_2XY substitution contains an appreciable odd component (this operator will be strictly odd if $\Delta\alpha_X = -\Delta\alpha_Y$). The odd component of the H_S operator for the $ML_4 \longrightarrow$ *cis*-ML_2XY substitution at $\Delta\alpha_X = -\Delta\alpha_Y$ will be smaller, and the *trans* form of ML_2XY will therefore be the more stable.

The $\Delta\mathcal{E}_{t\rightarrow c}$ *trans–cis* isomerization energy given by (4.23) bilinearly depends on the differences of the orbital ionization potentials of ligands X and L ($\Delta\alpha_X$) and Y and L ($\Delta\alpha_Y$) and is large if both $\Delta\alpha_X$ and $\Delta\alpha_Y$ are fairly large simultaneously. At the same time, ligand static *trans*-effect values are linearly related to $\Delta\alpha_X$ and $\Delta\alpha_Y$. Large differences of isomer energies should therefore be expected for the ML_2XY complexes with X and Y in the opposite ends of the series of substituent *trans*-influence on L. Stabilization of *cis*-ML_2XY should be maximum when X and Y are situated close to each other. It follows that in $Pt^{II}L_2XY$- and $Pd^{II}L_2XY$-type complexes, "covalent" interactions between the central atom and ligands can cause stabilization of the *trans* isomer regardless of electrostatic repulsion of the ligands. This conclusion is opposite to the one made above in considering ML_2X_2 complexes, in which, in the absence of nonbonded interactions, *cis*-isomers are stabilized.

2. Before we turn to comparing theory and experiment, recall that in $Pt^{II}Cl_3X$ complexes, the $P-Cl_{trans}$ bond experiences a maximum elongation when X is H, alkyl (R), or SiR_3. The Δl value is then

estimated at 0.12–0.11 Å (Table 13). The *trans*-effect of PR_3 and AsR_3 ligands amounts to half this value, $\Delta l \approx 0.06$ Å. It follows that *trans*-$Pt^{II}L_2XY$ isomers should be noticeably stabilized when L = PR_3 or AsR_3, Y = H, alkyl, or SiR_3, and X = Cl or another ligand with a static *trans*-effect value close to that of Cl.

The conclusion of predominant stabilization of *trans*-$Pt^{II}(PR_3)_2HX$ isomers is substantiated by the experimental data summarized in the review [136]. The author of this review notes that, except when phosphorus-containing ligands form chelate rings, almost all $Pt^{II}(PR_3)_2HX$ complexes have the *trans* configuration. Experimental proofs of the *trans* structure were obtained for about 40 complexes of this type with X = Cl, Br, I, NCO, NCS, RCOO, NO_2, NO_3, SR, SeR, and CN and R = Et or Ph. Similar Ni(II) and Pd(II) complexes and $Pt(AsR_3)_2HX$ compounds with X = Cl, Br, I, and NCS also have *trans* conformations [141]. The $Pt(PPh_3)_2HSi(CF_3C_6H_4)_3$ complex with two substituents, H and SiR_3, characterized by strong *trans*-influence has the *cis* configuration; as mentioned, the model that we use also predicts stabilization of *cis* isomers for such compounds.

Theoretical predictions are also in agreement with the data, again almost exclusively structural, on alkyl-substituted $Pt^{II}(PR'_3)_2RX$ complexes [192]. Approximately 20 such compounds with R and R′ = Me, Et, and Pr and X = Cl, Br, I, NO_2, NO_3, and SCN have *trans* configurations. Among these compounds, the existence of both *cis* and *trans* isomers was only established for $Pt(PMe_3)_2MeCl$ [192]. Both isomers were also observed for $Pt(PMe_{3-n}Ph_n)_2MeCl$ (n = 1, 2), where the role of L is played by bulky phosphine groups [193], and, consequently, steric interactions may make the major contribution to the energy difference between the isomers. For one of these compounds, $Pt(PMe_2Ph)_2MeCl$, the conclusion that the *trans* isomer had the higher stability was drawn, based on the differential scanning calorimetry data [194]. It was found that *cis* to *trans* isomerization of this compound was accompanied by heat evolution.

The $Pd(PR_3)_2(SnCl_3)Cl$ and $PtL_2(SnCl_3)Cl$ complexes with strong-*trans*-influence ligands $SnCl_3$ and L = $PPh_{3-n}Et_n$ (n = 0, 1, and 2) are characterized by thermodynamic stability of *trans* isomers [194] in conformity with the predictions based on the model that we use. (In $Pt(PEt_3)_2(MeC_6H_4)Cl$ and $(2,4,6\text{-}Me_3C_6H_2)Pt(PEt_3)_2Br$, where steric interactions should play an important role, *cis* isomers were only observed [193].) The heats of isomerization reported in [194] evidence, also in agreement with theoretical predictions, that the *cis* isomers of $Pt(PR_3)XCl_2$, where X is diethyl sulfide and diisopropyl sulfide, are the more stable. To end the discussion of PtL_2XY-type complexes, note that the conclusions that follow from the theory considered above are in agreement with the data on the structure of Pt(II) complexes obtained by the method of ^{95}Pt NMR spectroscopy [195]. Among complexes of the other metals, a fairly numerous class of $Au^{III}Me_2XY$ (X, Y = Hal, Py, SCN, and PMe_3) d^8 complexes merits interest. In agreement with theory, these complexes have *cis* conformations [176].

3. To conclude this section, consider attempts at solving problems related to mutual influence of ligands and relative isomer stabilities by purely computational *ab initio* methods. Typical examples are works [196, 197] concerned with quasi-square-planar Pt(II) complexes, namely, *cis*- and *trans*-$PtH_2(PMe_3)_2$ (I-c and I-t), *cis*- and *trans*-$PtHCl(PMe_3)_2$ (II-c and II-t), and *cis*- and *trans*-$PtCl_2(PMe_3)_2$ (III-c and III-t). Calculations for these complexes were performed by the MO LCAO SCF method with the use of effective core potentials. In [197], a relativistic core potential was employed, and the basis set of Gaussian orbitals was more flexible than in [196].

The first thing to be mentioned is that calculations reported in [197] lead to qualitatively incorrect predictions for the geometry of complexes III. A comparison with the X-ray data showed [197] that the calculated Pt—Cl distances differed from the experimental ones by 0.06–0.07 Å, and the trend of P—Cl bond length variations depending on the spatial structure of the compound was opposite to that observed in reality. According to the calculations, Pt—Cl bond lengths in III-t are longer than in III-c by 0.04 Å, whereas according to the X-ray data, they are shorter by 0.07–0.09 Å (Table 13). These results show that even high-level *ab initio* calculations do not guarantee a correct description of substitution effects in such complex inorganic systems as compounds of heavy transition metals.

Similarly, the data on the relative stabilities of the isomers of complexes I–III obtained in [197] cannot be considered reliable. Indeed, according to the calculations, the *trans* isomers of I-III should be the more stable. This is in agreement with model considerations and the experimental data for complexes II, but at variance with both for complexes III. As to complexes I, for which the model that we use predicts the higher stability of the *cis* isomer (measurements of the heat of isomerization of I are impeded by the decomposition of I with evolution of hydrogen [198]), the difference of the total energies of I-c and I-t is very small according to [197] (2.1 kcal/mol) and can easily change sign if relaxation of Pt—P bonds is taken into account in geometry optimization (in [197], Pt—P bond lengths were not optimized).

4.5. Thermal Isomerization of Magnus Salt-Type Compounds

1. When the relative stabilities of two isomers are compared, the transition from a homoligand complex to its heteroligand derivative, $ML_n \longrightarrow ML_{n-k}XY...Z$, is a speculative intermediate stage of the analysis. Magnus salt-type compounds (Fig. 16) $[M^{II}L_4][M^{II}Hal_4]$ (M = Pt, Pd) studied in detail experimentally [175] are of interest because heating them is accompanied by the occurrence of solid-state reactions with real transformation of $[ML_4]$ and $[MX_4]$ homoligand complexes into mixed heteroligand $[ML_2X_2]$ compounds by the reaction

$$[ML_4][MX_4] \xrightarrow{kT} 2[ML_2X_2]. \tag{4.24}$$

These transformations are similar to isomerization reactions in that they involve rearrangement of ligands but leave the composition unchanged. Equation (2.57) can therefore be used to describe the energy characteristics of these transformations.

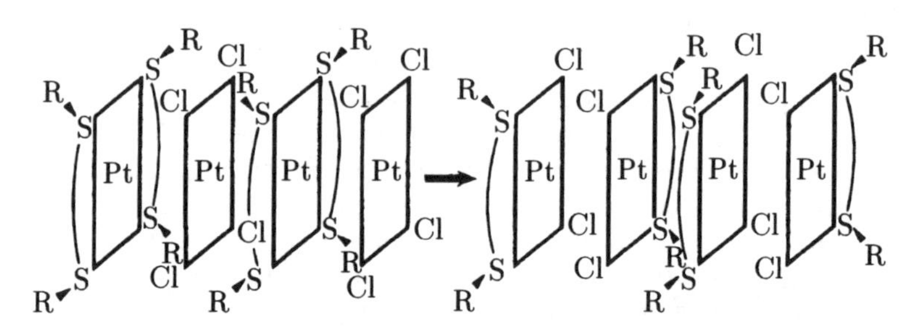

FIG. 16. STRUCTURAL $[PtL_4][PtCl_4] \rightarrow [PtL_2Cl_2]$ TRANSFORMATION ACCORDING TO [175].

The theory described in Section 2.4 gives an explanation of the experimental data [175] for reaction (4.24). According to (2.57), twice the total energy of an $[ML_2Hal_2]$ complex differs from the sum of the total energies of the $[ML_4]$ and $[MHal_4]$ homoligand complexes by two components. The first is the sum of the energies of electronic and geometric relaxation accompanying the $ML_4 \longrightarrow ML_2Hal_2$ and $MHal_4 \longrightarrow ML_2Hal_2$ transformations; it is determined as a second-order perturbation theory correction and contributes to the exothermic effect of reaction (4.24). The second component, which is a first-order perturbation theory correction, is the sum of two energy contributions of the $ML_4 \longrightarrow ML_2Hal_2$ and $MHal_4 \longrightarrow ML_2Hal_2$ substitutions without taking into account electronic and geometric relaxation. This component has the form

$$\begin{aligned}&\langle\Psi_0^{\mathrm{ML_4}}|\mathcal{H}_S(2\mathrm{L}\to 2\mathrm{X})|\Psi_0^{\mathrm{ML_4}}\rangle+\langle\Psi_0^{\mathrm{MX_4}}|\mathcal{H}_S(2\mathrm{X}\to 2\mathrm{L})|\Psi_0^{\mathrm{MX_4}}\rangle\\ =&\langle\Psi_0^{\mathrm{ML_4}}|\mathcal{H}_S(2\mathrm{L}\to 2\mathrm{X})|\Psi_0^{\mathrm{ML_4}}\rangle-\langle\Psi_0^{\mathrm{MX_4}}|\mathcal{H}_S(2\mathrm{L}\to 2\mathrm{X})|\Psi_0^{\mathrm{MX_4}}\rangle,\end{aligned} \tag{4.25}$$

where Ψ_0 is the wave function of the homoligand complex in the ground state, and $\mathcal{H}_S$ is, as usual, the substitution operator. To estimate the difference on the right-hand side of (4.25) with the required accuracy (to second-order perturbation terms), let us write $\Psi_0^{\mathrm{MX_4}}$ in terms of the wave functions of the ML_4 complex as

$$\Psi_0^{\mathrm{MX_4}}=\Psi_0^{\mathrm{ML_4}}-\sum_{j\neq 0}\frac{\langle\Psi_0^{\mathrm{ML_4}}|\mathcal{H}_S(4\mathrm{L}\to 4\mathrm{X})|\Psi_j^{\mathrm{ML_4}}\rangle}{\omega_j}\Psi_j^{\mathrm{ML_4}}. \tag{4.26}$$

Taking into account the scheme and the composition of the MOs of $Pt^{II}L_4$- and $Pd^{II}L_4$-type square-planar complexes (Fig. 13), we see that, in the approximation of frontier MOs, the second term in (4.26) and, accordingly, the difference in (4.25) vanishes. Indeed, the $ML_4 \longrightarrow MHal_4$ transition is described by a totally symmetric substitution operator, whereas the b_{1g}^* LUMO is even, while the e_u HOMO is odd. (The difference in (4.25) remains zero also in the approximation taking into account the $1a_{1g} \to b_{1g}^*$ transition, next in energy to $e_u \to b_{1g}^*$.) At the same time, the relaxation components of the total energy are nonzero in the approximation of frontier MOs, which predicts evolution of heat in reaction (4.24).

2. That reaction (4.24) is exothermic is confirmed by the thermochemical data [175, 200, 201]. The heats of reactions (4.24) were measured for thirty $[ML_4][MX_4]$ complexes (M = Pt and Pd, X = Cl and Br, and L_2 are bidentate ligands linked with the metals through P, S, or N atoms). All reactions were found to be exothermic [175]. The thermochemical measurements also showed that the heat effect of (4.24) for phosphine complexes was much higher than for their sulfide and amine counterparts (25–31, 3–11, and 1–6 kcal/mol, respectively; the data reported in [175] were recalculated for 1 mole of $[ML_4][MX_4]$). This dependence of the heat effect of reaction (4.24) on the ligand composition of complexes is easy to explain in terms of the approximation of frontier MOs, which relates this effect to electronic and geometric relaxation. We have already noted that according to (2.57), relaxation contributions to the stability of $ML_{n-k}X_k$ substituted complexes should sharply depend on the position of substituent X in the series of static *trans*-influence on L. Along this series, *trans*-effect values decrease rapidly from R_3P to R_2S and further to H_3N (Table 13).

The vibronic theory of heteroligand systems also gives an explanation of the experimental dependence [175] of the heat of reaction (4.24) for complexes with bidentate ligands $L_2 = RS(CH_2)_nSR$ on number n of methylene groups separating the sulfur atoms ($n = 2$ or 3, and R is an alkyl group). As mentioned, the heat of reaction (4.24) is in part due to geometric relaxation of $M^{II}L_2Hal_2$ complexes. Clearly, shorter and therefore more rigid methylene chains may hinder geometric relaxation and reduce its contribution to the total energy. The heat effect of the reaction should then decrease.

Lastly, we note that for complexes with *bis*-phosphine and *bis*-sulfide ligands L_2, the heat effect of (4.24) leading to *cis*-$M^{II}L_2Hal_2$ [175] exceeds the heat effects of *trans–cis* isomerization in reactions involving a pair of similar complexes with monodentate PR_3 or SR_2 ligands. This is also easy to explain. Suppose reaction (4.24) involves the formation of *trans*-$M^{II}L_2Hal_2$ followed by *trans–cis* isomerization. The total heat effect of (4.24) will then differ from that of 2*trans*-$M^{II}L_2X_2 \longrightarrow$ 2*cis*-$M^{II}L_2X_2$ isomerization by the energy of electronic and geometric relaxation accompanying the formation of two molecules of the *trans* isomer (Fig. 17). However, the heat of the $[M^{II}M_4][M^{II}X_4] \longrightarrow 2($*trans*-$M^{II}L_2X_2)$ reaction is only zero in the approximation of frontier MOs and becomes positive when the transition between occupied and unoccupied orbitals next in energy to the transition between the frontier MOs is taken into account. Note also that the qualitative considerations given above, which are based on the assumption of an intramolecular nature of stabilization of mixed complexes in Magnus salt-type compounds, also give an explanation of experimental heat effect patterns [199].

4.6. Planar Transition Metal d^9 Complexes

Although the theory discussed in this chapter was developed for heteroligand derivatives of closed-shell complexes, it can be extended to open-shell systems without orbital degeneracy, such as $Cu(gly)_2$ (gly stands for the glycinate anion) with a square-planar geometry of the nearest environment of Cu(II). Consider a square-planar d^9 ML_4 complex with one electron on the b_{1g}^* level, all lower levels being fully occupied (Fig. 13). Orbital degeneracy is absent, and the total energy of the corresponding ML_2X_2 heteroligand complex can be calculated with the use of the procedure for nondegenerate states described in Section 2.4 [202].

The frontier orbitals will be the b_{1g}^* and a_{1g}^* MOs, which are mixed by a substitution operator of B_{1g} symmetry. It follows from Table 4 that the $H_S(ML_2X_2)$ operator includes a component of B_{1g} symme-

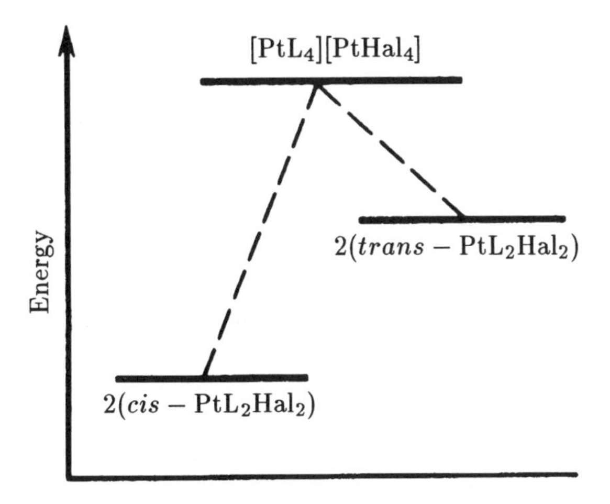

FIG. 17. RELATIVE STABILITIES OF $[PtL_4][PtX_4]$ AND $2[PtL_2X_2]$ COMPLEXES.

try only when the substitution leads to the formation of the *trans* configuration. Therefore, $S^{trans}_{b^*_{1g}a^*_{1g}} \neq 0$ and $S^{cis}_{b^*_{1g}a^*_{1g}} = 0$, that is, the *trans* isomer should be the more stable. It follows that passing from $Pt^{II}L_2X_2$-type complexes with the d^8 configuration to $Cu^{II}L_2X_2$-type complexes with the d^9 configuration is accompanied by the inversion of the relative stabilities of geometric isomers, at least if the energy gap between the b^*_{1g} and a^*_{1g} antibonding orbitals is fairly narrow. Note also that the $Cu^{II}L_4 \longrightarrow$ *trans*-$Cu^{II}L_2X_2$ substitution mixes the frontier MOs, which results in HOMO energy lowering and LUMO energy rise. For this reason, the energy of optical transitions from the HOMO to the higher MOs should be higher for the *trans* isomer.

Both conclusions are in agreement with the data for the heat of isomerization and optical spectra of $Cu(gly)_2$, where two N and two O atoms form an almost planar quadrangle with metal in the center [203, 204]. It was found in [203] that the *cis* $\rightarrow$ *trans* isomerization of this complex was an exothermic reaction (2 ± 0.2 kcal/mol). In addition, the wavelength corresponding to the first electronic transition was found to be 630 nm for the *cis* conformer and only 600 nm for the *trans* [204]. The higher stability of *trans*-$Cu^{II}L_2X_2$ isomers also follows from the observation that in the crystalline state, square-planar $Cu^{II}L_2X_2$ complexes as a rule have the *trans* configuration. It was noted in review [205] that such complexes exist as two isomers only rarely and only if formed by chelating ligands; complexes with simple monodentate ligands have the *trans* configuration.

4.7. Quasi-Tetrahedral Molecules and Complexes

1. For purely geometric reasons, there is no problem of the relative stability of *cis* and *trans* isomers or differences in the behavior of *cis* and *trans* bonds in $ML_{4-k}X_k$ quasi-tetrahedral molecules and complexes. There, however, remains the problem of substitution-induced changes in M — L bond lengths and bond angles. We will discuss this problem in terms of the general theory described in Section 2.4 taking into account numerous experimental data including data for organic compounds.

2. First, consider nontransition element compounds. A typical scheme of σ MOs constructed from ligand σ AOs and *ns* and *np* central atom AOs for tetrahedral ML_4 molecules and complexes is shown in Fig. 18. The vacant *nd* AOs of the central atom are not included because their contribution to bonding is small and has no effect on the final conclusions. For simplicity, the discussion will be conducted at the level of the approximation of frontier MOs; the basis set of all MOs including d(M) AOs is used in treatment of quasi-tetrahedral complexes in [206].

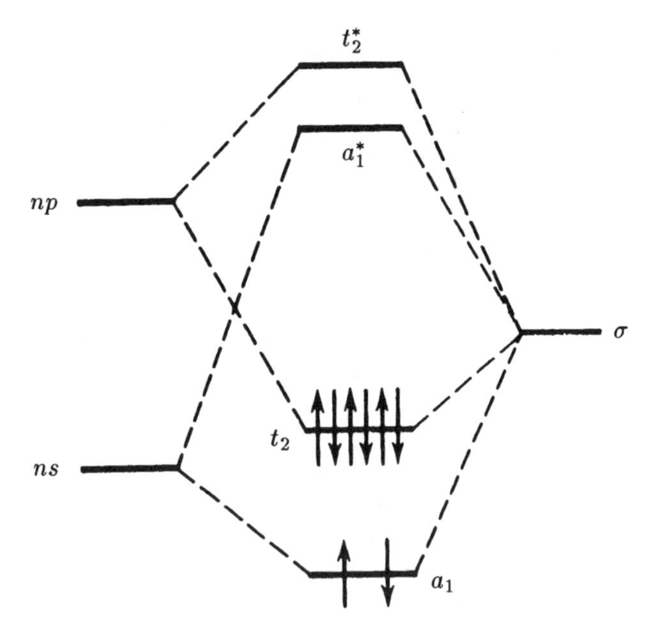

FIG. 18. QUALITATIVE SCHEME OF σ MOs FOR TETRAHEDRAL ML_4 MOLECULES, WHERE M IS A NONTRANSITION ELEMENT.

In the molecular systems under consideration, the HOMO is the t_2 MOs. At first sight, identifying the LUMO poses difficulties, because although in most of ML_4 molecules, vacant MOs are ordered as in Fig. 18, the a_1^* MO can be situated somewhat above the t_2^* one. This ambiguity is removed if, in addition to energy denominators ω_{nm}, we take into account numerator values in (2.47) and (2.49) or (2.55) and (2.56). The ns levels of M atoms in typical ML_4 molecules lie much lower than the np levels, and the contribution of ligand AOs to the a_1^* MO is larger than to the t_2^* ones. For this reason, the $|S_{ta^*}| > |S_{tt^*}|$ inequality should hold for the matrix elements of the H_S operator. A similar inequality should be satisfied for the orbital vibronic coupling constants because of a more localized character of nontransition element ns AOs in comparison with its np AOs (a similar situation was analyzed in Section 3.2). It follows that in (2.55) and (2.56), the largest contribution is made by the t_2–a_1^* transition, which allows us to assume that the role of the effective LUMO is played by the a_1^* MO. The frontier MOs can then be written as (Table 1 and Fig. 19)

$$\text{HOMO: } t_2, \begin{cases} \psi_x = c_t p_x + (l_t/2)(\sigma_1 - \sigma_2 + \sigma_3 - \sigma_4), \\ \psi_y = c_t p_y + (l_t/2)(\sigma_1 - \sigma_2 - \sigma_3 + \sigma_4), \\ \psi_z = c_t p_z + (l_t/2)(\sigma_1 + \sigma_2 - \sigma_3 - \sigma_4), \end{cases} \tag{4.27}$$
$$\text{LUMO: } a_1^*, \quad \psi_a^* = l_a s - (c_a/2)(\sigma_1 + \sigma_2 + \sigma_3 + \sigma_4).$$

Here, c_i and l_i are the coefficients of the ith AO of the central atom and the corresponding (by symmetry) group AO of the ligands. According to selection rules (2.25), the frontier MOs can only be mixed by T_2-symmetry modes, because $T_2 \times A_1 = T_2$. These are T_2' stretching and T_2'' bending vibrations (Table 2), of which only T_2' (Q_4, Q_5, and Q_6, Fig. 20) involve bond length changes.

The behavior of the Q_4, Q_5, and Q_6 modes and, therefore, the $(\partial H/\partial Q_i)_0$ (i = 4, 5, and 6) derivatives under the T_d group symmetry operations are the same as that of the x, y, and z central atom coordinates, respectively. Therefore of the orbital vibronic coupling constants, only the three: $A^4_{xa^*} = A^5_{ya^*} = A^6_{za^*}$ are nonzero. Bearing this in mind, let us apply equation (2.56), which, in the approximation of frontier MOs, yields

$$\begin{aligned} Q_4^f &= 4S_{xa^*}A^4_{xa^*}/K\omega, \\ Q_5^f &= 4S_{ya^*}A^5_{ya^*}/K\omega, \\ Q_6^f &= 4S_{za^*}A^6_{za^*}/K\omega, \end{aligned} \tag{4.28}$$

where K is the force constant for stretching vibrations of the ML_4 system, and ω is the energy gap between the t_2 HOMO and the a_1^* LUMO.

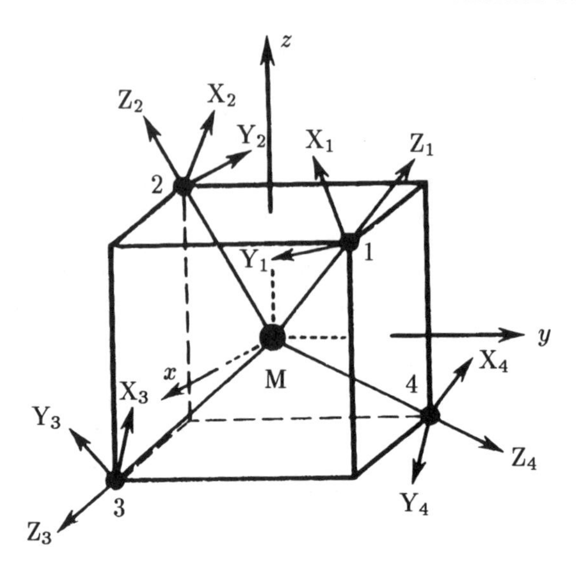

FIG. 19. FRAME OF REFERENCE AND NUMBERING OF LIGANDS FOR TETRAHEDRAL ML_4 MOLECULES.

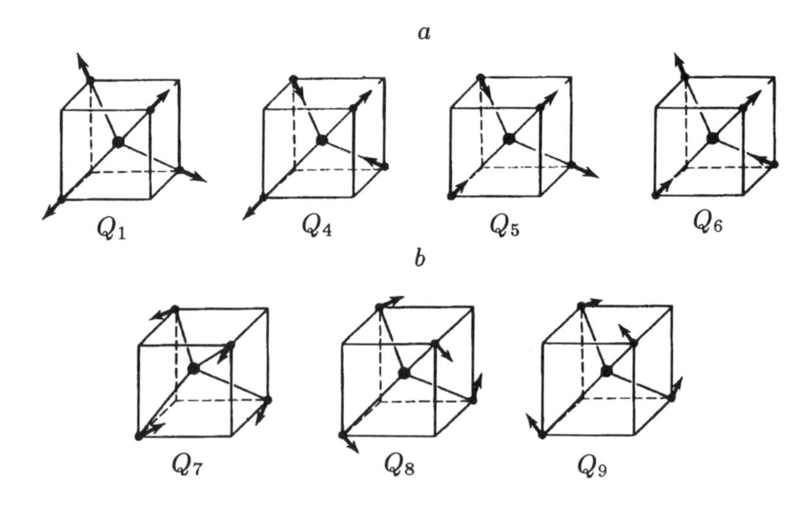

FIG. 20. NORMAL (SYMMETRIZED) STRETCHING (a) AND BENDING (b) MODES OF A_1 AND T_2 SYMMETRY TYPES FOR TETRAHEDRAL ML_4 MOLECULES.

Next, consider the $ML_4 \longrightarrow ML_3X$ substitution at position 1 and find the SA products in (4.28). It suffices to find only one such product, e. g., $S_{za^*} A^6_{za^*}$, because the x, y, and z directions are equivalent.

Applying (4.27) yields [206]

$$S_{za^\bullet} A^6_{za^\bullet} = \left\{ -\frac{c_a l_a l_t^2}{4} \left[\frac{\partial \beta(\sigma, s)}{\partial R} \right]_0 + \frac{c_a^2 c_t l_t}{4\sqrt{3}} \left[\frac{\partial \beta(\sigma, p)}{\partial R} \right]_0 \right\} \Delta\alpha. \quad (4.29)$$

This result is simple to analyze. Because the np AOs of the central atom are more diffuse than the ns AO, the derivative of $\beta(\sigma, p)$ is smaller than that of $\beta(\sigma, s)$. In addition, the coefficient of the former is also smaller (by a factor of $\sqrt{3}$). On the other hand, for Group III, IV, and V nontransition element derivatives considered below, the valence ns(M) AOs differ only insignificantly in energy from the σ AOs of typical ligands, whereas the np(M) AOs have higher energies than the σ(L) AOs. It follows that $c_a \approx l_a$ and $c_t < l_t$, and the $c_i l_i$ product in the second term in (4.29) is smaller than in the first. This means that the coefficient of $\Delta\alpha$ in (4.29) is negative, and the $S_{za^\bullet} A^6_{za^\bullet}$ product is opposite in sign to $\Delta\alpha$.

Certain comments need be made concerning first-period atoms M such as carbon. For carbon compounds, we have $c_t \lesssim l_t$ rather than $c_t < l_t$, because the orbital electronegativity of the $2p$(C) AO is more or less close to that of the σ AOs of some ligands such as H and CH_3. At the same time, the $2s$(M) AO, where M is a first-period element, is lower in energy than typical σ(L) AOs, and therefore $c_a > l_a$. This, however, does not change the sign of the right-hand side of (4.29). For such a change to occur, it is necessary that the c_a coefficient be appreciably larger than $\sqrt{3}\, l_a$, which is hardly probable for ML_4 molecules (e.g., see the calculations for CH_4 [10, 207]).

3. The conclusion on the signs of the SA product and $\Delta\alpha$ made above also refers to $S_{xa^\bullet} A^4_{xa^\bullet}$ and $S_{ya^\bullet} A^5_{ya^\bullet}$. According to (4.28), all Q^f_ν (ν = 4, 5, and 6) modes are negative at $\Delta\alpha > 0$ and positive at $\Delta\alpha < 0$. Considering the corresponding atomic displacements, the conclusion can be drawn that the $ML_4 \longrightarrow ML_3X$ replacement of L in ML_4 by a more electropositive ligand X causes elongation of the M—L bonds in comparison with ML_4 and shortening of the M—X bond in comparison with its length in MX_4. The replacement of L by a more electronegative ligand X has the opposite effect. Both effects are approximately proportional to $|\Delta\alpha|$. Lastly, the effect of substituting two or three ligands in ML_4 ($ML_4 \longrightarrow ML_2X_2$ and $ML_4 \longrightarrow MLX_3$) should be the sum of partial effects of the replacement of separate ligands (equation (3.12)).

4. It remains to discuss tetrahedral bond angle deformations caused by substitution. For this purpose, consider the Q_7, Q_8, and Q_9 bending modes (Fig. 20 and Table 2). In the approximation of frontier MOs, the substitution at position 1 is described by the obvious relations of type (4.28)

Table 14. Geometry of fluoromethanes and fluorosilanes according to experimental data and *ab initio* calculations (in parentheses)

Molecule	M—F	M—H	F—M—F	F—M—H	H—M—H
	Å		degrees		
CF_4	1.320	—	109.5	—	—
	(1.302)	—	(109.5)	—	—
CF_3H	1.332	1.098	108.8	110.1	—
	(1.317)	(1.074)	(108.5)	(110.4)	—
CF_2H_2	1.357	1.093	108.3	108.7	113.7
	(1.338)	(1.078)	(108.6)	(108.9)	(112.5)
CFH_3	1.383	1.100	—	108.3	110.6
	(1.365)	(1.082)	—	(109.1)	(109.8)
CH_4	—	1.094	—	—	109.5
	—	(1.084)	—	—	(109.5)
SiF_4	1.552	—	109.5	—	—
	(1.557)	—	(109.5)	—	—
SiF_3H	1.562	1.447	108.3	110.6	—
	(1.569)	(1.449)	(108.0)	(110.9)	—
SiF_2H_2	1.577	1.471	107.9	109.2	112.0
	(1.581)	(1.461)	(107.6)	(108.9)	(113.4)
$SiFH_3$	1.593	1.486	—	108.5	110.4
	(1.594)	(1.470)	—	(108.8)	(110.1)
SiH_4	—	1.481	—	—	109.5
	—	(1.475)	—	—	(109.5)

$$\begin{aligned} Q_7^f &= 4S_{xa^\bullet} A_{xa^\bullet}^7/(K\omega), \\ Q_8^f &= 4S_{ya^\bullet} A_{ya^\bullet}^8/(K\omega), \\ Q_9^f &= 4S_{za^\bullet} A_{za^\bullet}^9/(K\omega). \end{aligned} \tag{4.30}$$

Here, K is the force constant for bending vibrations. Following the same line of reasoning as in Section 3.7, we obtain

$$S_{za^\bullet} A_{za^\bullet}^9 = (1/2\sqrt{6})c_a^2 c_t l_t (\beta(\sigma, p)/R)_0 \Delta\alpha. \tag{4.31}$$

As $\beta(\sigma, p) < 0$, it follows from (4.31) that the sign of Q_9^f is again opposite to that of $\Delta\alpha$, and the same is true of the Q_7^f and Q_8^f coordinates.

Table 15. Geometry of quasi-tetrahedral nontransition element molecular ions $MF_{4-k}H_k^{\pm}$ according to *ab initio* calculations [208a]

Ion	M—F	M—H	F—M—F	F—M—H	H—M—H
	Å		degrees		
BH_4^-	—	1.243	—	—	109.5
BFH_3^-	1.440	1.248	—	110.3	108.6
$BF_2H_2^-$	1.421	1.246	108.4	109.5	110.4
BF_3H^-	1.406	1.236	108.6	110.4	—
BF_4^-	1.394	—	109.5	—	—
NH_4^+	—	1.013	—	—	109.5
NFH_3^+	1.333	1.015	—	107.8	111.1
$NF_2H_2^+$	1.308	1.018	109.0	108.4	114.2
NF_3H^+	1.289	1.023	108.9	110.1	—
NF_4^+	1.280	—	109.5	—	—
AlH_4^-	—	1.651	—	—	109.5
$AlFH_3^-$	1.700	1.650	—	109.2	109.7
$AlF_2H_2^-$	1.692	1.643	107.9	109.4	111.5
AlF_3H^-	1.685	1.629	108.4	110.5	—
AlF_4^-	1.677	—	109.5	—	—
PH_4^+	—	1.380	—	—	109.5
PFH_3^+	1.523	1.375	—	108.3	110.6
$PF_2H_2^+$	1.506	1.370	108.0	108.4	115.0
PF_3H^+	1.491	1.364	108.0	110.9	—
PF_4^+	1.480	—	109.5	—	—

That is, the $ML_4 \longrightarrow ML_3X$ substitution at position 1 causes widening of the $X{-}M{-}L_i$ ($i = 2$, 3, and 4) bond angles from tetrahedral values if X is a more electropositive ligand than L and narrowing of these angles if L is more electropositive. The $L_i{-}M{-}L_j$ (i, $j = 2$, 3, and 4) bond angles change differently; they decrease if X is more electropositive and increase if X is more electronegative than L. We hardly need to explain why angle deformations caused by multiple substitution obey the same additivity rule as bond length changes.

5. The data on neutral molecules and charged complexes (Tables 14 and 15, respectively [208]) and the more recent data [208] on the series

$MCl_{4-k}(CH_3)_k$ (M = C, Si, Ge, Sn, and Pb and k = 0–4) show that the theoretical conclusions conform with experiment and the data of reliable *ab initio* calculations. Theory correctly reproduces the trends of geometry changes in series of related substituted molecules and complexes, although these changes are often fairly insignificant. Note that a comparison with the computation results [208] on species not studied experimentally is justified, because these computations correctly reproduce the available experimental data.

6. With minor modifications, the theory described above can be extended to σ-bonded transition metal complexes such as $TiL_{4-k}X_k$. The frontier MOs of these complexes in the basis set of metal $(n-1)d$ and ns AOs and ligand σ AOs will again be the t_2 MOs (HOMO) and the a_1^* MO (LUMO). The LUMO has form (4.27), and the HOMO is obtained from (4.27) by replacing p_x, p_y, and p_z AOs with d_{yz}, d_{xz}, and d_{xy}. This makes it easy to write equations similar to (4.28)–(4.31). The steric consequences of the $ML_4 \longrightarrow ML_3X$ substitution in TiL_4-type complexes will be the same as for nontransition element compounds. This conclusion is in agreement with the experimental data [208b] and detailed *ab initio* calculations on the $TiCl_{4-k}(CH_3)_k$ (k = 0–4) series.

CHAPTER 5

QUASI-OCTAHEDRAL $Cu^{II}L_{6-k}X_k$ d^9 COMPLEXES

In this chapter, we consider heteroligand derivatives of homoligand open-shell molecular systems with orbital degeneracy. Specifically, we discuss one of the simplest systems of this type, namely, quasi-octahedral copper(II) complexes [51, 52, 54, 209, 210]. It seems pertinent to mention here that six-coordinate Cu(II) complexes are among favorite objects of investigations by vibronic theory methods and are still extensively studied [11, 14, 15, 211].

5.1. Homoligand d^9 Complexes: E-e Problem

1. As mentioned in Chapter 2, for heteroligand derivatives of open-shell complexes, we cannot write an explicit formula of type (2.47) for finding the lower sheet of the adiabatic potential energy surface. Instead, we must solve secular equation (2.46), the roots of which depend on the form of the equation and should be found for every separate system. For this reason, we will first remind the reader how $Cu^{II}L_6$ complexes are treated traditionally [11, 14, 15] (here, L are ligands usually assumed to be monoatomic; if they are not, we will confine the discussion to the first coordination sphere of Cu). The scheme of MOs of such a complex of O_h symmetry is shown in Figs. 1 and 21. The d^9 complex has an open (e_g^{*3}) electronic shell and occurs in an orbitally doubly degenerate state; its ground state term is ^2E_g.

FIG. 21. ELECTRONS ON THE e_g^* MOs IN A d^9 COMPLEX.

To determine the adiabatic potential energy surface of such a system, all electrons of occupied MOs up to e_g^* are considered internal; it is assumed that these electrons together with repulsion of atomic cores contribute to the first (harmonic) component of the adiabatic potential (see (2.15)). The second adiabatic potential component is found by solving secular equation (2.16). This equation is written in terms of wave functions Ψ_1 and Ψ_2 formed from the $\psi^*_{z^2}$ and $\psi^*_{x^2-y^2}$ (e^*_g) MOs. These MOs transform under O_h point symmetry group operations as metal d_{z^2} and $d_{x^2-y^2}$ AOs. The Ψ_1 and Ψ_2 functions can be written explicitly in terms of the $\psi^*_{z^2}$ and $\psi^*_{x^2-y^2}$ MOs as

$$\begin{aligned}\Psi_1 &= (1/\sqrt{3!})\mathrm{Det}\|\psi^*_{z^2}\alpha\psi^*_{x^2-y^2}\alpha\psi^*_{x^2-y^2}\beta\|,\\ \Psi_2 &= (1/\sqrt{3!})\mathrm{Det}\|\psi^*_{z^2}\alpha\psi^*_{z^2}\beta\psi^*_{x^2-y^2}\alpha\|\end{aligned} \tag{5.1}$$

(the same can be done for unpaired electron spin β). The corresponding secular equation of type (2.16) takes the form

$$\mathrm{Det}\begin{Vmatrix} -AQ_3+A'(Q_2^2-Q_3^2)-\mathcal{E}' & AQ_2-2A'Q_2Q_3 \\ AQ_2-2A'Q_2Q_3 & AQ_3+A'(Q_2^2-Q_3^2)-\mathcal{E}' \end{Vmatrix} = 0, \tag{5.2}$$

where A and A' are the linear and quadratic vibronic coupling constants

$$\begin{aligned}A &= \langle\Psi_2|(\partial\mathcal{H}/\partial Q_3)_0|\Psi_2\rangle,\\ A' &= (1/2)\langle\Psi_1|(\partial^2\mathcal{H}/\partial Q_2^2)_0|\Psi_1\rangle.\end{aligned} \tag{5.3}$$

Here, Q_2 and Q_3 are the normal coordinates (Fig. 4 and Table 2). These coordinates of E_g symmetry and the totally symmetric Q_1 coordinate are the only ones allowed by selection rules (2.13), because $[E_g^2] = A_{1g} + E_g$. Coordinate Q_1 can, however, be excluded, for it is assumed that bond lengths in the complex have already been optimized with respect to this coordinate. Solving (5.2) to find roots $\mathcal{E}' = \mathcal{E}'_{1,2}$ and adding the harmonic term $(1/2)K(Q_2^2 + Q_3^2)$ yields adiabatic potential energy surface $\mathcal{E}$ as a function of Q_2 and Q_3,

$$\mathcal{E} = (1/2)K(Q_2^2 + Q_3^2) + \mathcal{E}'. \tag{5.4}$$

This equation can conveniently be transformed by transition to polar coordinates ρ and φ in the Q_2, Q_3 plane, that is

$$Q_3 = \rho\cos\varphi, \qquad Q_2 = \rho\sin\varphi. \tag{5.5}$$

Written in these coordinates, (5.4) becomes

$$\mathcal{E} = (1/2)K\rho^2 \pm \rho\sqrt{A^2 + A'^2\rho^2 + 2AA'\rho\cos 3\varphi}. \tag{5.6}$$

It is worth while to analyze the last equation in more detail.

2. In the simplest approximation, quadratic vibronic coupling constant A' is ignored (the linear approximation). Equation (5.6) then becomes

$$\mathcal{E} = (1/2)K\rho^2 \pm |A|\rho. \tag{5.7}$$

Function (5.7) describes the surface known as Mexican hat (Fig. 22a); this surface, consisting of two sheets, has cylindrical symmetry and forms a circular trough of radius $\rho_0 = |A|/K$ passing around the origin. At the bottom of the trough, $\mathcal{E}$ is minimum. This means that in the approximation linear with respect to vibronic coupling, the adiabatic potential has an infinite number of identical minima, that is, there is a continuum of sets of Q_2, Q_3 coordinates describing an infinite number of distorted forms of the complex, each such form corresponding to the same minimum energy. The $\mathcal{E}(0) - \mathcal{E}(\rho_0)$ difference between the energy at the origin ($\rho = 0$) and the energy at the bottom of the trough ($\rho = \rho_0$) is called Jahn–Teller stabilization energy and equals

$$\mathcal{E}_{\text{J-T}} = \mathcal{E}(0) - \mathcal{E}(\rho_0) = A^2/2K, \qquad \rho_0 = |A|/K. \tag{5.8}$$

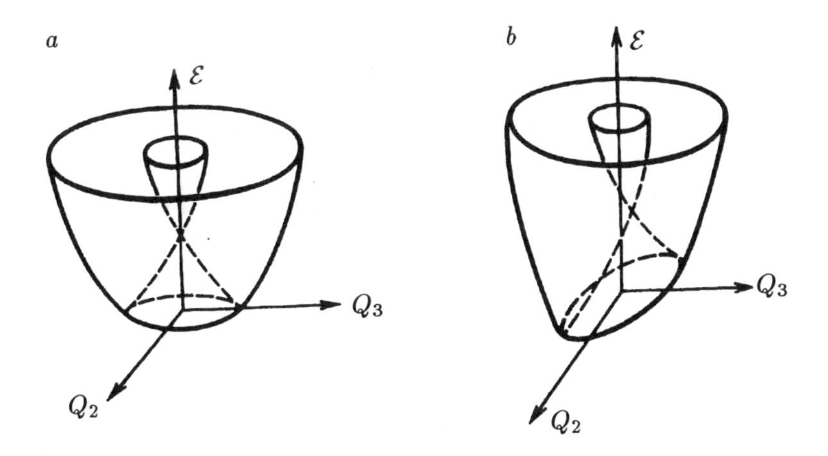

FIG. 22. ADIABATIC POTENTIAL FOR $Cu^{II}L_6$-TYPE COMPLEXES, E-e PROBLEM (a), AND $Cu^{II}L_{6-k}X_k$-TYPE COMPLEXES, E-e-S PROBLEM IN THE APPROXIMATION LINEAR IN VIBRONIC COUPLING (b).

Taking into account constant A' complicates the adiabatic potential enrgy surface. At the bottom of the trough, there appear three equivalent minima separated by barriers. The Jahn–Teller stabilization energy defined as the difference of the $\mathcal{E}(\rho, \varphi)$ values at the origin and at the bottom of a minimum then takes the form

$$\mathcal{E}_{\text{J-T}} = A^2/2(K - 2A'), \tag{5.9}$$

and the minimum height of the barriers separating the minima is given by

$$\Delta\mathcal{E} = 4A'\mathcal{E}_{\text{J-T}}/(K - 2A') = 2A^2A'/(K - 2A')^2. \tag{5.10}$$

The local minima are situated at $\varphi = (0, 2\pi/3, -2\pi/3)$ or $(\pi, \pi/3, -\pi/3)$ depending on the sign of the product AA' of vibronic coupling constants. The positions of the minima correspond to the first set of φ values if $AA' > 0$ and to the second set otherwise. Physically, this means that at $AA' > 0$, adiabatic potential energy surface has minima corresponding to distorted ML_6 octahedra of D_{4h} symmetry elongated along the z, x, or y axis, whereas at $AA' < 0$, these octahedra are contracted in the same directions. Direct calculations of vibronic coupling constants and experimental data show that Cu(II) complexes exist largely as elongated octahedra.

Table 16. Axial and equatorial Cu—F bond lengths (Å) in $Cu^{II}F_6$ complexes of D_{4h} symmetry in various crystals [246]

Compound	R_{ax}	R_{eq}
Na_2CuF_4	2.37	1.91
$NaCuF_3$	2.26	1.93
$KCuF_3$	2.26	1.92
	2.25	1.93
	1.96	2.07
K_2CuF_4	2.22	1.92
	1.95	2.08
Ba_2CuF_6	2.33	1.89
	2.32	1.87
$CuF_2(s)$	2.27	1.93

Jahn–Teller instability of the high-symmetry O_h configuration for $Cu^{II}L_6$-type complexes and the conclusion of preferred formation of elongated octahedra are illustrated by Table 16. The table also shows that in crystals, equilibrium bond lengths are influenced by environmental effects.

5.2. Linear *E-e-S* Problem

1. To pass from the adiabatic potential of a homoligand open-shell complex to the adiabatic potential of its heteroligand derivative, we must complement secular determinant (2.16) by the matrix elements of substitution operator $\mathcal{H}_S$ to obtain (2.46). As with the *E-e* problem considered above, solving the *E-e-S* (*S* stands for substitution) problem for d^9 complexes requires the use of functions (5.1). In the approximation linear with respect to vibronic coupling ($A' = 0$), equation (2.46) takes the form

$$\mathrm{Det}\begin{Vmatrix} -AQ_3 + S_{11} - \mathcal{E}' & AQ_2 + S_{12} \\ AQ_2 + S_{12} & AQ_3 + S_{22} - \mathcal{E}' \end{Vmatrix} = 0, \tag{5.11}$$

where A is the linear vibronic coupling constant, and S_{ij} $(i, j = 1, 2)$ are the matrix elements of the $\mathcal{H}_S$ operator in the basis set of functions Ψ_1 and Ψ_2, that is, $S_{ij} = S_{ji} = \langle \Psi_i | \mathcal{H}_S | \Psi_j \rangle$.

Equation (5.11) describes the adiabatic potential energy surface of a heteroligand complex in terms of the electronic structure of its homoligand predecessor. This equation can also be interpreted as written in terms of the electronic structure of the heteroligand system. Let $\widetilde{\Psi}_1$ and $\widetilde{\Psi}_2$ be linear combinations of Ψ_1 and Ψ_2 such that form the basis in which the S_{ij} matrix is diagonal. Equation (5.11) then describes the special case of the pseudo-Jahn–Teller effect [15] in the $ML_{6-k}XY...Z$ system having the high-symmetry geometry of the ML_6 complex. In such a system, the *E* term is already split as a result of ligand substitution. As distinguished from the situation considered in Section 2.1.5, we here have two modes rather than one. One of these ("mixing") mixes the split components of the *E* term, and the other ("tuning") changes the energy gap between these components. Table 17 shows that for all $ML_{6-k}X_k$ complexes except *trans*-ML_3X_3, the $\widetilde{\Psi}_1$, $\widetilde{\Psi}_2$ basis set coincides with the Ψ_1, Ψ_2 one, that is, equation (5.11) can be treated in terms of the pseudo-Jahn–Teller effect for $ML_{6-k}X_k$, where Q_2 and Q_3 are the mixing and tuning modes, respectively.

Solving (5.11) yields the roots

$$\mathcal{E}'_{1,2} = \frac{S_{11} + S_{22}}{2} \pm \left\{ \left(AQ_3 + \frac{S_{11} - S_{22}}{2} \right)^2 + (S_{12} + AQ_2)^2 \right\}^{1/2}, \tag{5.12}$$

and, as is usual in the theory of the Jahn–Teller effect (see Section 2.4), the adiabatic potential is written as the sum of $\mathcal{E}'$ and the harmonic term in the form

$$\mathcal{E}(Q_2, Q_3) = (1/2)K(Q_2^2 + Q_3^2) + \mathcal{E}'_{1,2}(Q_2, Q_3). \tag{5.13}$$

Table 17. Matrix elements of $\mathcal{H}_S$ operator between E-term wave functions and extrastabilization energy $\Delta\mathcal{E}_E$

Complex	S_{11}	S_{22}	S_{12}	$\Delta\mathcal{E}_E$
ML_5X	$\frac{1}{\sqrt{6}}S_A - \frac{1}{\sqrt{3}}S_E$	$\frac{1}{\sqrt{6}}S_A + \frac{1}{\sqrt{3}}S_E$	0	$\frac{1}{\sqrt{3}}\lvert S_E\rvert$
trans-ML_4X_2	$\frac{2}{\sqrt{6}}S_A - \frac{2}{\sqrt{3}}S_E$	$\frac{2}{\sqrt{6}}S_A + \frac{2}{\sqrt{3}}S_E$	0	$\frac{2}{\sqrt{3}}\lvert S_E\rvert$
cis-ML_4X_2	$\frac{2}{\sqrt{6}}S_A + \frac{1}{\sqrt{3}}S_E$	$\frac{2}{\sqrt{6}}S_A - \frac{1}{\sqrt{3}}S_E$	0	$\frac{1}{\sqrt{3}}\lvert S_E\rvert$
trans-ML_3X_3	$\frac{3}{\sqrt{6}}S_A - \frac{\sqrt{3}}{2}S_E$	$\frac{3}{\sqrt{6}}S_A + \frac{\sqrt{3}}{2}S_E$	$\frac{1}{2}S_E$	$\lvert S_E\rvert$
cis-ML_3X_3	$\frac{3}{\sqrt{6}}S_A$	$\frac{3}{\sqrt{6}}S_A$	0	0
trans-ML_2X_4	$\frac{4}{\sqrt{6}}S_A + \frac{2}{\sqrt{3}}S_E$	$\frac{4}{\sqrt{6}}S_A - \frac{2}{\sqrt{3}}S_E$	0	$\frac{2}{\sqrt{3}}\lvert S_E\rvert$
cis-ML_2X_4	$\frac{4}{\sqrt{6}}S_A - \frac{1}{\sqrt{3}}S_E$	$\frac{4}{\sqrt{6}}S_A + \frac{1}{\sqrt{3}}S_E$	0	$\frac{1}{\sqrt{3}}\lvert S_E\rvert$
MLX_5	$\frac{5}{\sqrt{6}}S_A + \frac{1}{\sqrt{3}}S_E$	$\frac{5}{\sqrt{6}}S_A - \frac{1}{\sqrt{3}}S_E$	0	$\frac{1}{\sqrt{3}}\lvert S_E\rvert$
MX_6	$\sqrt{6}\,S_A$	$\sqrt{6}\,S_A$	0	0

Note: See Table 4 for substituent positions; numbering of ligands is given in Fig. 4.

To find a minimum of the lower sheet of the adiabatic potential energy surface, let us rewrite (5.13) in polar coordinates, which can conveniently be introduced as

$$\begin{aligned} Q_3 &= \rho\cos\varphi + (S_{11} - S_{22})/2A, \\ Q_2 &= \rho\sin\varphi - S_{12}/A. \end{aligned} \tag{5.14}$$

In these coordinates, the adiabatic potential takes the form

$$\begin{aligned} \mathcal{E}(\rho,\varphi) = \frac{K}{2}\bigg\{\rho^2 &+ \rho\frac{S_{11} - S_{22}}{A}\cos\varphi - 2\rho\frac{S_{12}}{A}\sin\varphi \\ &+ \left(\frac{S_{11} - S_{22}}{2A}\right)^2 + \left(\frac{S_{12}}{A}\right)^2\bigg\} + \frac{S_{11} + S_{22}}{2} \pm |A|\rho. \end{aligned} \tag{5.15}$$

The wave functions corresponding to two adiabatic potential sheets are

$$\begin{aligned} \Psi_+ &= \sin(\varphi/2)\Psi_1 + \cos(\varphi/2)\Psi_2, \\ \Psi_- &= \cos(\varphi/2)\Psi_1 - \sin(\varphi/2)\Psi_2. \end{aligned} \tag{5.16}$$

Selecting sign minus corresponding to the lower energy surface sheet in (5.15) and applying the usual extremum conditions allows us to locate the minimum of $\mathcal{E}(\rho, \varphi)$. Its coordinates are

$$\cos\varphi_0 = \frac{A(S_{22} - S_{11})}{|A|R}, \quad \sin\varphi_0 = \frac{2AS_{12}}{|A|R}, \quad \rho_0 = \frac{|A|}{K} + \frac{R}{|A|}, \quad (5.17)$$

where $R = \{(S_{11} - S_{22})^2 + 4S_{12}^2\}^{1/2}$ is the splitting (usually, on the order of 1 eV, see below) of the E term that would be observed if the $ML_{6-k}XY...Z$ complex retained octahedral symmetry of ML_6.

It follows from these considerations that even in the approximation linear in vibronic coupling, heteroligand d^9 $ML_{6-k}XY...Z$ complexes, unlike homoligand ML_6 ones, are characterized by the presence of a single adiabatic potential minimum rather than a continuum of minima in the form of a circular trough. Substituting (5.17) into (5.15), we easily obtain the minimum energy value

$$\mathcal{E}_{\text{min}} = (1/2)(S_{11}+S_{22}) - \{(1/4)(S_{11}-S_{22})^2+S_{12}^2\}^{1/2} - A^2/2K. \quad (5.18)$$

2. Consider equation (5.18). It follows from (5.12) that the first term in (5.18), $(1/2)(S_{11}+S_{22})$, describes the change in the electronic energy of ML_6 caused by ligand substitution in the hypothetical state with an ideal octahedral geometry without taking into account splitting of the E term. The energy of this state equals half the sum of the energies of the levels formed by splitting the doubly degenerate Jahn–Teller state as a result of substitution; this energy includes the averaged shift of the originally doubly degenerate term along the energy axis. It will be shown that this contribution to $\mathcal{E}_{\text{min}}$ is the same for *cis*- and *trans*-$ML_{6-k}X_k$ isomers.

It follows from (5.11) that the second term in (5.18),

$$\Delta\mathcal{E}_E = -\{1/4(S_{11} - S_{22})^2 + S_{12}^2\}^{1/2}, \quad (5.19)$$

describes energy lowering caused by substitution-induced splitting of the E term at $Q_2 = Q_3 = 0$, that is, under the condition that the geometry of the initial ML_6 complex remains unchanged. By analogy with the terminology of crystal field theory, $\Delta\mathcal{E}_E$ can be called E-term extrastabilization energy. (In crystal field theory, extrastabilization is the difference of the energy of the unsplit d level and the energy of the ground state in the crystal field [11, 53]). Equation (5.19) describes relaxation of the electronic subsystem as a result of substitution. The $\Delta\mathcal{E}_E$ values, which are different for different isomers, will be our special concern, for in the linear E-e-S problem, it determines the relative stabilities of *cis* and *trans* isomers, because the third term in (5.18) is independent of S_{ij}.

The last term in (5.18), $A^2/2K$, coincides with energy lowering in the unsubstituted ML_6 complex due to the Jahn–Teller effect (Jahn–Teller stabilization energy, see (5.8)) and equals energy gain from geometry optimization, when $ML_{6-k}XY...Z$ relaxes from the octahedral configuration of ML_6 to its equilibrium geometry.

3. It should be stressed that the independence of the geometric relaxation energy from the matrix elements of the substitution operator is determined by the conical shape of the $\mathcal{E}'(\rho, \varphi)$ surface given by (5.11). A geometric interpretation of this statement was suggested in [52]. According to (5.7), adiabatic potential $\mathcal{E}(\rho, \varphi)$ of the unsubstituted complex (Mexican hat) treated as a function of two variables is the sum of two functions. One of these, $\mathcal{E} = K\rho^2/2$, is a paraboloid of revolution expanding upward, and the other, $\mathcal{E} = -|A|\rho$, is the surface of a circular cone expanding downward (Fig. 23a). According to (5.14), the $ML_6 \longrightarrow ML_{6-k}XY...Z$ transformation can, accurate to an unimportant constant in the equation for the energy, be treated as a shift of the cone from the origin parallel to the (Q_2, Q_3) plane toward the $Q_2^0 = -S_{12}/A$, $Q_3^0 = (S_{11} - S_{22})/2A$ point (Fig. 23b). The new adiabatic potential has the form of a deformed Mexican hat. As was shown analytically above and is clear from geometric considerations, this potential energy surface has a single minimum situated along the straight line directed from the origin to the (Q_2^0, Q_3^0) point. Next, consider the cross section passing through this straight line and the $\mathcal{E}$ axis. Let Q (Fig. 24) denote the coordinate along the straight line counted from the origin. The specified adiabatic potential cross section then satisfies the equation

$$\mathcal{E}(Q) = KQ^2/2 - |A||Q - Q^0| = KQ^2/2 \mp |A|Q + |A|Q^0. \qquad (5.20)$$

Here, the upper sign is selected if $Q - Q^0 > 0$, and the lower one, if $Q - Q^0 < 0$. The $\mathcal{E}(Q)$ function is plotted in Fig. 24 together with the corresponding cross section of an undistorted Mexican hat of the usual Jahn–Teller effect.

It follows from (5.20) and Fig. 24 that in the direction toward the minimum, the $\mathcal{E}(ML_{6-k}XY\ldots Z)$ function only differs from $\mathcal{E}(ML_6)$ by the $|A|Q^0$ constant, and the dependence of energy $\mathcal{E}$ on the S_{ij} matrix elements is fully accounted for by the $Q^0 = Q^0(S_{ij})$ dependence. For this reason, the extremum condition $(\partial\mathcal{E}/\partial Q) = 0$ does not contain S_{ij} and corresponds to the same Q and $\mathcal{E}(0) - \mathcal{E}_{\min}$ values for the substituted and unsubstituted systems. Note that this conclusion is not valid if the $\mathcal{E}'(Q)$ surface does not form a cone, as, e. g., when the A' constant is taken into account in the E-e-S problem (see Sections 5.4–5.6).

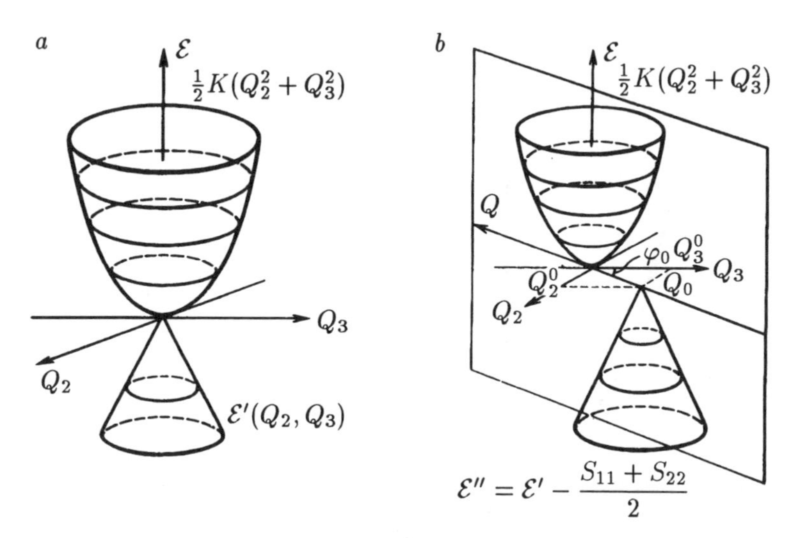

FIG. 23. TWO COMPONENTS OF THE LOWER ADIABATIC POTENTIAL SHEET (a) FOR HOMOLIGAND $Cu^{II}L_6$ COMPLEXES AND (b) FOR HETEROLIGAND $Cu^{II}L_{6-k}X_k$ COMPLEXES.

4. An interesting problem is that of barrier U to motion of nuclei along the adiabatic potential trough ($\rho \approx \rho_0$) in a d^9 heteroligand complex. This barrier is given by

$$U = R\{1 + RK/2A^2\}, \tag{5.21}$$

which shows that $U \geq R$. In ML_6, the energy of motion of nuclei along the Mexican hat trough is on the order of several cm^{-1} [11, 15], and therefore even small E term splitting R caused by the $ML_6 \longrightarrow ML_{6-k}XY...Z$ substitution suppresses free motion of nuclei along the trough. It follows that the distortion of a substituted complex is of a static character [11, 15], and, if the energy of zero-point oscillations of nuclei about the ρ_0, φ_0 point is ignored, the total energy of the $ML_{6-k}XY...Z$ complex equals $\mathcal{E}_{\min}$.

5.3. Linear *E-e-S* Problem: Stability of *cis* and *trans* Isomers

To estimate the relative stability of various isomers of heteroligand d^9 complexes, it suffices to show that the substitution operator is additive with respect to substituents (2.33). For this purpose, let us apply selection rules (2.37) taking into account that the symmetry group of the initial ML_6 homoligand complex is O_h, and the $\Psi_i = \Psi_1$ and

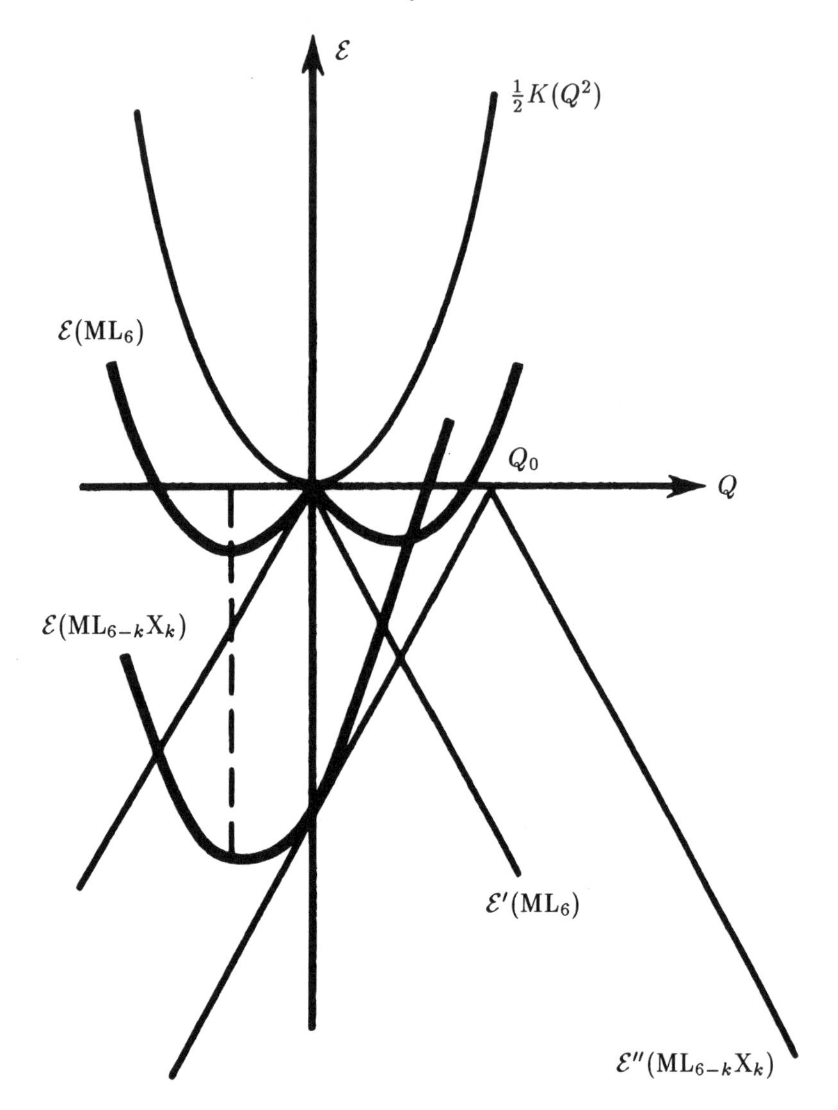

FIG. 24. THE FORM OF ADIABATIC POTENTIAL COMPONENTS SHOWN IN FIG. 23 AND THE RESULTING POTENTIAL SURFACES: CROSS SECTION ALONG THE $(Q, \mathcal{E})$ PLANE.

$\Psi_j = \Psi_2$ functions given by (5.1) belong to the same E term. It follows from (2.37) that the $\langle\Psi_1|\mathcal{H}_\Gamma^\gamma|\Psi_2\rangle$ matrix element can only be nonzero if Γ is either A_{1g} or E_g: $[E_g] = A_{1g} + E_g$. It is, however, fairly clear that the even Ψ_1 and Ψ_2 functions are not mixed by the T_{1u} components of

$\mathcal{H}_S$. For this reason, for the $ML_{6-k}X_k$ complex, the S_{ij} $(i, j = 1, 2)$ matrix elements in (5.17) and (5.18) do not depend on the T_{1u} component of the $\mathcal{H}_S$ operator and only depend on its $\mathcal{H}_{A_{1g}}$, $\mathcal{H}^1_{E_g}$, and $\mathcal{H}^2_{E_g}$ components. Taking into account the Clebsch–Gordan coefficients for the octahedral group [53b], we can write the nonzero matrix elements of these components in the basis formed by the Ψ_1 and Ψ_2 functions as

$$\langle\Psi_1|\mathcal{H}_{A_{1g}}|\Psi_1\rangle = \langle\Psi_2|\mathcal{H}_{A_{1g}}|\Psi_2\rangle = S_A,$$
$$\langle\Psi_2|\mathcal{H}^1_{E_g}|\Psi_2\rangle = \langle\Psi_1|\mathcal{H}^2_{E_g}|\Psi_2\rangle = -\langle\Psi_1|\mathcal{H}^1_{E_g}|\Psi_1\rangle = S_E. \qquad (5.22)$$

The S_{ij} matrix elements of the $\mathcal{H}_S$ operator can then be represented in terms of S_A and S_E as in Table 17.

2. We can now prove the independence of the first term in (5.18) from the mutual arrangement of substituents. According to Table 17, the $S_{11} + S_{22}$ sum in (5.18) only depends on S_A and is proportional to the contribution of the totally symmetric component $a_A\mathcal{H}_A$ for separation (2.36) of the substitution operator. From Table 3, we see that this contribution depends on the number rather than the arrangement of substituents X in $ML_{6-k}X_k$. The dependence of a_A on the number of substituents is linear, and the $(1/2)(S_{11} + S_{22})$ sum has the same value for all isomers. This value can be positive or negative depending on ligands L and X. We will see that if X is more electropositive than L, then the $(1/2)(S_{11} + S_{22})$ term is positive. Conversely, if X is more electronegative than L, then $(1/2)(S_{11} + S_{22}) < 0$.

3. The second term in (5.18) is determined by the $|S_E|$ parameter and the contribution of symmetrized operators of E_g symmetry to $\mathcal{H}_S$. For this reason, the $\Delta\mathcal{E}_E$ term is structurally sensitive and, as is seen from Table 17, depends not only on the number but also on the mutual arrangement of substituents X in $ML_{6-k}X_k$. According to (5.18), this term is always negative (or zero) and favors stabilization of substituted complexes in comparison with unsubstituted ones. We conclude that the most stable isomers of $ML_{6-k}X_k$ are those for which this term is larger in magnitude. According to Table 3, the coefficients of $\mathcal{H}^1_{E_g}$ and $\mathcal{H}^2_{E_g}$ are smaller in magnitude for *cis* isomers. It follows that *trans* isomers should be stabler than *cis* ones. The energies of *cis* $\to$ *trans* isomerization are given by

$$\Delta\mathcal{E}_{c\to t}(ML_4X_2) = |S_E|/\sqrt{3}, \quad \Delta\mathcal{E}_{c\to t}(ML_3X_3) = |S_E|. \qquad (5.23)$$

Note that $\Delta\mathcal{E}_E$ only equals zero for *cis*-ML_3X_3 (the $\mathcal{H}_S$ operator then does not contain the $\mathcal{H}^1_E$ and $\mathcal{H}^2_E$ contributions) and MX_6 (the transition to this complex is described by the $\mathcal{H}_S = \sqrt{6}\mathcal{H}_A$ totally symmetric substitution operator). As a result, the energy of the *cis*-ML_3X_3

complex is described by a simple additive scheme, whereas the other $ML_{6-k}X_k$ mixed complexes should have higher thermodynamic stabilities than those predicted by the additive scheme (Fig. 25). The energies of all $ML_{6-k}X_k$ complexes are expressed in terms of the energies of the ML_6 and MX_6 complexes and the S_E value. The magnitudes of the deviations from additive total energy values equal $|S_E|/\sqrt{3}$ for ML_5X, MX_5L, *cis*-ML_4X_2, and *cis*-ML_2X_4; $2|S_E|/\sqrt{3}$ for *trans*-ML_4X_2 and *trans*-ML_2X_4, and $|S_E|$ for *trans*-ML_3X_3.

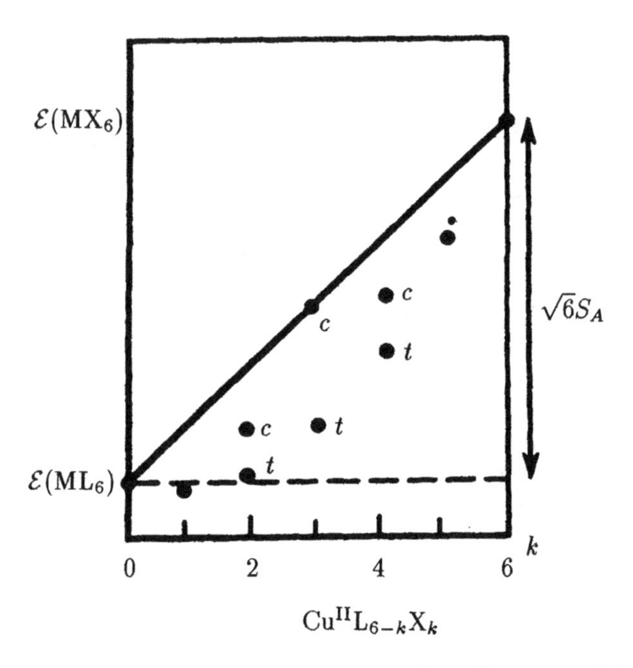

FIG. 25. DEPENDENCE OF THE ENERGY OF d^9 $Cu^{II}L_{6-k}X_k$-TYPE COMPLEXES ON THEIR COMPOSITION AND STRUCTURE IN THE APPROXIMATION LINEAR IN VIBRONIC COUPLING.

Lastly, note that the conclusion of predominant stabilization of *trans* isomers of heteroligand quasi-octahedral d^9 transition metal complexes has a simple physical meaning. As the wave functions of the E term are even, a purely antisymmetric (odd) substitution operator should not affect the E_g term at all. The substitution operator corresponding to the asymmetric substitution with the formation of *cis*-ML_4X_2 includes both centrally symmetric (even) and antisymmetric components, of which only the former participates in splitting of the E_g term. On the contrary, the substitution operator leading to *trans*-ML_4X_2

is centrally symmetric as a whole, and its contribution to splitting of E_g is therefore larger. The same reasoning applies to the formation of *trans*-ML_3X_3. Although the substitution operator then contains an antisymmetric component, its contribution is smaller than that of the antisymmetric component of the *cis*-substitution operator; accordingly, the contribution of the centrally symmetric component of the *trans*-substitution operator is larger than that of the corresponding component of the operator leading to *cis*-ML_3X_3.

So far, we have discussed d^9 complexes. The conclusion of the predominant stabilization of *trans* isomers is, however, also valid for low-spin d^7 and high-spin d^4 complexes with the $t_{2g}^6 e_g^{*\,1}$ and $t_{2g}^3 e_g^{*\,1}$ electronic configurations, respectively. The latter complexes contain one electron in the e_g^* level, whereas the d^9 complexes, one hole (three electrons). Their t_{2g} levels are either fully filled or half-filled. It follows that unexcited low-spin d^7 and high-spin d^4 ML_6 complexes also occur in states with double orbital degeneracy, and their ground state term is also E_g; its wave functions have the same spatial symmetry properties as functions (5.1) of d^9 complexes. Recall that the derivation of equations (5.23) for the isomerization energy was based solely on these symmetry properties.

4. Some problems require more specific expressions for numerically estimating the matrix elements of the $\mathcal{H}_S$ operator and the corresponding $\Delta\mathcal{E}_{c\to t}$ value. As in Chapters 3 and 4, such expressions and estimates will be obtained in terms of the MO LCAO method on the assumption that the difference between ligands X and L is described by the $\Delta\alpha = \alpha(X) - \alpha(L)$ value, where α is the Coulomb integral for the ligand σ AO. We will use the explicit expressions

$$\begin{aligned} \psi_{z^2}^* &= l_e d_{z^2} - (c_e/2\sqrt{3})(2\sigma_1 + 2\sigma_2 - \sigma_3 - \sigma_4 - \sigma_5 - \sigma_6), \\ \psi_{x^2-y^2}^* &= l_e d_{x^2-y^2} - (c_e/2)(\sigma_3 + \sigma_4 - \sigma_5 - \sigma_6). \end{aligned} \tag{5.24}$$

Here, σ_i are the ligand σ AOs, and d_{z^2} and $d_{x^2-y^2}$ are the transition metal AOs. (Overlap integrals between different AOs are, as previously, ignored; therefore, $c_e^2 + l_e^2 = 1$.)

Alongside substitution operator $\mathcal{H}_S$, consider the corresponding one-electron operator H_S (see Section 2.3). It follows from (5.24) and form (2.39) of the matrix elements of the H_S operator that, for both $e_g^{*\,3}$ and $e_g^{*\,1}$ configurations, S_E parameter (5.22) can be written in the form

$$S_E = \langle \psi_{z^2}^* | H_E^2 | \psi_{x^2-y^2}^* \rangle = -(c_e/2\sqrt{3})\Delta\alpha,$$

where c_e appears in equations (5.24) for the MOs. The isomerization energy then equals

$$\begin{aligned} \Delta\mathcal{E}_{c\to t}(\mathrm{ML_4X_2}) &= c^2|\Delta\alpha|/6, \\ \Delta\mathcal{E}_{c\to t}(\mathrm{ML_3X_3}) &= c^2|\Delta\alpha|/2\sqrt{3}. \end{aligned} \tag{5.25}$$

Equations (5.25) can be used to approximately estimate the $\Delta\mathcal{E}_{c\to t}$ value. Assuming that $\Delta\alpha$ varies from one to several eV and $c^2 \approx 1/2$, we obtain estimates ranging from several to 20 kcal/mol.

5. The conclusion of the predominant stabilization of *trans* isomers of quasi-octahedral d^9 and d^7 complexes is in agreement with the available experimental data [205, 212–225]. Examples of coexistence of *cis* and *trans* isomers of the same complex are few and largely provided by complexes with chelating ligands. None of the several dozens of copper(II) complexes with monodentate ligands studied by diffraction methods [215–220] had the *cis* conformation. Typical examples are the $\mathrm{Cu^{II}L_4(py)_2^{2-}}$ (L = Cl or Br) complexes, in which the pseudooctahedral environment of copper is characterized by the presence of the center of inversion [215–220]. Note also complex I with three pairs of different ligands, in which identical ligands are positioned *trans* to each other:

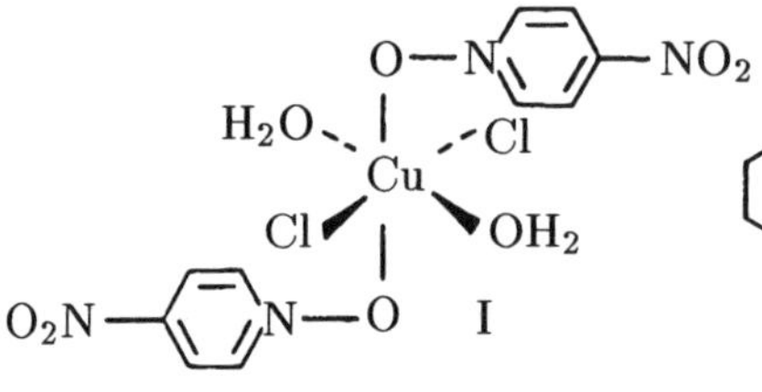

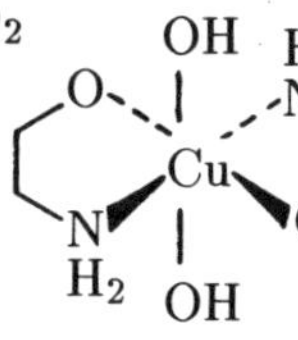

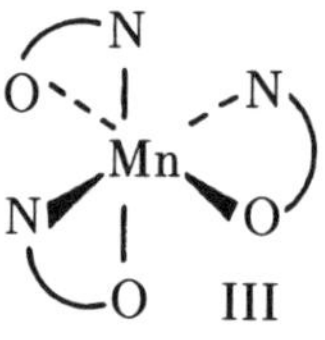

There are structural data [215–220] on several chelate Cu(II) compounds with bidentate amines, for instance $\mathrm{Cu(en)_2(SCN)_2}$, where four nitrogen atoms form a square-planar environment of copper, and thiocyanate ligands are situated above and blow this plane. In complexes with bidentate ligands containing nitrogen and oxygen, like atoms occupy *trans* positions (e. g., in complex II above). The same is true of Cu(II) complexes with ligands containing nitrogen and sulfur, perhaps, except one compound where both *cis* and *trans* isomers were observed. Structural data exist for d^4 manganese(III) complexes with bidentate organic ligands linked with Mn through N and O [226, 227]. These complexes have the *trans* structure III drawn above, in agreement with theoretical expectations for $\mathrm{MnL_3X_3}$ complexes with $\mathrm{Mn^{3+}}$ in the high-spin state.

Lastly, consider direct thermochemical measurements of the heats of *cis–trans* isomerization for Cu(II) complexes. The corresponding data are exceedingly scarce, because *cis* isomers were only observed in few complexes, all formed by chelating ligands [212–225]. It was, however, found that *cis* → *trans* isomerization of Cu^{II} complexes with 1,2-diaminocyclohexane and H_2O was accompanied by heat release (1–2 kcal/mol) [228]. Similarly, complex V with the *trans* structure was found to be 2 kcal/mol more stable than complex IV with the *cis* structure, the formation of which was determined by the character of the chelating ligand (tren stands for 2,2,2-triaminotriethylamine, and trien, for triethylenetetramine) [228]:

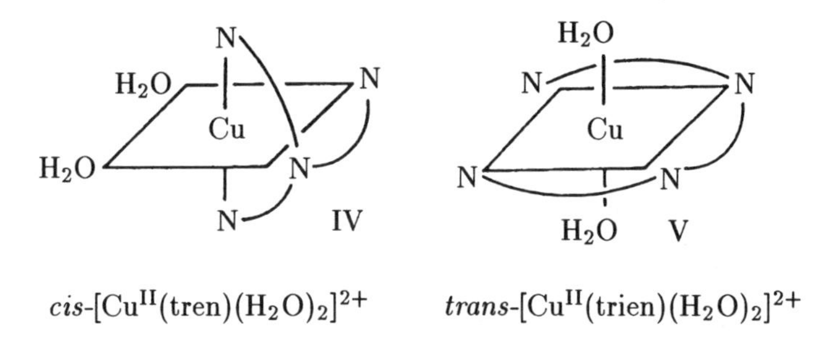

cis-$[Cu^{II}(tren)(H_2O)_2]^{2+}$ *trans*-$[Cu^{II}(trien)(H_2O)_2]^{2+}$

5.4. Quadratic *E-e-S* Problem

1. In the preceding sections, the adiabatic potential energy surfaces of $Cu^{II}L_{6-k}X_k$ complexes were found by only taking into account the linear vibronic coupling constant. In this section, we include into consideration the quadratic vibronic coupling constant [209, 210]. Such an analysis is necessary, because the deformation of the bottom of the Mexican hat due to quadratic vibronic coupling may amount to several kcal/mol [14], that is, be of the same order as the deformation due to ligand replacement. We cannot therefore *a priori* disregard quadratic effects on the relative stability of different isomers (it will, however, be shown below that taking into account quadratic vibronic coupling does not change the conclusion of predominant stabilization of *trans* isomers in comparison with *cis* ones). In addition, including this constant will allow us to explain the nature and symmetry of so-called distortion isomers of chain $CuHal_2(NH_3)_2$ complexes exclusively in terms of intramolecular vibronic interactions, without involving intermolecular forces.

As a zero approximation, we, as previously, use an ML_6 complex with an ideal octahedral structure. The sought adiabatic potential then again has form (5.13). However, if quadratic vibronic coupling is included, the anharmonic adiabatic potential component, $\mathcal{E}'_{1,2}$, should be calculated from the secular equation more complex than (5.11), *viz.*,

$$\mathrm{Det}\left\| \begin{matrix} -AQ_3+A'(Q_2^2-Q_3^2)+S_{11}-\mathcal{E}' & AQ_2-2A'Q_2Q_3+S_{12} \\ AQ_2-2A'Q_2Q_3+S_{12} & AQ_3-A'(Q_2^2-Q_3^2)+S_{22}-\mathcal{E}' \end{matrix} \right\| = 0. \quad (5.26)$$

2. First, consider doubly substituted ML_4X_2 complexes. Note that the S_{12} matrix element in (5.26) vanishes for both isomers of such complexes. Indeed, the Ψ_1 and Ψ_2 functions are mixed by an $\mathcal{H}_E^2$-type perturbation, but the $\mathcal{H}_S$(*trans*-ML_4X_2) and $\mathcal{H}_S$(*cis*-ML_4X_2) operators do not contain $\mathcal{H}_E^2$ components (it is assumed that the substitution occurs at coordinate z in the *trans* isomer and in the xy plane in the *cis* one). The contribution of the totally symmetric component of $\mathcal{H}_S$ is the same for the two isomers and, therefore, does not affect the energy of isomerization. This contribution can be excluded from consideration after shifting the origin of the energy scale by $(S_{11}+S_{22})/2$.

For the quadratic E-e-S problem, the use of coordinates (5.14) does not simplify the search for the minima of adiabatic potential energy surfaces. We will therefore use polar coordinates independent of the matrix elements of $\mathcal{H}_S$, namely, $Q_2 = \rho\sin\varphi$ and $Q_3 = \rho\cos\varphi$. In these coordinates, the lower sheet of the sought adiabatic potential surface is given by

$$\begin{aligned} \mathcal{E} = K\rho^2/2 - \{A^2\rho^2 + A'^2\rho^4 + 2AA'\rho^3\cos 3\varphi \\ + S(A\rho\cos\varphi + A'\rho^2\cos 2\varphi) + S^2/4\}^{1/2}. \end{aligned} \quad (5.27)$$

Here, $S = S_{22} - S_{11}$; $S = -(2/\sqrt{3})S_E$ for the *cis* isomer of ML_4X_2 and $+(4/\sqrt{3})S_E$ for its *trans* counterpart. Note the following properties of adiabatic potential (5.27): simultaneous reversal of the signs of A and A' is equivalent to the reversal of the sign of S; the reversal of the sign of A is equivalent to the substitution $\varphi \to \varphi + \pi$; the reversal of the sign of A' is equivalent to the simultaneous substitutions $\varphi \to \varphi + \pi$ and $S \to -S$; lastly, the adiabatic potential is invariant with respect to the substitution $\varphi \to -\varphi$. It follows from these properties that it suffices to analyze (5.27) in the range $0 \le \varphi \le \pi$ at arbitrary signs of A and A'.

3. Surface (5.27) is difficult to study analytically. It can more conveniently be analyzed by computations [209]. Applying this approach to $Cu^{II}L_4X_2$ with the use of various reduced substitution constant S_E values and the typical vibronic coupling and force constant values $A = -0.05$ and $A' = -0.004$ au and $K = 0.08$ au [229] gave the polar coordinates of the local minima of surface (5.27) shown in Fig. 26; the energy values at the minima are plotted in Fig. 27. (For the specified A, A', and K values, the barrier between adiabatic potential energy surface minima in the homoligand complex, tetragonal

distortion parameter ρ_0, and the Jahn–Teller stabilization energy have typical values of 600 cm^{-1}, 1/3 Å, and 3000 cm^{-1}, respectively.) According to Fig. 27, the lowest lines in the diagram correspond to the *trans* isomer. It follows that no matter what the S value is taken, the *trans* form is the more stable, as when linear vibronic constant is only taken into account.

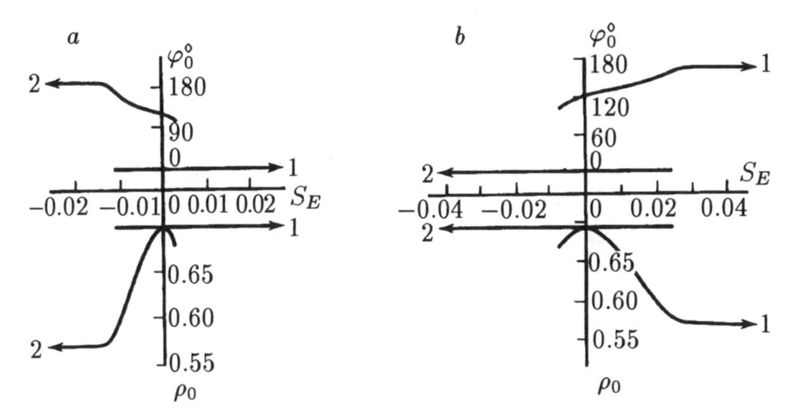

FIG. 26. DEPENDENCES OF THE ρ_0 AND φ_0 COORDINATES OF ADIABATIC POTENTIAL MINIMA ON THE S_E CONSTANT FOR $Cu^{II}L_4X_2$ COMPLEXES; THE DEPENDENCES WERE OBTAINED TAKING INTO ACCOUNT LINEAR AND QUADRATIC VIBRONIC COUPLING TERMS AT TYPICAL K, A, AND A' CONSTANT VALUES (φ_0 IN DEGREES AND ρ_0 AND S_E IN ATOMIC UNITS): (*a*) *trans* ISOMER, (*b*) *cis* ISOMER; *1* AND *2* ARE THE NUMBERS OF MINIMA. TWO PLOTS ARE SHOWN, FOR φ_0 IN THE TOP AND FOR ρ_0 IN THE BOTTOM PART.

We wish to stress that this conclusion is independent of specific A, A', and K parameter values. First, by virtue of the symmetry properties of adiabatic potential energy surface (5.27) specified above, this surface is independent of vibronic constant signs. To show that the conclusion for *trans* isomer stabilization is also independent of the magnitudes of these model parameters, we again turn to calculations. Let us make the substitutions

$$L = K(A'/A^2), \qquad M = A'/A, \qquad N = S/A'$$

and rewrite (5.27) in the form

$$(A'/A^2)\mathcal{E} = (1/2)L\rho^2 - \{M^2\rho^2 + M^4\rho^4 + 2M^3\rho^3\cos 3\varphi + M^3N\rho\cos\varphi + M^4N\rho^2\cos 2\varphi + M^4N^2/4\}^{1/2}. \tag{5.28}$$

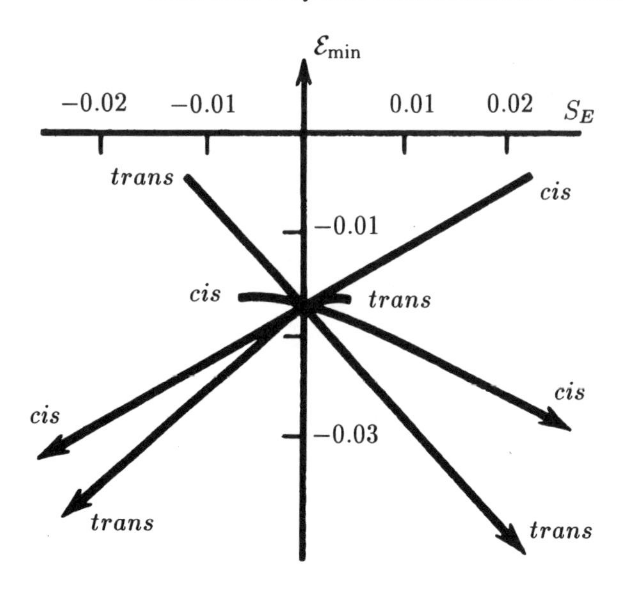

FIG. 27. DEPENDENCE OF $\mathcal{E}_{\min}$ ENERGY AT ADIABATIC POTENTIAL MINIMA ON S_E FOR $Cu^{II}L_4X_2$ COMPLEXES TAKING INTO ACCOUNT LINEAR AND QUADRATIC VIBRONIC COUPLING TERMS; ($\mathcal{E}_{\min}$ AND S_E IN ATOMIC UNITS).

The right-hand side of (5.28) contains three parameters, L, M, and N. The L and M parameters are conformation-independent, whereas $N_{cis} = -1/2\, N_{trans}$. The minima of (5.28) for the *cis* and *trans* isomers were calculated for different (L_i, M_i, N_i) points. The $M = (A'/A)$ parameter was varied from zero to 0.5 au, and L, from zero to $0.5M + 2M^2$ (this value was found from the condition that the minima of the adiabatic potential of the unsubstituted complex must be situated in the physically realistic $\rho_0 \leq 2$ au region). The range of N variations (in reality, $-1 \leq N \leq 1$ au) was chosen to meet the requirement that at its boundary, the φ coordinates of two ML_4X_2 adiabatic potential minima situated symmetrically with respect to the Q_3 axis should attain their limiting values, 0 or π (Fig. 26). The calculations showed that the *trans* isomer of ML_4X_2 was more stable than its *cis* counterpart across the whole range of parameter values.

It remains to consider the situation when the minima of the adiabatic potential energy surface occur on the Q_3 axis. This situation is simpler and can be analyzed analytically [209]. Assume for definiteness that $A < 0$ and $A' < 0$ (it was shown above that the other combinations of signs reduce to this). It will also be taken that $K > 2|A'|$;

this assumption is physically warranted and is usually introduced in studies of the Jahn–Teller effect for homoligand complexes [14, 15]. The results of the analysis can then be summarized as follows. At $S < 0$, the arrangement of the minima along the Q_3 axis corresponds to the condition $|S| > 36A^2|A'|/(K^2 - 4A'^2)$; at typical A, A', and K constant values, this condition gives $|S| > 1.6$ eV, that is, the difference of the orbital electronegativities of ligands L and X should be $\Delta\alpha \geq 5$ eV. At $S > 0$, the minima can be situated at the Q_3 axis if $|S| > 18A^2|A'|/(K^2 - 4A'^2)$, that is, at $|S| > 0.8$ eV and $|\Delta\alpha| \geq 2.5$ eV. Whether $S > 0$ or $S < 0$, the adiabatic potential energy surface of each isomer has only one minimum. The minimum corresponding to *trans*-ML_4X_2 is deeper than that of the *cis* isomer, and the isomerization energy is given by

$$\Delta\mathcal{E}_{c\to t} = \frac{|S_E|}{\sqrt{3}} \mp \frac{2A^2|A'|}{K - 4A'^2}, \tag{5.29}$$

as is easy to see by substituting $\varphi = 0$ and $\varphi = \pi$ into (5.27). Signs minus and plus in (5.29) are selected at $S_E > 0$ and < 0, respectively, and the first and second terms have the meaning of electronic and geometric relaxation contributions to isomerization energy, respectively. Recall that in the linear E-e-S problem (at $A' = 0$), geometric relaxation makes no contribution to isomerization energy. Note also that here, this contribution appears even in the region of S_E values within which quadratic effects are suppressed by low-symmetry substitution.

4. The character of the difference between the adiabatic potentials calculated taking into account quadratic vibronic coupling for *cis*- and *trans*-ML_3X_3 is fairly obvious. For the *cis* isomer, the $\mathcal{H}_S$ operator does not contain a tetragonal component (Table 3), whereas for the *trans* isomers, the contributions of $\mathcal{H}_E^1$ and $\mathcal{H}_E^2$ are nonzero. It follows that in *cis*-ML_3X_3, the adiabatic potential surface only shifts as a whole along the energy axis in comparison with ML_6. The shift value equals $(S_{11}+S_{22})/2$ and is determined by the totally symmetric component of substitution operator $\mathcal{H}_S$. In *trans*-ML_3X_3, this shift is accompanied by lowering of one side of the Mexican hat with respect to the other. As a result, the potential energy surface of the *trans* isomer always contains one or two minima deeper than those characteristic of the *cis* isomers (*cis* isomer minima are equal in energy). The relative stabilities of the *cis* and *trans* forms of ML_3X_3 are illustrated by Fig. 28, where the dependence of $\mathcal{E}_{\text{min}}$ on S_E is shown for the specified K, A, and A' parameter values; the coordinates of the minima are given in Fig. 29.

It follows from the above reasoning that the inclusion of quadratic vibronic coupling does not change the conclusion of predominant thermodynamic stabilization of *trans* isomers of d^9 and d^7 complexes. Note

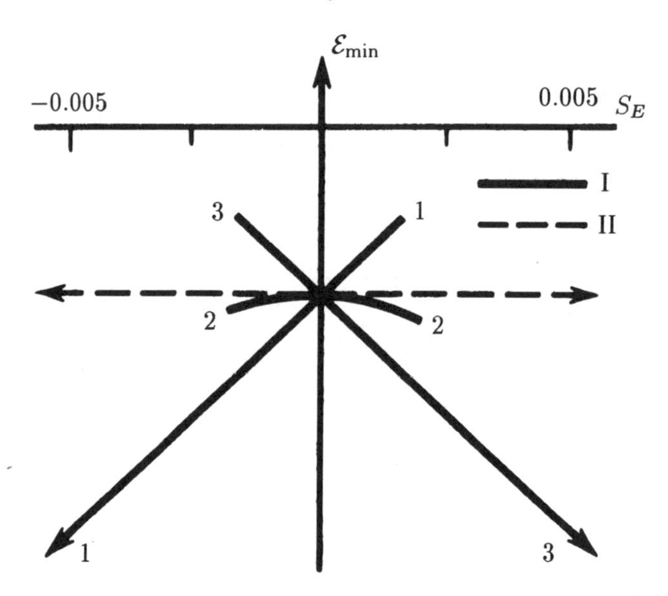

FIG. 28. DEPENDENCE OF $\mathcal{E}_{\min}$ ENERGY AT ADIABATIC POTENTIAL MINIMA ON S_E FOR $Cu^{II}L_3X_3$ COMPLEXES ($\mathcal{E}_{\min}$ AND S_E IN ATOMIC UNITS): I, *trans* ISOMER AND II, *cis* ISOMER; *1–3* ARE THE NUMBERS OF MINIMA.

that in $Cu^{II}L_6 \longrightarrow Cu^{II}L_4X_2$ substitution reactions, kinetic factors also favor larger yields of the *trans* isomer. Indeed, homoligand $Cu^{II}L_6$ complexes usually have the geometry of elongated octahedra and are characterized by the presence of two longer and therefore weaker Cu—L bonds positioned *trans* with respect to each other [211a]. Note in conclusion that, as is seen from Fig. 26, the inclusion of quadratic vibronic coupling reveals a new feature of the structure of heteroligand $Cu^{II}L_{6-k}X_k$ complexes, *viz.*, the possibility of the existence of their specific nonequivalent forms called distortion isomers.

5.5. Distortion Isomers of Chain $Cu^{II}(NH_3)_2Hal_2$ Complexes

The term distortion isomers is applied to geometric forms of a complex characterized by the same mutual arrangement of ligands in the first coordination sphere of the metal but differing in the shape of the coordination polyhedron, e. g., having different central atom–ligand bond lengths. Distortion isomers of Cu(II) complexes were experimentally observed in copper compounds of the composition $Cu(NH_3)_2Hal_2$ (Hal = Cl or Br) having chain structures [230–231].

Fragments of $Cu(NH_3)_2Hal_2$ chains corresponding to two isomers (α

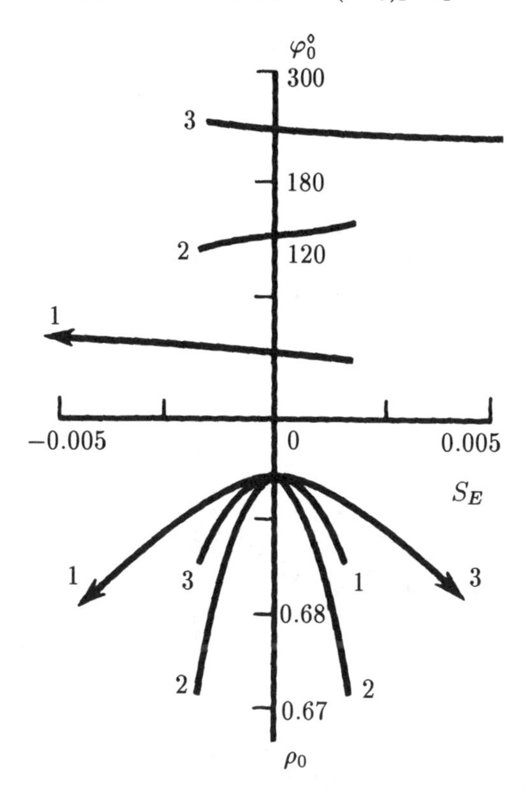

FIG. 29. DEPENDENCE OF THE COORDINATES OF ADIABATIC POTENTIAL MINIMA ON S_E FOR *trans*-CuL_3X_3 COMPLEXES; *1–3* ARE THE NUMBERS OF MINIMA; FOR *cis* ISOMERS, THE COORDINATES ARE THE SAME AS FOR HOMOLIGAND COMPLEXES (φ_0 IN DEGREES AND ρ_0 AND S_E IN ATOMIC UNITS).

and β) are depicted in Fig. 30a, which shows that both isomers have the *trans* arrangement of NH_3 ligands but different geometries of the first coordination sphere of Cu (to be specific, let Hal = Br). In one of the isomers, four Cu—Br bond lengths are identical (the D_{4h} form, or the β isomer), whereas in the other, the lengths of the Cu—Br bonds lying opposite to each other coincide but differ from the lengths of the other two Cu—Br bonds (the D_{2h} form, or the α isomer). Distortion isomers were also observed for some isolated (nonchain) complexes with more complex ligands including organic ones [205, 230–237], *viz.*, $Cu(NH_3)_2(SCN)_2$ [232], Cu(salicylaldehyde)$_2$ [234], CuL_2Cl_2 (L_2 = 3-picoline-N-oxide, 4-picoline-N-oxide, 2,6-lutidine-N-oxide [235, 236], or perchlorate-*bis*(1,3-diaminobutane) [231]), Cu(2-picoline)$_2(NO_3)_2$ [231], and $Cu(NH_3)_4(NO_3)_2$ [237]. The theoretical problems of distortion isomerism are, however, usually discussed for $Cu(NH_3)_2Hal_2$ compounds, which contain simple ligands and are easiest to be treated theoretically.

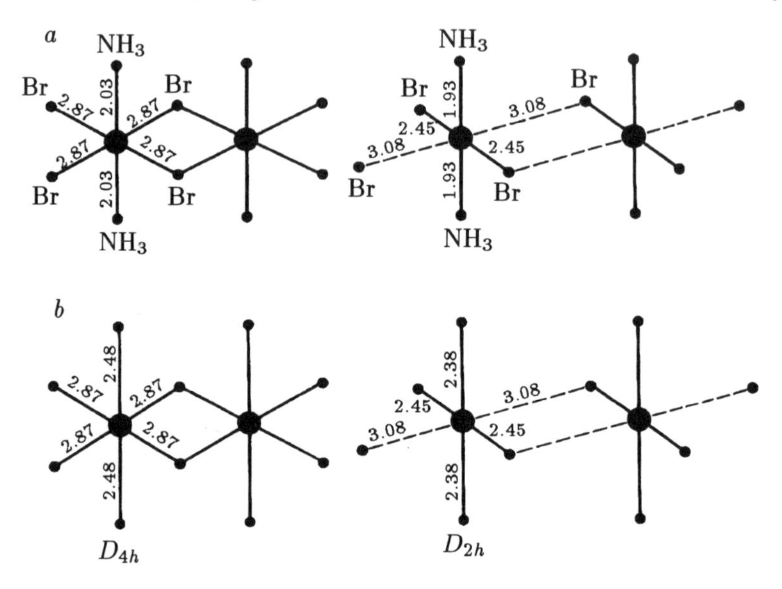

FIG. 30. STRUCTURE OF DISTORTION ISOMERS OF $Cu(NH_3)_2Br_2$ CHAIN COMPOUNDS: (*a*) EXPERIMENTAL INTERATOMIC DISTANCES AND (*b*) INTERATOMIC DISTANCES RECALCULATED TAKING INTO ACCOUNT THE DIFFERENCE OF THE ATOMIC RADII OF NITROGEN AND BROMINE.

Immediately after the discovery of distortion isomerism, a suggestion was made that the phenomenon was related to Jahn–Teller instability of octahedral $Cu^{II}L_6$ complexes. As mentioned in [238], homologinad $Cu^{II}L_6$ complexes typically have D_{4h} structures with two long and four short Cu—L bonds. The transition to *trans*-$Cu^{II}L_4X_2$ can therefore involve the replacement of two L ligands situated either on the fourfold symmetry axis of the elongated octahedron with preservation of D_{4h} symmetry or in the plane normal to this axis, and then symmetry lowers to D_{2h}.

In [238], these conclusions were not substantiated analytically or computationally. This was achieved in [239] in terms of the theory of Jahn–Teller effect; in that work, ligands were modeled by point charges. It was shown that the approximation linear with respect to vibronic coupling predicts the formation of the D_{4h} form of *trans*-$Cu^{II}L_4X_2$. Later, the problem of distortion isomers was tackled in [14, 15, 232] taking into account the pseudo-Jahn–Teller effect. An analysis of the form of the adiabatic potential of such a model (again in the linear approximation) showed that because of intramolecular vibronic coupling, *trans*-$Cu^{II}L_4X_2$ can exist as either the high-symmetry (D_{4h}) or the low-symmetry (D_{2h}) form, but not both simultaneously.

The approach suggested as a way of solving the problem took into account the chain structure of $Cu(NH_3)_2Hal_2$ complexes in crystals [11, 14, 15]. Distortional isomerism of these complexes was assumed to be a consequence of a strong pseudo-Jahn–Teller effect for each quasi-octahedral structural unit in the chain and interchain interaction in the crystalline state.

Let the chain be assigned a D_{4h} structure with each copper atom having a tetragonal bipyramidal environment formed by four equatorial Hal atoms and two axial NH_3 groups. Because ligands Hal and NH_3 are different, the ^2E_g ground term characteristic of octahedral Cu(II) complexes splits into the $^2A_{1g}$ and $^2B_{1g}$ components, which can be mixed by mode Q of symmetry B_{1g}. As a result of the strong pseudo-Jahn–Teller effect, this mixing causes the appearance of two equivalent minima on the lower sheet of the adiabatic potential energy surface (see Section 2.2). At both minima, the tetragonal bipyramid around Cu is distorted, and the square formed by equatorial Hal atoms turns into a rhombus. This results in the formation of the more stable α isomer of the $Cu(NH_3)_2Hal_2$ chain.

Interchain interaction, however, tends to maintain the D_{4h} configuration of chain units [11, 14, 15], and a third shallower minimum appears on the lower sheet of the adiabatic potential energy surface. This minimum corresponds to $Q = 0$, which explains the existence of the less stable β isomer of the chain $Cu(NH_3)_2Hal_2$ complex.

2. The problem of distortion isomerism can, however, be given a different solution [209, 210] based on a purely one-center intramolecular mechanism. The approach taken in [209, 210] is close to that employed in [239], but it goes beyond the approximation linear in vibronic coupling and, in addition to the Jahn–Teller e_g^* MOs, takes into account the a_{1g}^* unoccupied level.

The problem of possible symmetry (no geometry) of $Cu(NH_3)_2Hal_2$ complexes has the simplest solution in the limit of weak low-symmetry perturbation [209]. According to (5.6), adiabatic potential energy surface $\mathcal{E}(\rho, \varphi)$ of a homoligand $Cu^{II}L_6$ complex has three equivalent minima at $\varphi = 0$ and $\pm 2\pi/3$ or at $\varphi = \pi$ and $\pm\pi/3$ depending on the signs of vibronic coupling constants A and A', that is, depending on whether the octahedron is tetragonally elongated or tetragonally contracted (it is assumed that the fourfold symmetry axis coincides with the x, y, or z axis). Accoring to (5.27), in the presence of substituents on axis z, that is, at $S \neq 0$, only the $\varphi = \pm\pi/3$ and $\varphi = \pm 2\pi/3$ minima can remain equivalent, because the $\mathcal{E}(\rho, \varphi)$ function does not change with the reversal of the sign of φ.

The question of whether or not the minima exist in reality requires special consideration. It follows from the continuity of $\mathcal{E}$ as a function of ρ and φ that at fairly small $|S_E|$, there exist all three minima, and their coordinates are close to the coordinates of the minima of the unsubstituted complex. This means that if the difference between ligands L and X is small, two forms of the *trans*-$Cu^{II}L_4X_2$ complex can exist, one with D_{4h} symmetry ($\varphi = 0$ or $\varphi = \pi$) and the other with D_{2h} symmetry ($\varphi = \pm\pi/3$ or $\varphi = \pm 2\pi/3$). Earlier, this possibility was discussed in [240a]; also see [240b]. At not very small $|S_E|$, the minima of adiabatic potential (5.27) can, however, only be found by computations.

Figure 26 shows that at S_E values small in magnitude, there are two nonequivalent minima corresponding to the high-symmetry D_{4h} and the low-symmetry D_{2h} isomers. This conclusion applies to complexes with $|S_E|$ of about 0.01 au, that is, realistic difference values, $\Delta\alpha = \alpha_X - \alpha_L \approx 1.5$ eV, between the Coulomb integrals of the σ AOs of ligands L and X. This leads us to conclude that in principle, Jahn–Teller effect considerations correctly predict the existence and possible symmetry of distortion isomers at realistic substitution parameter and vibronic coupling constant values if linear and quadratic vibronic couplings and differences between ligands in the first coordination sphere of the metal are taken into account.

3. The treatment given above is, however, insufficient, for it does not explain the observed bond lengths in the D_{4h} and D_{2h} forms of $Cu(NH_3)_2Hal_2$, see Fig. 30a. Difficulties arise because for adiabatic potential (5.27) in the region of the existence of distortion isomers ($|S_E| < 0.01$ au), the quasi-octahedral structural unit of the D_{4h} isomer should be elongated along the fourfold symmetry axis and contracted in the equatorial plane, whereas a similar unit of the D_{2h} isomer should be elongated along a twofold axis in the equatorial plane and contracted in two other directions (x and z or y and z).

Let us compare these predictions with the structural data (Fig. 30a). First, it should be taken into account that the observed Cu—Br and Cu—NH_3 distances in $Cu(NH_3)_2Br_2$ depend on vibronic effects and the effect of substitution on the electronic subsystem on the one hand and the difference of the atomic radii r of bromine and nitrogen on the other. (According to various data, r(Br) is on average larger than r(N) by 0.45 Å.) To introduce corrections for atomic radii, we may, e. g., replace r(N) by r(Br). The interatomic distances recalculated in this way are shown in Fig. 30b. These corrected distances only depend on Jahn–Teller and substitution-induced distortions of the $Cu^{II}Br_6$ regular octahedron, which we use as a zeroth-order approximation in adiabatic potential calculations. The experimental distortion values are obtained by comparing the distances given in Fig. 30b with the interatomic distances in the ideal octahedron. These distances can be estimated in a number of ways. For instance, averaging the distances observed in crystalline $CuBr_2$ yields 2.66 Å for the mean Cu—Br bond length. Using the tabulated ionic radii [241], we obtain a fairly close value of 2.76 Å. In our treatment, we used an approximate estimate of 2.7 Å.

These corrections and estimates make it clear (Fig. 30) that the quasi-octahedral structural unit of $Cu(NH_3)_2Br_2$ (D_{4h} form) should be considered tetragonally contracted rather than elongated in the z direction. Likewise, a similar structural unit in the D_{2h} form is an octahedron tetragonally elongated along the x (or y) axis. It follows that reconciling theory and experiment would require that all three adiabatic potential minima of this structural unit be situated in the left half-plane of the Q_2, Q_3 plane (in the region of negative Q_3 values). Agreement with experiment is attained when the regular $Cu^{II}Br_6$ octahedron with 2.7 Å Cu—Br bond lengths is largely distorted by E_g symmetry deformations, the contribution of the totally symmetric Q_1 deformation being several times smaller.

The same conclusion can be reached in another way, through considering deformations transforming the bond lengths experimentally observed for the D_{4h} isomer into the bond lengths of the D_{2h} chain. The contribution of deformation Q_1 to such a transformation is 2.5 times smaller than that of E_g deformations and can safely be ignored. Computations, however, show that the required arrangement of adiabatic potential (5.27) minima does not correspond to any reasonable A, A', and S parameter values. As in homoligand $Cu^{II}L_6$ complexes, all three minima can never be situated in the same half-plane.

4. This result leads us to consider a more general adiabatic potential to describe the isomeric forms of $Cu(NH_3)_2Br_2$. This adiabatic potential is obtained by including substitution-induced and vibronic coupling-induced mixing of the E term states with the states produced in the transfer of one electron from the e_g^* MO to the unoccupied a_{1g}^* MO separated from the former by a $\omega \approx 2$ eV energy gap. (Mixing of these states can be treated as ds mixing; the importance of including ds mixing in consideration of a number of properties of copper complexes was discussed in, e. g., [242, 243].) To simplify the analysis, the problem can conveniently be treated in terms of MOs that experience mixing under deformations Q_2 and Q_3 and the action of the substitu-

tion operator [210]. This leads to the secular equation

$$\mathrm{Det}\left\| \begin{matrix} -aQ_3+a'(Q_2^2-Q_3^2)+s_{11}-\varepsilon & aQ_2-2a'Q_2Q_3 & bQ_3+s_{1a} \\ aQ_2-2a'Q_2Q_3 & aQ_3-a'(Q_2^2-Q_3^2)+s_{22}-\varepsilon & bQ_2 \\ bQ_3+s_{1a} & bQ_2 & \omega+s_a-\varepsilon \end{matrix} \right\| = 0. \tag{5.30}$$

(Here, orbital vibronic coupling constants and the matrix elements of the one-electron substitution operator in the basis set of MOs are, to avoid confusion, denoted by small letters.) The adiabatic potential then takes the form

$$\begin{aligned} \mathcal{E}(Q_2,Q_3) &= (1/2)K(Q_2^2+Q_3^2)+\mathcal{E}'(Q_2,Q_3), \\ \mathcal{E}'(Q_2,Q_3) &= 2\varepsilon_1(Q_2,Q_3)+\varepsilon_2(Q_2,Q_3), \end{aligned} \tag{5.31}$$

where ε_1 is the smallest and ε_2 second to smallest roots of secular equation (5.30).

In comparison with (5.27), adiabatic potential (5.30) includes additional parameters, namely, energy gap ω, vibronic constant b describing interaction between the e_g^* and a_{1g}^* MOs, and substitution operator matrix elements s_{1a} and s_a. The number of model parameters can be reduced through using the LCAO method; this method also makes it possible to identify chemically acceptable regions of parameter values. To follow this course, we must, in addition to MO (5.24), consider the a_{1g} MO given by

$$\psi_a^* = l_a s - (c_a\sqrt{6})(\sigma_1+\sigma_2+\sigma_3+\sigma_4+\sigma_5+\sigma_6). \tag{5.32}$$

Using Slater determinants (5.1) to describe many-electron E-term wave functions, we can show that for d^9 complexes, orbital vibronic coupling constants a and a', except for sign, coincide with constants A and A' in (5.27), *viz.*, $a=-A$ (0.05 au) and $a'=-A'$ (0.004 au). An approximate estimate of b can be obtained from the a value in the usual way, by applying the technique for estimating vibronic constants in terms of the derivatives of resonance integrals with respect to interatomic distances (see Capters 3 and 4). For a and b, this approach yields

$$\begin{aligned} a &= \frac{2}{\sqrt{3}}c_e l_e\left[\frac{\partial\beta(x^2-y^2,\sigma)}{\partial R}\right]_0, \\ b &= -\frac{2}{\sqrt{6}}c_a l_e\left[\frac{\partial\beta(x^2-y^2,\sigma)}{\partial R}\right]_0 - c_e l_a\left[\frac{\partial\beta(s,\sigma)}{\partial R}\right]_0. \end{aligned} \tag{5.33}$$

Here, β are the resonance integrals corresponding to interactions between the central atom AOs specified in (5.33) and the σ AOs of ligands.

Consider equations (5.33) in more detail. Clearly, if the derivative of the resonance integral for the metal 4*s* AO with respect to R was not taken into account, (5.33) would give $b = -c_a a/\sqrt{2}\,c_e$, where the c_a/c_e ratio can be estimated based on the quantum-chemical calculations [244] for Cu(II) complexes. In the *sd* basis set, the ratio between the populations of the 4*s* and 3*d* AOs, that is, c_a^2/c_e^2, can be estimated at 0.03–0.06, which gives $b \approx -0.01$ au.

In reality (if the second term is taken into account), b should be smaller. In [210], b was set equal to -0.02, and the energy gap was assumed to be $\omega = 2$ eV, which approximately coincides with the difference of the one-electron 4*s* and 3*d* levels in the isolated copper atom. Substitution operator matrix elements s_{11}, s_{22}, and s_{1a} for the ψ_a^*, $\psi_{z^2}^*$, and $\psi_{x^2-y^2}^*$ MOs were treated as free model parameters (the s_a parameter is discussed below), and for various values of these parameters, adiabatic potential surfaces were calculated. The computations show that the three adiabatic potential minima can indeed all be situated in the left half-plane ($Q_3 < 0$) within some region of s_{11}, s_{22}, and s_{1a} variations near $s_{11} = 6.9$, $s_{22} = 3.1$, and $s_{1a} = 1.27$ eV. The form of adiabatic potential energy surface (5.31) at these parameter values, which are fairly acceptable from the chemical bonding point of view [210], is shown in Fig. 31 for the chain compounds under consideration.

To summarize, the conclusion can be drawn that not only symmetry but also actual interatomic distances in the distortion isomers of $Cu(NH_3)_2Br_2$ compounds can be explained in terms of only intramolecular interactions at realistic values of vibronic coupling constants, substitution operator matrix elements, and energy gap width between the e_g^* and a_{1g}^* orbitals.

This treatment also explains the relative stabilities of the α and β isomers. The calculations [210] show that, on the potential energy surface shown in Fig. 31, the minimum at $Q_2 = 0$ corresponding to the D_{4h} configuration of $Cu(NH_3)_2Hal_2$ chains is less deep. This implies that the β isomer (the D_{4h} structure) should be less stable, which agrees with experimental observations.

To avoid ambiguity, note that the one-center intramolecular mechanism [209, 210] explaining the existence of the α and β forms of $Cu(NH_3)_2Hal_2$ chain complexes neither presupposes nor rules out cooperative binding of chain units in a chain and chains in a crystal. This mechanism is in agreement with the observed stereochemistry of chains. Neither does it lead to any distortions of the crystal as a whole (which would be at variance with experiment), because, as mentioned, this mechanism only includes those deformations of the initial geome-

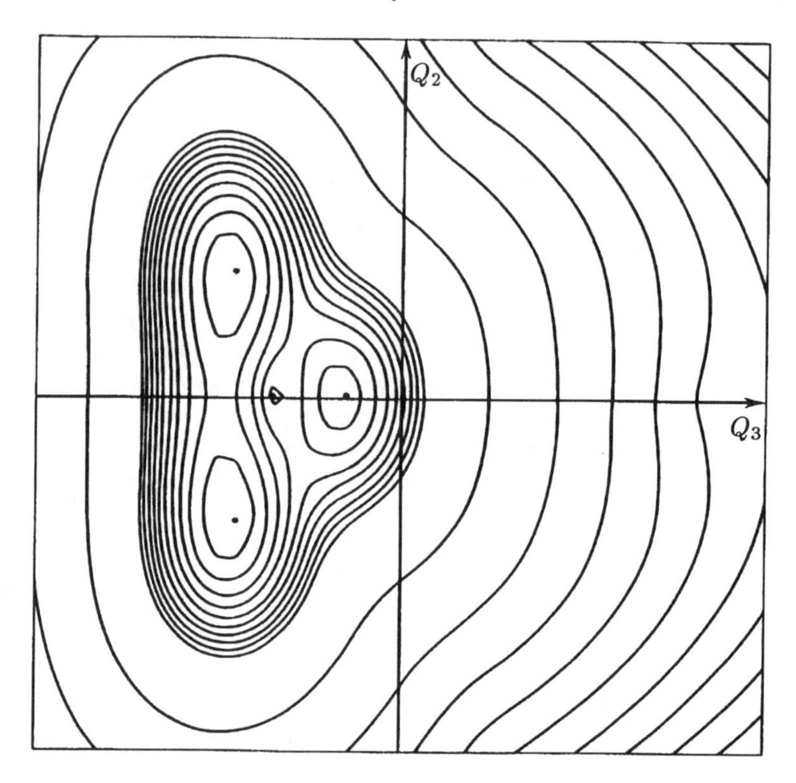

FIG. 31. CONTOUR PLOT OF LOWER ADIABATIC POTENTIAL SHEET FOR $Cu(NH_3)_2Hal_2$ CHAIN UNITS CALCULATED BY (5.30) AND (5.31).

try of structural chain units that lead to actually observed interatomic distances in both (α and β) isomers.

5.6. Distortion Isomers of *cis*-$Cu^{II}L_4X_2$ Complexes

1. It was shown in Sections 5.3 and 5.4 that $Cu^{II}L_4X_2$-type d^9 complexes largely form *trans* isomers, which are thermodynamically more stable than *cis*. Stable *cis*-$Cu^{II}L_4X_2$ complexes can, nevertheless, be prepared by specially selecting bi- or polydentate ligands. Such complexes are formed with L_2 = phenanthroline (phen), bipyridine (bipy), and bipyridylamine (bipyam), which are bonded by their nitrogen atoms, and $X_2 = NO_2$, NO_3, HCOO, and CH_3COO, which form Cu—O bonds,

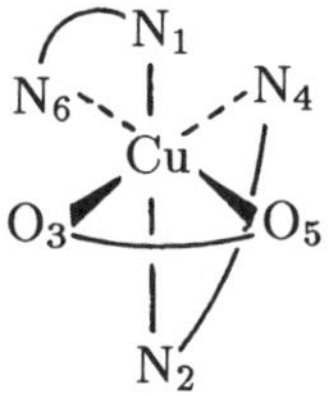

Quite a number of such compounds containing the *cis*-$Cu^{II}N_4O_2$ chromophore have been studied by X-ray diffraction (Table 18). The available experimental data show that in some of these complexes, transitions between distortion isomers occur. For instance, the temperature dependences of the EPR spectra of $[Cu(phen)_2O_2CMe]A$ with $A = BF_4 \cdot 2H_2O$, $ClO_4 \cdot 2H_2O$, ClO_4, and BF_4 evidence that the *cis*-$Cu^{II}N_4O_2$ chromophore does not have a rigid structure [245–247]. These results are interpreted based on the assumption that at low temperatures, we observe the more stable isomer, whereas at elevated temperatures, a part of complexes occurs in a less stable form, which explains why crystals contain mixtures of isomers. Abnormally large-amplitude atomic vibrations along metal–ligand bonds detected in X-ray structure studies of the compounds and correlation between these amplitudes and the experimental Cu—N and Cu—O bond length values [245] substantiate the suggestion that compounds containing the *cis*-$Cu^{II}N_4O_2$ chromophore form distortion isomers.

2. In this section, we will discuss the geometric structure and distortion isomerism of *cis*-$Cu^{II}L_4X_2$ complexes.

Consider the adiabatic potential energy surface of a d^9 ML_4X_2 complex given by (5.27) for the quadratic E-e-S problem. To describe the *cis* isomer, we must substitute $S = (4/\sqrt{3})S_E$ into this equation. In the LCAO approximation (5.24), $S_E = -c_e^2\Delta\alpha/2\sqrt{3}$, where $\Delta\alpha = \alpha_X - \alpha_L$. The ρ_0, φ_0 coordinates of adiabatic potential minima calculated for the *cis* form at various S_E parameter values and orbital vibronic coupling constant and force constant values of $a = 0.05$, $a' = 0.004$, and $K = 0.09$ au typical of $Cu^{II}L_6$ complexes are represented in Fig. 26b. According to this figure, at $S_E \gtrsim 0.025$ and $S_E \lesssim -0.01$ au, that is, when ligands L and X are fairly different, the adiabatic potential has a single minimum, and *cis*-ML_4X_2 can only exist in the form of C_{2v} symmetry. On the other hand, at substitution parameters fairly small in magnitude ($-0.01 \lesssim S_E \lesssim 0.025$), the $\mathcal{E}(\rho, \varphi)$ surface has three minima at $\varphi_0 = 0$, $180° \geq \varphi_0 \geq 90°$, and (because $\mathcal{E}$ is invariant with respect to the reversal of the sign of φ) $-90° \geq \varphi_0 \geq -180°$ (see Fig. 26b). These minima correspond to two nonequivalent *cis*-ML_4X_2 forms of C_{2v} ($\varphi_0 = 0$) and C_s symmetry.

Table 18. Cu—O and Cu—N bond lengths (Å) in compounds containing the *cis*-$Cu^{II}N_4O_2$ chromophore at room temperature [245–247]

Compound	Cu—O_3	Cu—O_5	Cu—N_4	Cu—N_6	Cu—N_1	Cu—N_2
$[Cu(bipy)_2NO_2]BF_4$	2.1169	2.4623	2.0523	2.1416	1.9900	2.0043
$[Cu(bipy)_2NO_2]NO_3$	2.2382	2.3289	2.0650	2.1001	1.9810	2.0053
$[Cu(phen)_2NO_2]BF_4$	2.0714	2.5974	2.0494	2.1673	1.9987	2.0195
$[Cu(bipyam)_2NO_2]BF_4$	2.1116	2.5513	2.0217	2.1393	1.9845	2.0131
$[Cu(bipyam)_2NO_2]NO_3$	2.0735	2.5504	2.0951	2.0951	2.0079	2.0079
$[Cu(bipy)_2NO_3]NO_3 \cdot H_2O$	2.2984	2.8178	2.0219	2.0449	1.9841	1.9826
$[Cu(bipy)_2(NO_3)]PF_6$	2.1525	2.7460	2.0345	2.1030	1.9813	1.9766
$[Cu(bipy)_2CH_3COO]BF_4$	1.9794	2.7842	2.0333	2.2088	1.9948	2.0167
$[Cu(bipy)_2CH_3COO]ClO_4 \cdot H_2O$	2.0317	2.6476	2.0564	2.1674	1.9712	1.9933
$[Cu(phen)_2CH_3COO]BF_4$	1.9953	2.6706	2.0625	2.2189	2.0095	2.0254
$[Cu(phen)_2CH_3COO]ClO_4$	2.2171	2.4204	2.0965	2.1315	1.9905	2.0055
$[Cu(phen)_2CH_3COO]ClO_4 \cdot 2H_2O$	2.2520	2.2520	2.1225	2.1225	1.9985	1.9985
$[Cu(phen)_2CH_3COO]BF_4 \cdot 2H_2O$	2.2616	2.2615	2.1233	2.1233	2.0004	2.0004
$[Cu(phen)_2CH_3COO]NO_2 \cdot 2H_2O$	2.1237	2.4478	2.0823	2.1712	2.0003	2.0192
$[Cu(bipyam)_2CH_3COO]NO_3$	2.0331	2.6726	2.0314	2.1607	2.0082	2.0099
$[Cu(bipy)_2HCOO]BF_4 \cdot 0.5H_2O$	2.0234	2.8691	2.0613	2.1578	1.9777	2.0022
$[Cu(phen)_2HCOO]BF_4$	2.363	2.363	2.111	2.111	1.990	1.990
$[Cu(bipyam)_2HCOO]BF_4$	2.0015	2.8755	2.0232	2.1663	1.9999	2.0155

This leads us to conclude that in the specified range of S_E parameter values, there can exist distortion isomers. For the covalent model of the corresponding ML_6 homoligand complex with $c_e^2 \approx 1/2$, the region of the existence of distortion isomers corresponds to quite realistic differences of the Coulomb integrals of ligands L and X, *viz.*, $-4 \lesssim \Delta\alpha \lesssim 1.5$ eV. Note also that the range of $\Delta\alpha$ values at which distortion isomers can exist is two times larger for *cis* complexes than for *trans*, and it may be expected that distortion isomers of the *cis* forms of d^9 ML_4X_2 complexes should be more common. Note also that for complexes with $\Delta\alpha < 0$, the region of the existence of distortion isomers is broader than for complexes with $\Delta\alpha > 0$.

3. With reference to complexes containing the *cis*-$Cu^{II}N_4O_2$ chromophore, the replacement of nitrogen-containing ligands by those containing oxygen should be treated as the introduction of a more electronegative ligand with $\Delta\alpha = \alpha_O - \alpha_N < 0$: $S_E = -c_e^2\Delta\alpha/2\sqrt{3} > 0$. This means that adiabatic potential energy surface minima determining the geometry of complexes with the specified chromophore are situated in Fig. 26b at positive S_E parameter values.

According to Fig. 26b, at fairly large ($\geq$ 0.03 au) S_E values, the adiabatic potential energy surface has a minimum at $\varphi_0 = \pi$, and the equilibrium configuration of *cis*-$Cu^{II}N_4O_2$ corresponds to octahedron deformation along the Q_3 mode taken with a minus sign, which means contraction along the N_1-Cu-N_2 axis and expansion in the equatorial plane normal to this axis. It follows that the axial $Cu-N$ bonds should be shorter than the equatorial ones, and two $Cu-O$ and also two $Cu-N$ bonds in the equatorial plane should have equal lengths. According to Table 18, such a geometry is characteristic (at room temperature) of the nearest environment of Cu in $[Cu(phen)_2CH_3COO]ClO_4 \cdot 2H_2O$, $[Cu(phen)_2CH_3COO]BF_4 \cdot 2H_2O$, and $[Cu(phen)_2HCOO]BF_4$.

4. It also follows from Fig. 26b that quadratic vibronic coupling is suppressed by the substitution effect in the $S_E \geq 0.03$ au region and is not at $0 < S_E \leq 0.3$ au. At such S_E values at the point of the global adiabatic potential energy surface minimum (Fig. 26b), $|\varphi_0|$ varies in the range 120–180°; therefore, $Q_2 \neq 0$ and $Q_3 < 0$. The equilibrium configuration of *cis*-$Cu^{II}N_4O_2$ is then characterized by the presence of two pairs of nonequivalent bonds $Cu-N_4$, $Cu-N_6$ and $Cu-O_3$, $Cu-O_5$ in the equatorial plane as a result of Q_2 deformation. The negative Q_3 mode value means that the axial $Cu-N_1$ and $Cu-N_2$ bonds are shorter than the equatorial ones. The lengths of the axial $Cu-N$ bonds become equal to the lengths of the short equatorial $Cu-N$ bonds in the limit of $\varphi_0 = 120°$.

Such a structure is characteristic of most of the complexes included in Table 18. According to this table, the Cu—N_1 and Cu—N_2 bond lengths vary from 1.97 to 2.01 Å, whereas the Cu—N_4 and Cu—N_6 bonds positioned *trans* with respect to oxygen are 0.01–0.21 Å longer. The equatorial plane contains both long Cu—N and Cu—O bonds positioned *trans* to each other and shorter Cu—N and Cu—O bonds, also *trans* to each other.

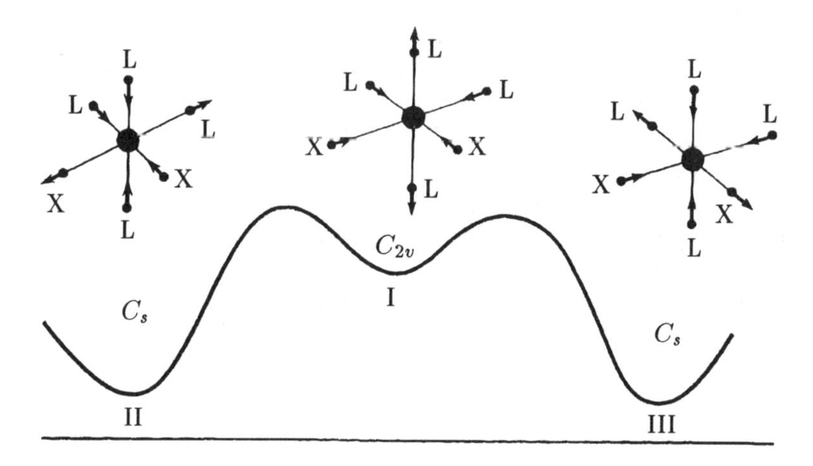

FIG. 32. DEFORMATIONS AND RELATIVE STABILITIES OF DISTORTION ISOMERS FOR *cis*-$Cu^{II}L_4X_2$-TYPE COMPLEXES.

In the range $0 < S_E \leq 0.025$ au, there exists not only the relatively deep adiabatic potential energy surface minimum considered above but also a shallower minimum at $\varphi_0 = 0$ (Fig. 32). The corresponding *cis*-$Cu^{II}N_4O_2$ isomer formed as a result of deformation along the $Q_3 > 0$ mode has C_{2v} symmetry. Accordingly, the complex has long Cu—L_1 and Cu—L_2 axial bonds and short Cu—L_4 and Cu—L_6 equatorial bonds of equal lengths.

The data on the temperature dependence of the Cu—N and Cu—O interatomic distances in $[Cu(bipy)_2NO_2]NO_3$ (Table 19), which were measured from 20 K to room temperature, give a fairly convincing substantiation of the conclusions drawn on the structure of the distortional isomers of complexes with *cis*-$Cu^{II}N_4O_2$-type chromophores. (It appears that in the solid phase, the compound is a mixture of two different *cis* forms, and the observed interatomic distances are average values [247].) According to calculations, the low-symmetry C_s isomer possesses the higher stability, and temperature lowering is therefore accompanied by an increase in the differences of equatorial bond lengths: at 20 K, the difference of Cu—O bond lengths amounts to 0.48 Å,

Table 19. Temperature dependence of bond lengths (Å) in the $[Cu(bipy)_2O_2N]^+$ complex [247]

T, K	Cu—O_3	Cu—O_5	Cu—N_4	Cu—N_6	Cu—N_1	Cu—N_2
298	2.299	2.320	2.073	2.084	1.979	1.988
165	2.204	2.351	2.070	2.098	1.984	1.989
100	2.156	2.414	2.060	2.110	1.987	1.992
20	2.052	2.536		2.142	1.982	1.997

Note: Numbering of ligands is given in Fig. 4.

whereas at 298 K, it decreases to 0.02 Å. At the same temperatures, the difference of Cu—N bond lengths equals 0.11 and 0.02 Å, respectively.

5. It should, however, be noted that one point in Table 19 cannot be explained in terms of simple treatment of the E-e-S problem based on equation (5.27). According to the experimental data, axial Cu—N bond lengths remain almost unchanged in the temperature range of measurements. This would be easy to explain on the assumption that the less stable C_{2v} form of *cis*-CuL_4X_2 is, like the C_s one, contracted along the L_1—Cu—L_2 direction. The transition between these two forms then would not be accompanied by so appreciable changes in axial Cu—L bond lengths as the transition from the axially elongated configuration of the C_{2v} isomer to the axially contacted configuration of the C_s one. Such an interpretation, however, requires using a more general approach to the E-e-S problem and, as in the preceding section, can be substantiated through taking into account mixing of the partially occupied e_g^* MOs with the unoccupied a_{1g}^* one.

6. The distortional isomerism of *cis*-$Cu^{II}L_4X_2$ and *cis*-$Cu^{II}L_2X_4$ complexes can be given a different but equivalent treatment based on the pseudo-Jahn–Teller effect [211g].

In considering the linear E-e-S problem (Section 5.2), we noted that, at $S_{12} = 0$, secular equation (5.11) can be treated as a secular equation describing the pseudo-Jahn–Teller effect in a heteroligand but yet undeformed complex. We can adopt a similar approach to secular equation (5.26) for the quadratic E-e-S problem, for instance, for *cis*-$Cu^{II}L_4X_2$ or *cis*-$Cu^{II}L_2X_4$ complexes. Indeed, Table 17 shows that the S_{12} matrix element then vanishes. It follows that (5.26) describes vibronic mixing of the ^2A and ^2B terms, into which the ^2E term of the initial $Cu^{II}L_6$ octahedral complex splits in the *cis* complexes under consideration. The mixing is effected by the Q_a and Q_b modes, which originate from Q_3 and Q_2, respectively, when symmetry lowers from O_h to C_s.

This reduces the problem of ditortional isomerism of *cis* complexes to examining the $(A + B) - (a + b)$ problem. Such an approach was taken in [211g], where the adiabatic potential was augmented by the introduction of a third-order anharmonic term and a linear term for the purpose of a more detailed description of distortion isomers.

CHAPTER 6

MUTUAL INFLUENCE OF LIGANDS AND ISOMERISM IN QUASI-OCTAHEDRAL COMPLEXES WITH MULTIPLY BONDED SUBSTITUENTS

The stereochemistry of quasi-octahedral transition metal d^0, d^1, and d^2 complexes with one or several multiply bonded ligands has been extensively studied by the X-ray diffraction and spectral (IR and NMR) methods. The structures of dozens of such complexes have been determined to arrive at two general conclusions concerning their geometry [248–251], *viz.*, (i) independent of the electronic configuration of the central atom (d^0, d^1, or d^2), the introduction of a multiply bonded ligand (O$\equiv$ or N$\equiv$) into a σ-bonded complex results in a substantial elongation of the bond positioned *trans* to this ligand and (ii) while the steric effect of the introduction of one multiply bonded substituent is independent of the electronic configuration of the central atom, the mutual arrangement of several such substituents depends on the number of d electrons.

In d^0 complexes, multiply bonded ligands occupy positions *cis* with respect to each other. According to reviews [248–251a], such an arrangement of two or three multiply bonded ligands is characteristic of V(V) and Nb(V), Mo(VI) and W(VI), and Tc(VII) and Re(VII) complexes. The complete list of d^0 complexes with two or three multiply bonded substituents X that have been studied (some 30 compounds) is given in [250]. It is stressed in [250, 251a] that the rule of the *cis* arrangement of multiply bonded ligands in d^0 complexes has no exclusions. Conversely, multiple bonds in d^2 complexes are positioned *trans* with respect to each other. This dependence of the mutual arrangement of multiply bonded ligands on the electronic configuration of the central atom seems unexpected from the point of view of empirical correlations between *trans*-influence and stabilization of *cis* isomers, because the introduction of one multiply bonded ligand has more or less the same stereochemical consequences (strong *trans*-influence [248–251a]) in d^0, d^1, and d^2 complexes (although the effect appears to be somewhat weaker in d^1 and d^2 complexes than for d^0). The *trans* arrangement of multiple bonds was observed in Mo(IV) and W(IV), Tc(V) and Re(V), and Ru(VI) and Os(VI) complexes.

In this chapter, we discuss quasi-octahedral transition metal complexes with multiply bonded substituents and factors determining their structure. The theory described in the preceding chapters will be generalized to a certain degree by combining the results obtained for closed-shell (d^0) and open-shell complexes. Together with transition metal complexes, quasi-octahedral nontransition element compounds with multiply bonded substituents will shortly be described. At the end of the chapter, we will discuss an important although insufficiently studied problem of mutual influence of ligands and the relative stability of *cis* and *trans* isomers in quasi-octahedral complexes of actinides.

6.1. Influence of Multiply Bonded Ligand: Bond Lengths

1. First, consider the mutual influences of ligands in quasi-octahedral transition metal d^0 complexes ML_5X with one multiply bonded ligand X and five σ-bonded ligands L [252]. We will use the general approach to substitution effects in octahedral closed-shell complexes described in Chapters 3 and 4. There will, however, be one difference. When a σ-bonded ligand L is replaced by multiply bonded X, the substitution operator H_S cannot be described in terms of only the $\Delta\alpha = \alpha(X) - \alpha(L)$ value determining the change in the Coulomb integral for ligand σ AOs. We must also take into account the formation of an M$\overset{\leftarrow}{=}$X or M$\equiv$X multiple bond in place of the σ bond through the introduction of resonance integrals $\beta_\pi = \beta(\pi, d_\pi)$ describing interaction between the p_π AO of ligand X and the d_π AO of metal M. These integrals have previously been ignored, although in reality, they have nonzero, although small, values for σ-bonded ligands too. For generality, let all ligands be assigned certain $\beta(\pi, d_\pi)$ values; it will be assumed that for ligand X, this integral differs from that for ligand L by $\Delta\beta_\pi = \beta_\pi(X) - \beta_\pi(L)$. For the same purpose, consider similar $\Delta\beta_{\sigma,i} = \beta_{\sigma,i}(X) - \beta_{\sigma,i}(L)$ ($i = s,\ p,\ d$) values describing the difference in resonance integrals between M$-$X and M$-$L σ bonds. In addition to resonance integral changes involved in the $ML_6 \longrightarrow ML_5X$ substitution, we will take into account the difference of the Coulomb integrals $\Delta\alpha_\sigma = \alpha_\sigma(X) - \alpha_\sigma(L)$ and $\Delta\alpha_\pi = \alpha_\pi(X) - \alpha_\pi(L)$.

2. As previously, homoligand ML_6 complexes will be treated as initial systems, which implies the use of the corresponding schemes of MOs constructed from the valence *s*, *p*, and *d* AOs of the central atom and the σ and p_π AOs of the ligands. Typical schemes of homoligand system MOs for transition metal (MoF_6) and nontransition element (SF_6) compounds are shown in Fig. 33. They are based on the exper-

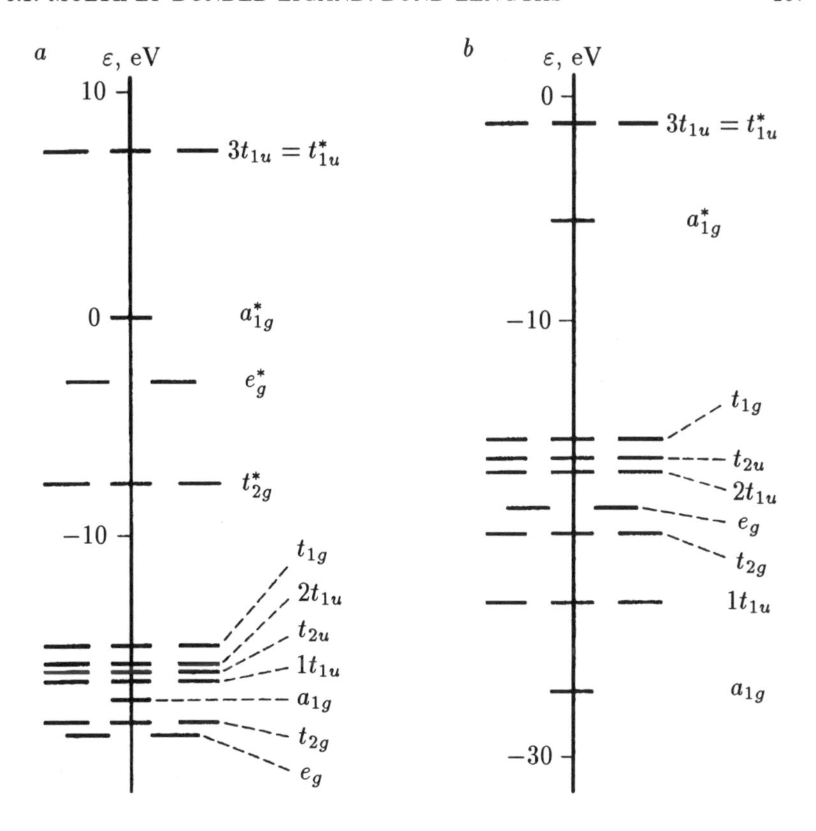

FIG. 33. ENERGY LEVEL DIAGRAM FOR MoF_6 (a) AND SF_6 (b) ACCORDING TO THE EXPERIMENTAL DATA [163, 253] AND THE DVM-X_α CALCULATION RESULTS [254, 255].

imental data [163, 253] on valence levels augmented by the results of the DVM-X_α calculations [254, 255] of unoccupied levels. The figure shows that in both molecular systems, the inclusion of π AOs causes the appearance of a band of closely spaced MOs. (Some of these levels, namely, those nonbonding, would have equal energies but for interligand interactions.) For this reason, the approximation of frontier MOs, that we for simplicity use as in Section 3.3, should be modified. The lowest unoccupied MOs (LUMO, t^*_{2g} in Fig. 33a and a^*_{1g} in Fig. 33b) are determined unambiguously, whereas the t_{1g} MOs can only formally be considered the highest occupied MOs (HOMO). It would be more natural to assume that the HOMO is represented by all closely spaced MOs or even all occupied MOs for transition metal complexes and all MOs between t_{1g} to t_{2g} for nontransition element compounds. In addition to the MOs specified above, we will need the e_g^* ones. This completes the list of the MOs that will be used in the further discussion.

In the basis set of the $(n-1)d$, ns, and np transition metal AOs or the ns, np, and nd nontransition element AOs and the σ and p_π ligand AOs, they take the form

$$
\begin{aligned}
\psi_{a_{1g}} &= c_a s + (l_a/\sqrt{6})(\sigma_1+\sigma_2+\sigma_3+\sigma_4+\sigma_5+\sigma_6),\\
\psi^*_{a_{1g}} &= l_a s - (c_a/\sqrt{6})(\sigma_1+\sigma_2+\sigma_3+\sigma_4+\sigma_5+\sigma_6),\\
\psi_{e_g,z^2} &= c_e d_{z^2} + (l_e/2\sqrt{3})(2\sigma_1+2\sigma_2-\sigma_3-\sigma_4-\sigma_5-\sigma_6),\\
\psi^*_{e_g,z^2} &= l_e d_{z^2} - (c_e/2\sqrt{3})(2\sigma_1+2\sigma_2-\sigma_3-\sigma_4-\sigma_5-\sigma_6),\\
\psi_{t_{1g},xz} &= (1/2)(-\pi_{1x}+\pi_{2x}+\pi_{3z}-\pi_{4z}),\\
\psi_{t_{2g},xz} &= c_\tau d_{xz} + (l_\tau/2)(\pi_{1x}-\pi_{2x}+\pi_{3z}-\pi_{4z}),\\
\psi^*_{t_{2g},xz} &= l_\tau d_{xz} - (c_\tau/2)(\pi_{1x}-\pi_{2x}+\pi_{3z}-\pi_{4z}),\\
\psi^{(T)}_{1t_{1u},x} &= (l_{t2}/\sqrt{2})(\sigma_3-\sigma_4) + (l_{t1}/2)(\pi_{1x}+\pi_{2x}+\pi_{5x}+\pi_{6x}),\\
\psi^{(T,N)}_{2t_{1u},x} &= (l_{t1}/\sqrt{2})(\sigma_3-\sigma_4) - (l_{t2}/2)(\pi_{1x}+\pi_{2x}+\pi_{5x}+\pi_{6x}),\\
\overline{\psi}^{(N)}_{2t_{1u},x} &= \overline{c}_t p_x + \sqrt{(1-\overline{c}_t^2)/2}\,\psi^{(N)}_{2t_{1u},x},\\
\psi_{2t_{2u},x} &= (1/2)(\pi_{1x}+\pi_{2x}-\pi_{5x}-\pi_{6x}).
\end{aligned}
\tag{6.1}
$$

Here, degenerate states are represented by one MO each, and the π AOs are written in the common coordinate system (Fig. 4). The c_i, $l_i = \sqrt{1-c_i^2}$, l_{t1}, and l_{t2} (but not $\overline{c}_t$) coefficients are positive, and indices T and N labeling the t_{1u} MOs are used to distinguish between transition metal (T) and nontransition element (N) complexes. (The $\overline{\psi}^{(N)}_{2t_{1u},x}$ MO can *a priori* be written with both $\overline{c}_t > 0$ and $\overline{c}_t < 0$, but the character of mutual ligand effects is independent of the sign of $\overline{c}_t$, because this coefficient appears in the corresponding equations as $\overline{c}_t^2$.) It is taken into account in (6.1) that in transition metal complexes, the unoccupied np central atom AOs are situated well above the valence AOs of typical ligands. They therefore make insignificant contributions to the occupied $1t_{1u}$ and $2t_{1u}$ MOs; this also follows from the DVM-X_α calculation results [254], according to which in transition metal complexes, the np AOs have low populations. These results also show that the np AOs of the central atom are virtually absent in the $2t_{1u}$ MOs for SF_6 [255] and SeF_6, although to be on the safe side, we wrote (6.1) with an expression for the $2t_{1u}$ MO involving nontransition element np AOs [256].

Equation (6.1) can conveniently be used to find the matrix elements of H_S and orbital vibronic coupling constants for transition metal complexes. These elements and constants take the form

$$
\begin{aligned}
S_{t_{1g}xz,t^{*}_{2g}xz} &= \frac{c_\tau}{4}\Delta\alpha_\pi - \frac{l_\tau}{2}\Delta\beta(\pi, d_\pi),\\
S_{t_{2g}xz,t^{*}_{2g}xz} &= -\frac{c_\tau l_\tau}{4}\Delta\alpha_\pi + (l_\tau^2 - c_\tau^2)\Delta\beta(\pi, d_\pi),\\
S_{1t_{1u}x,t^{*}_{2g}xz} &= -\frac{c_\tau l_{t1}}{4}\Delta\alpha_\pi + \frac{l_{t1} l_\tau}{2}\Delta\beta(\pi, d_\pi),\\
S_{t_{2u}x,t^{*}_{2g}xz} &= -\frac{c_\tau}{4}\Delta\alpha_\pi + \frac{l_\tau}{2}\Delta\beta(\pi, d_\pi),\\
S_{2t_{1u}x,t^{*}_{2g}xz} &= (c_\tau l_{t2}/4)\Delta\alpha_\pi - (l_\tau l_{t2}/2)\Delta\beta(\pi, d_\pi),\\
A^{3}_{t_{1g}xz,t^{*}_{2g}xz} &= -\frac{\sqrt{3}\,l_\tau}{2}\Big(\frac{\partial(\beta(\pi, d_\pi))}{\partial R}\Big)_0,\\
A^{1}_{t_{2g}xz,t^{*}_{2g}xz} &= \frac{\sqrt{6}}{3}(l_\tau^2 - c_t^2)\Big(\frac{\partial(\beta(\pi, d_\pi))}{\partial R}\Big)_0,\\
A^{3}_{t_{2g}xz,t^{*}_{2g}xz} &= \frac{\sqrt{3}}{6}(l_\tau^2 - c_t^2)\Big(\frac{\partial(\beta(\pi, d_\pi))}{\partial R}\Big)_0,\\
A^{12}_{1t_{1u}x,t^{*}_{2g}xz} &= \frac{\sqrt{2}}{2}l_\tau l_{t1}\Big(\frac{\partial(\beta(\pi, d_\pi))}{\partial R}\Big)_0,\\
A^{12}_{2t_{1u}x,t^{*}_{2g}xz} &= -\frac{\sqrt{2}}{2}l_\tau l_{t2}\Big(\frac{\partial(\beta(\pi, d_\pi))}{\partial R}\Big)_0,\\
A^{12}_{t_{2u}x,t^{*}_{2g}xz} &= \frac{\sqrt{2}}{2}l_\tau\Big(\frac{\partial(\beta(\pi, d_\pi))}{\partial R}\Big)_0.
\end{aligned}
\tag{6.2}
$$

Similarly, for nontransition element complexes, we obtain

$$
\begin{aligned}
S_{e_g z^2,a^{*}_{1g}} &= -(\sqrt{2}\,c_a l_e/6)\Delta\alpha_\sigma,\\
S_{2t_{1u}z,a^{*}_{1g}} &= -(\sqrt{3}\,c_a l_{t1}/6)\Delta\alpha_\sigma,\\
S_{\overline{2t_{1u}z},a^{*}_{1g}} &= -(\sqrt{3}\,c_a l_{t1}\sqrt{1-\overline{c}_t^2}\,/6)\Delta\alpha_\sigma,\\
A^{12}_{2t_{1u}z,a^{*}_{1g}} &= l_a l_{t1}\Big(\frac{\partial(\beta(\sigma, s))}{\partial R}\Big)_0,\\
A^{3}_{e_g z,a^{*}_{1g}} &= l_a l_e\Big(\frac{\partial(\beta(\sigma, s))}{\partial R}\Big)_0 - \frac{\sqrt{2}}{2}c_a c_e\Big(\frac{\partial(\beta(\sigma, d_\sigma))}{\partial R}\Big)_0,\\
A^{12}_{\overline{2t_{1u}z},a^{*}_{1g}} &= l_{t1} l_a\sqrt{1-\overline{c}_t^2}\Big(\frac{\partial(\beta(\sigma, s))}{\partial R}\Big)_0 - \frac{\sqrt{3}}{3}c_a\overline{c}_t\Big(\frac{\partial(\beta(\sigma, p_\sigma))}{\partial R}\Big)_0.
\end{aligned}
\tag{6.3}
$$

The further analysis of mutual influence of ligands can be performed with the use of Table 20, which, like Table 5, can easily be constructed by applying rules (2.25) (for definiteness, the substitution is assumed to occur at position 1, see Fig. 4). According to Table 20, mutual influence of ligands depends on the contributions of the Q_1, Q_3, and Q_{12} modes (Fig. 4), which can be represented through atomic Cartesian displacements as indicated in Table 2.

Table 20. Modes contributing geometry distortions in quasi-octahedral ML_5X complexes with σ and π bonds*

LUMO	HOMO	Modes mixing LUMO and HOMO	MOs mixed by operator H_S **		Active modes
a_{1g}^*	a_{1g}	A_{1g}; Q_1	a_{1g}^*	a_{1g}	Q_1
	e_g	E_g; Q_2, Q_3		$e_{g,\,z^2}$	Q_3
	t_{1u}	T_{1u}; Q_{10}, Q_{11}, Q_{12}		$t_{1u},\ z$	Q_{12}
e_g^*	a_{1g}	E_g; Q_2, Q_3	$e_{g,\,z^2}^*$	a_{1g}	Q_3
	e_g	A_{1g}; Q_1		$e_{g,\,z^2}$	Q_1
		E_g; Q_2, Q_3			Q_3
	t_{1u}	T_{1u}; Q_{10}, Q_{11}, Q_{12}		$t_{1u,z}$	Q_{12}
t_{2g}^*	t_{1g}	E_g; Q_2, Q_3	$t_{2g,xz}^*$; $t_{2g,yz}^*$	$t_{1g,xz}$; $t_{1g,yz}$	Q_3
	t_{2g}	A_{1g}; Q_1		$t_{2g,xz}$; $t_{2g,yz}$	Q_1
		E_g; Q_2, Q_3			Q_3
	t_{1u}	T_{1u}; Q_{10}, Q_{11}, Q_{12}		$t_{1u,x}$; $t_{1u,y}$	Q_{12}
	t_{2u}	T_{1u}; Q_{10}, Q_{11}, Q_{12}		$t_{2u,x}$; $t_{2u,y}$	Q_{12}

* Modes affecting bond lengths only are included.
** The z, z^2, xz, and yz indices of the t_{1u}, e_g, and t_{2g} MOs correspond to symmetry of central atom AOs. For the t_{2u} MOs, the x and y indices indicate the axes along which the p_π AOs of ligands are oriented, and for the t_{1g} MOs, the xz and yz indices specify the planes in which they lie.

3. Let us analyze mutual influence of ligands in transition metal d^0 complexes and find forces F_{cis} and F_{trans} that appear as a result of the $ML_6 \longrightarrow ML_5X$ substitution and act on the ligands positioned *cis* and *trans* with respect to X. The $\Delta F = F_{trans} - F_{cis}$ difference will also be considered (see Section 3.6). To do this, it should be taken into account that in the complexes under consideration, energy gaps between the MOs fully (t_{1g} and t_{2u}) or almost fully ($1t_{1u}$ and $2t_{1u}$)

built of ligand AOs are small, of ~0.5 eV on average, whereas the t_{2g} MOs are situated comparatively far from them (Fig. 33). Note also that the a_{1g} and e_g σ MOs can be excluded from calculations of the forces, because they do not interact with the t_{2g}^* LUMO. Approximating insignificantly different energy gaps $t_{1g}-t_{2g}^*$, ..., $1t_{1u}-t_{2g}^*$ by their mean $\omega_{\alpha T}^*$, we obtain

$$F_{cis}(\Delta\alpha) = \frac{c_T l_T}{2\omega_{\alpha T^*}}\left[1 - \frac{\omega_{\alpha T^*}(l_T^2 - c_T^2)}{\omega_{TT^*}}\right]\left(\frac{\partial\beta(\pi, d_\pi)}{\partial R}\right)_0 \Delta\alpha_\pi, \tag{6.4}$$

$$F_{cis}(\Delta\beta) = -\frac{l_T{}^2}{\omega_{\alpha T^*}}\left[1 - \frac{\omega_{\alpha T^*}(l_T^2 - c_T^2)}{\omega_{TT^*} l_T{}^2}\right]\left(\frac{\partial\beta(\pi, d_\pi)}{\partial R}\right)_0 \Delta\beta(\pi, d_\pi), \tag{6.5}$$

$$F_{trans}(\Delta\alpha) = 2F_{cis}(\Delta\alpha), \qquad \Delta F(\Delta\alpha) = F_{cis}(\Delta\alpha), \tag{6.6}$$

$$F_{trans}(\Delta\beta) = 2F_{cis}(\Delta\beta), \qquad \Delta F(\Delta\beta) = F_{cis}(\Delta\beta), \tag{6.7}$$

$$F_{cis,\,trans} = F_{cis,\,trans}(\Delta\alpha) + F_{cis,\,trans}(\Delta\beta), \tag{6.8}$$
$$\Delta F = \Delta F(\Delta\alpha) + \Delta F(\Delta\beta)$$

(the t_{2g}^* LUMO does not contain σ AOs, and the expressions for forces therefore do not include $\Delta\alpha_\sigma$ and $\Delta\beta_{\sigma,\,i}$).

For convenience, the total expressions for forces were divided into two contributions, one depending on $\Delta\alpha_\pi$ and the other, on $\Delta\beta(\pi, d_\pi)$. According to Fig. 33, $0 < (\omega_{\alpha T^*}/\omega_{TT^*}) < 1$, where $\omega_{\alpha T^*} = \varepsilon(t_{2g}^*) - \alpha_\pi(\mathrm{L})$ and $\omega_{TT^*} = \varepsilon(t_{2g}^*) - \varepsilon(t_{2g})$. At the same time, valence ligand orbitals are lower in energy than metal $(n-1)d$ orbitals in typical transition metal complexes. The coefficients of AOs in MOs should therefore satisfy the inequalities $0 < l_T^2 - c_T^2 < l_T^2 < 1$. For this reason, the values in square brackets in (6.4) and (6.5) are positive, and the derivative of the resonance integral between the p_π orbital of ligand L and the d_π orbital of atom M with respect to R is also so. To complete the analysis of (6.4) and (6.5), it remains to find the signs of $\Delta\alpha_\pi$ and $\Delta\beta(\pi, d_\pi)$.

As distinguished from σ-bonded ligands L, multiply bonded ligands X with covalent or donor-acceptor π bonds show a greater propensity to form delocalized systems of π electrons. In terms of the LCAO method, this presupposes fairly small $|\alpha_\pi|$ and (or) large $|\beta_\pi|$ parameter values. The class of multiply bonded ligands can therefore be characterized by the inequalities $\alpha_\pi^{\mathrm{X}} > \alpha_\pi^{\mathrm{L}}$ and $\beta_\pi^{\mathrm{X}} < \beta_\pi^{\mathrm{L}}$; this implies that $\Delta\alpha_\pi > 0$ and $\Delta\beta_\pi < 0$. In addition, the α_π and β_π parameters of multiply bonded ligands differ from those of σ-bonded ones much more strongly than the parameters of σ-bonded ligands differ from each other. As concerns σ orbital parameters (α_σ and $\beta_{\sigma,\,i}$), multiply bonded ligands may be close to σ-bonded ones (although, e. g., for $\mathrm{MoHal_5O^-}$, it is reasonable to assume that $\Delta\alpha_\sigma > 0$). To elucidate the character of mutual influence of ligands, we must substitute $\Delta\alpha_\pi > 0$ and $\Delta\beta(\pi, d_\pi) < 0$ into (6.4)–(6.8), which gives $F_{cis} > 0$ and $F_{trans} > F_{cis} > 0$. This means that the replacement of σ-bonded ligand L in ML_6 by a multiply bonded ligand X⩶ or X≡ should cause relative and absolute elongation of the *trans* bond, and the *trans*-influence of a multiply bonded substituent (X⩶ or X≡) should be stronger than the *trans*-influence of the σ-bonded (X−).

The conclusion that there is a strong *trans*-influence by multiply bonded ligands in d^0 complexes can, probably, be extended to d^1 and d^2 complexes. In these complexes, the t_{2g}^* MOs are only partially filled, and the mechanism of *trans*-influence considered above is still energetically favored. This mechanism involves admixing of the t_{2g}^* MOs to the lower fully occupied MOs as a result of the $ML_6 \longrightarrow ML_5X$ substitution of a multiply bonded ligand for a σ-bonded entity (although such a treatment is, of course, not impeccable, and a rigorous theory of d^1 and d^2 complexes requires consideration of mutual influence of ligands in open-shell systems).

4. Experimental data show that considerable elongation of the σ bond positioned *trans* with respect to the multiply bonded ligand is indeed characteristic of quasi-octahedral transition metal d^0, d^1, and d^2 complexes. Good examples are $MXHal_5$ complexes (X = O or N) such as the $NbOF_5^{2-}$ anion in K_2NbOF_5 crystals,

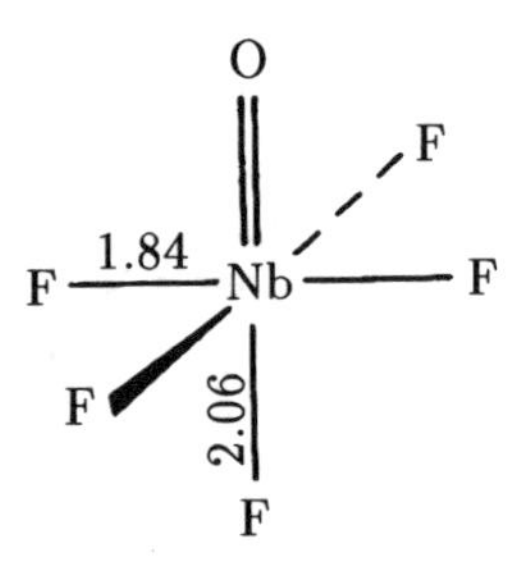

where the $Nb-F_{cis}$ bond lengths equal 1.84 Å, whereas the length of the $Nb-F_{trans}$ bond is 2.06 Å. Similar structures were observed for Mo and Re oxochlorides and the $NbO(NCS)_5^{2-}$ and $OsNCl_5^{2-}$ complexes. The difference of the $M-L_{trans}$ and $M-L_{cis}$ bond lengths in these compounds amounts to 0.16–0.46 Å (Table 21) and far exceeds similar differences in σ-bonded heteroligand complexes. *Trans*-influence of multiply bonded ligands is observed in complexes not only with monodentate ligands but also bidentate ones. Characteristic examples are [Tc^VO(2-methyl-8-quinolinol)$_2$Cl], [Tc^VO(salicylaldehyde)Nsalicylidene-D-glucoseamide], and [$Mo^{VI}O_2$ (2-hydroxyphenylbenzimidazole)$_2$] [251a],

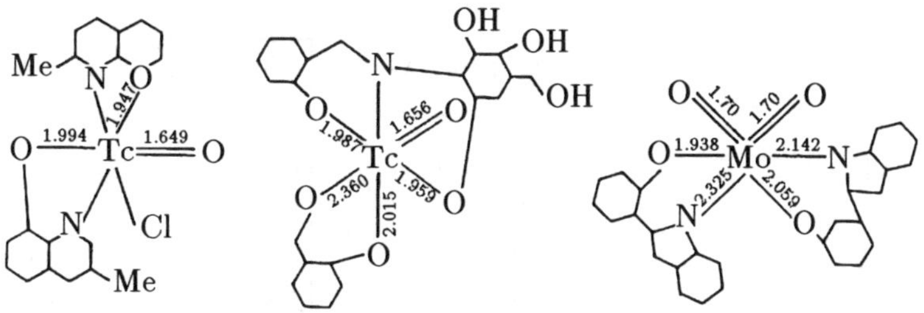

Other examples are provided by recently studied similar complexes of Re [257], Tc [258], and Mo [259]:

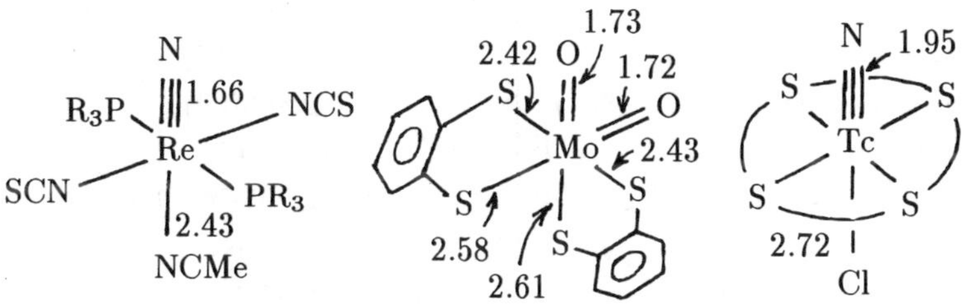

Lastly, given in Table 21 are the data on $MXCl_{5-m}(PR_3)_m$ complexes with multiply bonded ligand X and σ-bonded ligands of two kinds. In these complexes, positioned *trans* to the multiply bonded ligand is chlorine, and one or two more chlorine atoms occupy *cis* positions. The table shows that apart from complexes with an elongated *trans* M—Cl bond, this class of compounds includes complexes with *trans* M—Cl bonds shorter than the *cis* ones. This *trans*-shortening in $MXL_{5-m}L'_m$ complexes is discussed in detail in [257]. We will only note that in complexes with standard metal–oxygen bonds 1.66–1.70 Å long, elongation of *trans* bonds is observed, whereas *trans* bond shortening occurs when the multiple bond is abnormally long ($\gtrsim$ 1.80 Å) and, possibly, $\Delta\beta_\pi$ in (6.4)–(6.8) should be assigned a positive value.

5. For completeness, it remains to relate the "$\sigma + \pi$" version of the theory of mutual influence of ligands in transition metal complexes to the σ version considered in Chapter 3. Recall that in that chapter, σ MOs were only taken into account and the $ML_6 \longrightarrow ML_5X$ replacement of σ-bonded ligand L by ligand X, also σ-bonded but more electropositive or electronegative than L ($\Delta\alpha_\sigma > 0$ or $\Delta\alpha_\sigma < 0$), was considered. In this chapter, the results obtained previously will be re-

Table 21. Metal–ligand bond lengths (Å) in transition metal complexes with multiply bonded O and N ligands [250, 251a]

Complex	C*	$M-L_{trans}$	$M-L_{cis}$	$M-X$
$NbOF_5^{2-}$	d^0	2.06	1.84	1.68
$NbO(NCS)_5^{2-}$	d^0	2.27	2.09	1.70
$MoOCl_5^{2-}$	d^1	2.63	2.40	1.67
$MoOBr_5^{2-}$	d^1	2.83	2.56	1.86
$OsNCl_5^{2-}$	d^2	2.60	2.36	1.61–1.75
$ReOCl_5^{2-}$	d^2	2.47	2.39	1.66
trans-$ReOCl_3(PEt_2Ph)_2$	d^2	2.445(Cl)	2.40(Cl)	1.66
cis-$ReOCl_3(PEt_2Ph)_2$	d^2	2.294(Cl)	2.389(Cl)	1.869
cis-$MoOCl_2(PPhMe_2)_3$	d^2	2.551(Cl)	2.464(Cl)	1.676
cis-$MoOCl_2(PPhEt_2)_3$	d^2	2.424(Cl)	2.482(Cl)	1.801
cis-$MoOCl_2(PMePh_2)_3$	d^2	2.509(Cl)	2.466(Cl)	1.667
cis-$ReNCl_2(PEt_2Ph)_3$	d^2	2.56(Cl)	2.45(Cl)	1.79
cis-$MoO(NCO)_2(PPhEt_2)_3$	d^2	2.180(NCO)	2.095(NCO)	1.684
$WOCl_2$(N,N′,N″-trimethyl-	d^1	2.37(N)	2.24(N)	1.72
1,4,7,-triazacyclononane)		2.32(N)	2.25(N)	1.89

* Electronic configuration

vised taking into account not only σ but also π MOs. Naturally, the difference between L and X should then be described by both $\Delta\alpha_\sigma$ and $\Delta\alpha_\pi$ Coulomb integral differences ($\alpha_\sigma(X) = \Delta\alpha_\sigma(L) + \Delta\alpha_\sigma$ and $\alpha_\pi(X) = \Delta\alpha_\pi(L) + \Delta\alpha_\pi$, where the signs of $\Delta\alpha_\pi$ and $\Delta\alpha_\sigma$ coincide). At the same time, it is reasonable to set $\Delta\beta_{\sigma,\,i}$ and $\Delta\beta_\pi$ equal to zero, because σ-bonded ligands X and L having different orbital electronegativities may have only slightly different resonance integral values.

It should be expected that the σ version of the theory of mutual influence of ligands will be the limiting case of the $\sigma + \pi$ version corresponding to very weak π-bonding. To demonstrate this, we must go somewhat beyond the approximation of frontier MOs in the $\sigma + \pi$ version, because the t_{2g}^* LUMO do not contain σ AOs. It suffices to include the e_g^* σ MOs, which play the role of the LUMO in the σ version. As previously, the HOMO is assumed to include all occupied MOs from e_g to t_{1u}. Among the latter, a_{1g}, e_g, $1t_{1u}$, and $2t_{1u}$, which contain either σ or σ and π ligand AOs, interact with the e_g^* MOs. Using this approach, we obtain the total forces acting on *cis* and *trans* ligands and their difference in the form $F_{cis}+F'_{cis}$, $F_{trans}+F'_{trans}$, and $\Delta F+\Delta F'$, where as previously, F_{cis}, F_{trans}, and ΔF are given by (6.4)–(6.8), whereas the equations for F'_{cis}, F'_{trans}, and $\Delta F'$ can easily be obtained from (3.35) and (3.36),

$$F'_{cis} \approx \frac{c_e l_e}{3\sqrt{3}\,\omega_1}\left[(2c_e^2 - c_a^2)\left(\frac{\partial\beta(\sigma, d_\sigma)}{\partial R}\right)_0 - \frac{\sqrt{2}\,c_a c_e l_a}{l_e}\left(\frac{\partial\beta(\sigma, s)}{\partial R}\right)_0\right]\Delta\alpha_\sigma, \tag{6.9}$$

$$\Delta F' \approx \frac{c_e l_e}{\sqrt{3}}\left[\left(\frac{2}{\omega_{ae^*}} + \frac{c_e^2 - l_e^2}{\omega_{ee^*}} - \frac{l_a^2}{\omega_{ae^*}}\right)\left(\frac{\partial\beta(\sigma, d_\sigma)}{\partial R}\right)_0 + \frac{\sqrt{2}\,c_a c_e l_a}{\omega_{ae^*} l_e}\left(\frac{\partial\beta(\sigma, s)}{\partial R}\right)_0\right]\Delta\alpha_\sigma. \tag{6.10}$$

Here, $\omega_{ae^*} = \varepsilon(e_g^*) - \varepsilon(a_{1g})$, $\omega_{ee^*} = \varepsilon(e_g^*) - \varepsilon(e_g)$, and ω_1 denotes the average of the closely spaced ω_{ae^*} and ω_{ee^*} transition energies.

It was shown in the preceding section that at $\Delta\alpha_\pi > 0$, $F_{trans} > F_{cis} > 0$. The same conclusion follows from (6.9) and (6.10), that is, at $\Delta\alpha_\sigma > 0$, we have $F'_{trans} > F'_{cis} > 0$. Indeed, the populations of the *ns* orbitals in the complexes that we are considering are small in comparison with those of the d_σ AOs, that is, $c_a^2 \ll c_e^2$ and $l_e < l_a < 1$, and the right-hand sides of (6.9) and (6.10) are therefore positive. Equations (6.4)–(6.7), (6.9), and (6.10) considered together therefore show that the substitution of X for L in transition metal complexes ML_6 is accompanied by σ and π *trans*-effects of the same sign. It is now easy to correlate the $\sigma+\pi$ and σ versions of the theory of mutual influence of ligands. Weakening of π-bonding corresponds to the condition $\beta_\pi \approx 0$ and, therefore, $(\partial\beta_\pi/\partial R)_0 \approx 0$. According to (6.4)–(6.7), the F_{cis} and F_{trans} forces and ΔF then tend to zero. The F'_{cis} and F'_{trans} forces and $\Delta F'$, however, remain nonzero and positive, which corresponds to absolute and relative elongation of the *trans* bond by the purely σ mechanism in complete agreement with the conclusion obtained in the σ version of the theory. Shortening of the *trans* bond at $\Delta\alpha_\sigma < 0$ and $\Delta\alpha_\pi < 0$ is explained in the same way.

6. As concerns 12-electron quasi-octahedral nontransition element complexes, recall that their LUMO is the σ-type a_{1g}^* MO, which, of the group of MOs constituting the HOMO, only interacts with the $2t_{1u}$ and e_{g,z^2} ones. For the other HOMOs, all S_{n,a_{1g}^*} matrix elements equal zero. Because of the σ character of the LUMO, the $\Delta\alpha_\pi$ and $\Delta\beta_\pi$ values are absent from the equations for F_{cis} and ΔF. The effect of a multiply bonded substituent X therefore does not differ from that of

a σ-bonded one (in the approximation of frontier MOs). Nevertheless, the equations for F_{cis} and ΔF in the $\sigma+\pi$ and σ versions are different, because in the former, ligand σ AOs contribute not only to the $1t_{1u}$ MOs (as in the σ version) but also to the $2t_{1u}$ ones. Using (6.1) for MOs, we obtain

$$F_{cis}(\Delta\alpha) = \frac{c_a l_e}{3\sqrt{6}\,\omega_{ea^*}}\left[l_a l_e\left(\frac{\partial\beta(\sigma,s)}{\partial R}\right)_0 - \frac{c_a c_e}{\sqrt{2}}\left(\frac{\partial\beta(\sigma,d_\sigma)}{\partial R}\right)_0\right]\Delta\alpha_\sigma, \tag{6.11}$$

$$\Delta F(\Delta\alpha) = -\frac{2c_a l_e}{\sqrt{6}\,\omega_{ea^*}}\left[l_a l_e\left(1-\frac{l_{t1}^2\omega_{ea^*}}{l_e^2\omega_{aa^*}}\right)\left(\frac{\partial\beta(\sigma,s)}{\partial R}\right)_0 - \frac{c_a c_e}{\sqrt{2}}\left(\frac{\partial\beta(\sigma,d_\sigma)}{\partial R}\right)_0\right]\Delta\alpha_\sigma \tag{6.12}$$

(here, the $2t_{1u}$ MO is represented by the $\psi_{2t}^{(N)}$ function). We see that the inclusion of π orbitals results in the appearance in (6.12) of the $(1-l_{t1}^2\omega_{ea^*}/l_e^2\omega_{aa^*})<1$ additional multiplier, which decreases the first term in square brackets. The steric consequences of this decrease are as follows.

As mentioned in Section 3.5, complexes of nontransition elements in the higher oxidation states fall into two classes depending on the relative arrangement of the valence *ns* level of the central atom and ligand valence levels. In complexes of the first class, the *ns* AO of the central atom has energy higher than or approximately equal to that of ligand σ AOs, as is characteristic of typical complexes of elements situated in the left halves of nontransition element series. The second term in square brackets in (6.11) and (6.12) is then small in comparison with the first. Indeed, $c_a \lesssim l_a$ for the specified arrangement of levels and $c_e < l_e$ because of insignificance of contributions to bonding of unoccupied nontransition element *nd* AOs. Some decrease in the first term in (6.12) caused by taking into account π bonds is of little consequence. For example, at $\Delta\alpha_\sigma > 0$, we have $F_{cis} > 0$ and $\Delta F < 0$. It follows that the inclusion of π AOs does not change the conclusion that *cis* bonds experience elongation in compounds of the type of Ga(III) and Sn(IV) complexes when a more electropositive substituent is introduced. Conversely, if the *ns* AOs of the central atom are lower in energy than ligand σ AOs, that is, if $l_a < c_a$, the $(1-c^2\omega_{ea^*}/l_e^2\omega_{aa^*})$ multiplier changes the situation in favor of the second term in (6.12). The presence of π AOs is therefore another factor, additional to a decrease in l_a, which promotes transition from *cis* elongation to *trans* elongation observed in complexes of elements situated in the right halves of nontransition element series and discussed in Section 3.5.

Two points should be mentioned concerning the experimental data described in Section 3.5. First, it was noted above that multiply bonded ligand effects in nontransition element complexes with low-lying *ns* AOs do not differ significantly from σ-bonded ligand effects (in the approximation of frontier MOs). This conclusion is substantiated by a comparison of the lengths of M—Hal bonds in SF_5Cl-type and IOF_5 molecules. In both compounds, *trans* bond length changes are very small in comparison with *cis* bonds and only amount to several hundredths of Å. At the same time, in transition metal complexes, the difference between ligands of these two types is fairly significant (however, see the IR data [261]).

Secondly, there does not exist exact parallelism between variations of F_{cis} and ΔF, see (6.11) and (6.12). At $l_{t1} = 0$, that is, in the σ approximation, elongation of *trans* bonds should be accompanied by shortening of *cis* ones. However at $l_{t1} \neq 0$, there is a region within which the second term in square brackets is larger than the first term in (6.12) but smaller than the first term in (6.11). This is possible if $\Delta F(\Delta\alpha)$ is small. Examples are provided by $SbPhHal_5^-$ complexes in some Sb^V compounds [91], in which a small relative (and absolute) elongation of the *trans* bond is accompanied by an also small elongation of *cis* bonds. This effect is obtained in the σ version too, when all transitions between occupied and unoccupied MOs are taken into account. Including π orbitals, however, allows the effect to be explained in the approximation of frontier MOs.

Lastly, consider the effect of substituting $\overline{\psi}_{2t_{1u}}^{(N)}$ rather than $\psi_{2t_{1u}}^{(N)}$ (see (6.1)) for the $2t_{1u}$ MOs on the equations for F_{cis} and ΔF in 12-electronic nontransition element complexes with low-lying *ns* AOs. Clearly, this substitution does not affect the equation for F_{cis}, whereas ΔF takes the form

$$\Delta F(\Delta\alpha) = -\frac{2c_a l_e}{\sqrt{6}\,\omega_{ea^\bullet}}\Delta\alpha_\sigma \times \left[l_a l_e\left(1 - \frac{l_{t1}^2(1-\overline{c}_t^2)\omega_{ea^\bullet}}{l_e^2\omega_{aa^\bullet}}\right)\left(\frac{\partial\beta(\sigma,s)}{\partial R}\right)_0 - \frac{c_a c_e}{\sqrt{2}}\left(\frac{\partial\beta(\sigma,d_\sigma)}{\partial R}\right)_0 + \frac{\sqrt{2}\,l_{t1}c_a^2\overline{c}_t\sqrt{1-\overline{c}_t^2}}{3\omega_{aa^\bullet}}\left(\frac{\partial\beta(\sigma,p_\sigma)}{\partial R}\right)_0\right]. \tag{6.13}$$

A comparison of these results with (6.11) and (6.12) shows that an admixture of the *np* central atom AOs in the $2t_{1u}$ MOs does not change the conclusion of similar effects of multiply bonded and σ-bonded substituents X. This is also true of *cis*-influence in 12-electronic nontransition element ML_5X complexes with relatively high *ns*(M) AOs. In complexes with low-energy *ns*(M) AOs, the $(1 - \overline{c}_t^2)$ multiplier in square brackets together with the last term (for $\overline{c}_t > 0$) might prevent *cis*-influence from transforming into *trans*-influence. However for atoms in the middle of the right half of nontransition element series, this factor can hardly be of significance, because, as follows from the calculations on SF_6 [255] and similar calculations on SeF_6, the $\overline{c}_t$ coefficient is small.

The question of the relation between the $\sigma + \pi$ and σ versions of the theory of mutual influence of ligands for quasi-octahedral complexes of nontransition elements in the highest oxidation states does not require detailed consideration and can easily be answered by applying (6.11) and (6.12) or (6.11) and (6.13). If ligand π AOs do not participate in bonding, then the $1t_{1u}$ MOs only contain ligand σ AOs, and the $2t_{1u}$ MOs, ligand π AOs. The l_{t1} coefficient in (6.11), (6.12), and (6.13) and the $\overline{c}_t$ coefficient in (6.13) then vanish, and (6.11) and (6.12) or (6.11) and (6.13) are automatically transformed into the corresponding equations of the σ version.

6.2. Multiply Bonded Ligand: Bond Angles

1. In Section 3.6, we discussed angle deformations in quasi-octahedral σ-bonded complexes. Here, we turn to angle deformations in complexes with multiply bonded substituents. First, consider d^0 ML_5X complexes of the $MoHal_5O^-$ type with multiply bonded ligand X in position 1 (Fig. 4). Angle deformations then only depend on the Q_9 mode of T_{1u} symmetry (Fig. 7), which mixes the t^*_{2g} LUMO with the $2t_{1u}$ and $1t_{1u}$ MOs in homoligand ML_6 complexes. Consideration of the $1t_{1u}$ MOs does not change qualitative conclusions, and it suffices to take into account the $2t_{1u}$ HOMO. We will, however, include mixing of the $2t_{1u}$ and the unoccupied e^*_g MOs, which will not change conclusions concerning $MoHal_5O^-$-type complexes but will be of importance for analyzing nitroso compounds.

It is clear from symmetry considerations that the MOs that experience mixing due to the Q_9 mode can be divided into three pairs,

$$(2t_{1u,x},\, t^*_{2g,xz}), \quad (2t_{1u,y},\, t^*_{2g,yz}), \quad (2t_{1u,z},\, e^*_{g,z^2}),$$

such that mixing of MOs occurs within but not between the pairs. Applying (2.56) then yields

$$Q_9^f = \frac{4}{K_9}\left\{\frac{2S_{x,xz^*}A^9_{x,xz^*}}{\omega_{x,xz^*}} + \frac{S_{z,z^{2*}}A^9_{z,z^{2*}}}{\omega_{z,z^{2*}}}\right\}. \tag{6.15}$$

For shortness, we write x and xz^* in place of $t_{1u\,x}$ and $t^*_{2g,\,xz}$ etc. We also take into account that $S_{x,\,xz^*} = S_{y,\,yz^*}$ and $A^9_{y,\,yz^*} = A^9_{x,\,xz^*}$. Here, $S_{x,\,xz^*}$ is given by (6.2), and

$$S_{z,z^{2*}} = -(c_e l_{t1}/\sqrt{6})\Delta\alpha_\sigma + (l_e l_{t1}/\sqrt{2})\Delta\beta(\sigma, d_\sigma). \tag{6.16}$$

It follows from what has been said in Section 6.1 that for $MoHal_5O^-$-type complexes, $\Delta\alpha_\pi > 0$, $\Delta\alpha_\sigma > 0$, $\Delta\beta_\pi < 0$, and $\Delta\beta_\sigma \approx 0$. Therefore, $S_{x,\,xz^*} > 0$ and $S_{z,\,z^{2*}} < 0$. Applying (6.1) yields

$$\begin{aligned} A^9_{x,xz^*} &= \sqrt{2}\, l_{t1} l_\tau \langle d_{xz} | (\partial H/\partial Q_9)_0 | \sigma_3 \rangle, \\ A^9_{z,z^{2*}} &= -2 l_e l_{t2} \langle d_{z^2} | (\partial H/\partial Q_9)_0 | \pi_{3z} \rangle. \end{aligned} \tag{6.17}$$

(In the equation for $A^9_{x,\,xz^*}$, we ignore the $\langle d_{xz} | (\partial H/\partial Q_9)_0 | \pi_{5x} \rangle \sim \beta(\pi, d_\pi)$ integral, which is small in magnitude in comparison with the $\langle d_{xz} | (\partial H/\partial Q_9)_0 | \sigma_3 \rangle \sim \beta(\sigma, d_\sigma)$ one for σ-bonded ML_6 complexes.) The signs of vibronic coupling constants are easiest to estimate with the use of pictorial representations I and II

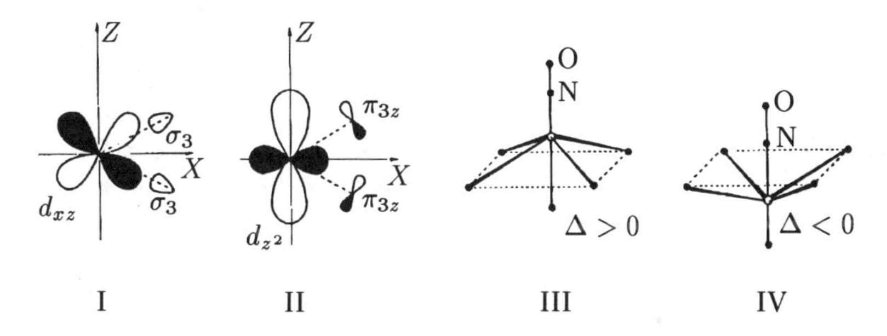

When Q_9 changes from negative to positive values, the $\langle d_{xz} | H | \sigma_3 \rangle$ integral changes sign from plus to minus, and its derivative with respect to Q_9 is negative. The sign of the $\langle d_{z^2} | H | \pi_{3z} \rangle$ integral is estimated similarly. We arrive at the conclusion that for ML_5X complexes of the $MoHal_5O^-$ type, $Q_9^f < 0$ by virtue of (6.15). This means that equatorial ligands are displaced from multiply bonded X toward the *trans* ligand (see III above); such deformations are indeed observed, e. g., in molybdenum(VI) oxofluorides [249]. It follows from rule (3.12) of superposition of deformations that in *cis*-ML_4X_2 complexes with multiply bonded ligands X in equatorial positions, axial σ-bonded ligands are displaced from the multiply bonded ligands toward two other equatorial ligands; this conclusion also corresponds with the structural data [249].

2. The behavior of angles in $ML_{6-k}X_k$ complexes with multiply bonded ligands X described above is considered a rule. However comparatively recently, exceptions from this rule were discovered. According to the X-ray structural data [262], in *trans*-$Os(NO)(NH_3)_4Hal^{2+}$ (Hal = Cl, Br, and I) complexes with linear M—NO groups (such complexes were observed in $[Os(NO)(NH_3)_4Hal]Cl_2$ crystals), equatorial NH_3 groups shift toward the Os$\equiv$NO multiple bond and experience

displacements Δ as large as 0.2 Å (structure IV above), although in similar $M(NO)Hal_5^{2-}$ and $[M(NO)(NH_3)_4F]^{2+}$ (M = Ru and Os) complexes, displacements Δ of equatorial ligands have usual values (structure III above, Table 22).

A comparison of the van der Waals radii of the ligands show that the anomalous behavior of bond angles in Os complexes (structure IV) cannot be explained by steric repulsion of NH_3 groups from the axial halogen atoms [268]; this leads us to apply the approach developed in the beginning of this Section [268–270].

The NO group is usually treated as a σ-donor and π^*-acceptor accepting electrons to its antibonding π^* MO, which corresponds with the traditional description of metal–nitroso group bonds as $M \leftrightharpoons NO$. Bonding π electrons of the NO group are usually not considered. This leaves 38 valence electrons in $M(NO)Hal_5^{2-}$- and $M(NO)(NH_3)_4Hal^{2+}$-type complexes with M = Ru and Os (d^6s^2 metal electrons, σ and π electrons of all ligands except NO, and σ and π^* electrons of the nitroso group). Accordingly, the heteroligand complexes under consideration should be treated as derivatives of 38-electronic ML_6 complexes of the $OsHal_6$ or $Os(NH_3)_6^{6+}$ type. Such ML_6 complexes have $t_{2g}^{*\,2}$ open electronic shells (Fig. 33), and formally, (2.56) cannot be applied to them. This equation can nevertheless be used taking into account the behavior of the MOs in transition from an ML_6 complex of O_h symmetry to $M(NO)L_5$ and *trans*-$M(NO)YL_4$ complexes of C_{4v} symmetry (the NO group occupies position 1, and Y, position 2, Fig. 4).

Symmetry lowering splits the t_{2g}^* MOs into the e^* ($t_{2g,\,xz}^*$ and $t_{2g,\,yz}^*$) and b_2^* ($t_{2g,\,xy}^*$) components. The energy of the b_2^* MO is lower than that of e^* and virtually independent of Q_9. Applying second-order perturbation theory immediately shows that the splitting of the t_{2g}^* MOs largely results from their interaction with the $2t_{1u}$ MOs of a lower energy under perturbation of C_{4v} symmetry, which accompanies the transformation of ML_6 into a C_{4v} heteroligand complex. If the splitting is fairly substantial, an analysis of the ensuing Q_9 deformation can therefore be performed on the assumption that the initial octahedral ML_6 complex has the $(t_{2g,\,xz}^*)^0(t_{2g,\,yz}^*)^0(t_{2g,\,xy}^*)^2$ configuration and can formally be treated as a closed-shell system.

Let us, as previously, use frontier MOs (6.14), from which (6.15) follows. In conformity with the special features of the NO ligand, we will use form (2.39) for the matrix elements of the substitution operator and, for generality, assume that in all nitroso complexes under consideration, substituents are situated in positions 1 and 2. (The substitution operator for $M(NO)Hal_5^{2-}$ may formally involve the ligand at position 2; the corresponding $\Delta\alpha_{\sigma 2}$ and $\Delta\alpha_{\pi 2}$ values are then set equal

Table 22. Displacements Δ (Å) and MNO and $L_{ax}ML_{eq}$ (deg) angles in osmium and ruthenium nitroso complexes

Compound	M = Os				M = Ru			
	MNO	Δ	$L_{ax}ML_{eq}$ (av)	Refs.	MNO	Δ	$L_{ax}ML_{eq}$ (av)	Refs.
$K_2[MNOF_5] \cdot H_2O$	180	0.212	83.9(5)	[263]	180	0.159	85.3(3)	[263]
$Rb_2[MNOF_5] \cdot H_2O$	180	0.178	84.9(3)	[263]	180	0.147	85.7(3)	[263]
$Cs_2[MNOF_5] \cdot H_2O$	180	0.184	84.7(2)	[263]				
$K_2[MNOCl_5]$	176.6	0.112	87.3(1)	[264]	176.8	0.083	88.0(4)	[267]
$K_2[MNOBr_5]$	178.7	0.089	88.0(1)	[264]	174.4	0.063	88.6(1)	[267]
$K_2[MNOI_5]$	177.9	0.070	88.5(1)	[264]	175.0	0.047	89.0(1)	[267]
$[MNO(NH_3)_4F][SiF_6]$	175.7	0.188	84.9(3)	[265]	175.5	0.143	86.1(2)	[265]
$[MNO(NH_3)_4NO_3]Cl_2 \cdot 0.5H_2O$	178.4	0.142	86.1(3)	[261]				
$[MNO(NH_3)_4OH]Br_2$	176.0	0.141	86.2(5)	[266]				
$[MNO(NH_3)_4Cl]Cl_2$	180	−0.022	90.6(4)	[262]				
$[MNO(NH_3)_4Br]Cl_2$	180	−0.188	95.1(4)	[262]				
$[MNO(NH_3)_4Br]Br_2$	180	0.020	89.5(4)	[262]				
$[MNO(NH_3)_4I]Cl_2$	180	−0.164	94.3(5)	[262]				

to zero.) The NO group will be treated as a ligand with two valence (σ-donor and π^*-acceptor) levels, whereas the π MOs of NO will not be taken into account. This yields

$$
\begin{aligned}
S_{x,xz^*} &= (c_\tau l_{t2}/4)(\Delta\alpha_{\pi 1} - \Delta\alpha_{\pi 2}), \\
S_{z,z^{2*}} &= -(c_e l_{t1}/\sqrt{6}\,)(\Delta\alpha_{\sigma 1} - \Delta\alpha_{\sigma 2}).
\end{aligned}
\tag{6.18}
$$

The vibronic coupling constants are as previously given by (6.17).

3. Next, let us estimate model parameters for Os and Ru complexes with bonds of different natures. First recall that in the corresponding ML_6 complexes, the $2t_{1u}$ MO largely includes ligand π orbitals, and the l_{t1} coefficient in (6.1) should therefore be smaller than l_{t2}. This refers to both σ-bonded complexes and complexes with σ and π bonds. The difference between complexes of these two types manifests itself by the difference in c_τ coefficient value for the t_{2g}^* MOs, which is small for σ-bonded complexes and commensurate with l_τ for complexes with σ and π bonds.

For the $\Delta\alpha_\sigma$ and $\Delta\alpha_\pi$ parameters, we can write

$$
\Delta\alpha_{\pi 1} - \Delta\alpha_{\pi 2} \approx \Delta\alpha_{\pi 1} \gg \Delta\alpha_{\sigma 1} - \Delta\alpha_{\sigma 2}. \tag{6.19}
$$

Indeed, it can be expected for the initial homoligand ML_6 (e. g., $MHal_6$) complex that the population of each $p_\pi(L)$ AO should be more or less close to 1.5–2.0 (the same is true of heteroligand complexes). However according to the X-ray [271] and NQR [272, 273] spectra, the population of each π^* orbital of the NO group is more likely to be 0.5–1.0. Similar estimates follow from the CNDO/2 calculations [274]. To describe one-electron charge redistribution at position 1, the $\Delta\alpha_{\pi 1}$ parameter should be assigned a large positive value. At the same time, the populations of halogen and NO σ AOs are close to each other and equal 1.5–1.7 [272, 273].

4. Prior to explaining the stereochemistry of nitroso complexes, consider initial unsubstituted ML_6 complexes of two types, one with σ and π bonds of appreciable strengths, and the other with only σ bonds. It follows from (6.2) and (6.17)–(6.19) and from what has been said above concerning the coefficients of AOs in the $2t_{1u}$ and t_{2g}^* MOs that if π bonds are fairly strong ($c_\tau \approx l_\tau$), the first term predominates in (6.15), whence

$$
\begin{aligned}
Q_9^f &= \frac{8S_{x,xz^*}A_{x,xz^*}^9}{K_9\omega_{x,xz^*}} \\
&\approx \frac{2\sqrt{2}\,c_\tau l_\tau l_{t1} l_{t2}}{K_9\omega_{x,xz^*}} \langle d_{xz}|(\partial H/\partial Q_9)_0|\sigma_3\rangle \Delta\alpha_{\pi 1},
\end{aligned}
\tag{6.20}
$$

where $\langle d_{xz}|(\partial H/\partial Q_9)_0|\sigma_3\rangle < 0$ and $\Delta\alpha_{\pi 1} > 0$. Therefore, for the complexes that we are considering, Q_9^f is negative and approximately proportional to $\Delta\alpha_{\pi 1} = \Delta\alpha_{\pi^*}(\mathrm{NO}) - \Delta\alpha_\pi(\mathrm{L})$. This leads us to conclude that in $\mathrm{M(NO)Hal_5^{2-}}$ complexes with fairly strong M — Hal π bonds, the Q_9^f values should be negative and increase in magnitude in the series I < Br < Cl, which is in agreement with experiment (Table 22).

For σ-bonded $\mathrm{ML_6}$ complexes, the c_τ coefficient in (6.1) is close to zero. According to (6.18), the first term in (6.15) is then small, and the second one predominates. Then

$$\begin{aligned} Q_9^f &= \frac{4S_{z,z^{2*}}A^9_{z,z^{2*}}}{K_9\omega_{z,z^{2*}}} \\ &= \frac{4c_e l_e l_{t1} l_{t2}}{\sqrt{6}\,K_9\omega_{z,z^{2*}}}\langle d_{z^2}|(\partial H/\partial Q_9)_0|\pi_{z3}\rangle(\Delta\alpha_{\sigma 1} - \Delta\alpha_{\sigma 2}), \end{aligned} \tag{6.21}$$

where $\langle d_{z^2}|(\partial H/\partial Q_9)_0|\pi_{z3}\rangle < 0$.

According to the X-ray photoelectron spectra [163], the mutual arrangement of the σ levels of the NO group and halogens is as follows:

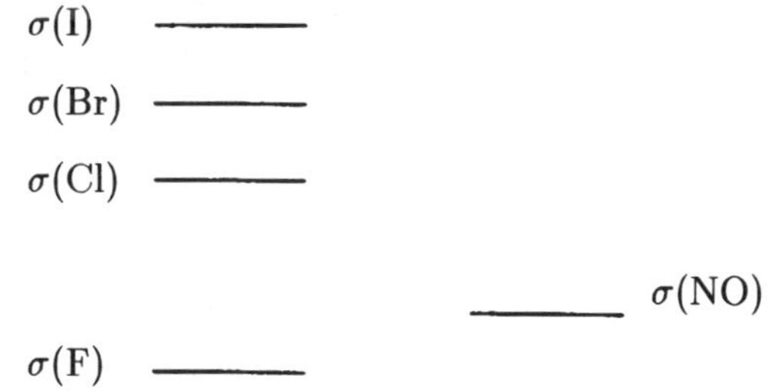

σ(I) ———

σ(Br) ———

σ(Cl) ———

——— σ(NO)

σ(F) ———

This scheme predicts that in $\mathrm{M(NO)(NH_3)_4Hal^{2+}}$ complexes with Cl, Br, and I for Hal, Q_9^f should be positive and increase from Cl to Br to I, which explains the anomalous behavior of bond angles observed experimentally (Table 22). According to experimental data, the sign of Q_9^f for *trans*-$\mathrm{Os(NO)(NH_3)_4F^{2+}}$ is opposite to that observed for the other *trans*-$\mathrm{Os(NO)(NH_3)_4Hal^{2+}}$ complexes, which can be explained if it is taken into account that the energy of the σ AO level of the NO group is lower than that of Cl, Br, and I but higher than the energy of the σ AOs of F.

The $\mathrm{M(NO)F_5^{2-}}$ complexes with M = Ru and Os should be considered separately, because their stereochemistry is independent of whether the corresponding $\mathrm{ML_6}$ complexes are considered systems with σ and π bonds or σ bonds only. Indeed, let π bonds be appreciably strong in the corresponding $\mathrm{ML_6}$ homoligand complex; $\mathrm{M(NO)F_5^{2-}}$ complexes are then described by (6.20), which gives $Q_9^f < 0$. The same conclusion is obtained if the ML_6 complexes are considered σ-bonded. On the assumption that $\alpha(\sigma\text{-F}) < \alpha(\sigma\text{-NO})$, we again find that $Q_9^f < 0$, as for $M(NO)(NH_3)_4F^{2+}$ complexes.

The conclusions drawn above follow from an analysis of complexes of two limiting types. In reality, π M—L bonds are not very strong in the complexes with halogens considered above, and the NH_3 group is also capable of forming weak π bonds with metals. For this reason, the difference in character between bonds in the $[Os(NO)Hal_5]^{2-}$ and $[Os(NO)(NH_3)_4Hal]^{2+}$ complexes is in reality not very substantial, and it cannot be ruled out that in certain situations (e. g., in the presence of certain outer-sphere anions etc.), $[Os(NO)(NH_3)_4Hal]^{2+}$ complexes will have a "normal" geometry. Such a geometry is observed for $[Os(NO)(NH_3)_4Br]^{2+}$ in $[Os(NO)(NH_3)_4Br]Br_2$ crystals, although the positive Δ value characterizing this structure, 0.02 Å, is small. An alternative explanation of the dual stereochemical behavior of the osmium complexes based on the concept of distortion isomers is also possible [275].

5. We already discussed distortion isomers of Cu(II) complexes in Sections 5.5 and 5.6. Such isomers are also called bond stretch isomers [276], which more closely corresponds with what was supposed [277] to be characteristic of *cis-mer*-$MoOCl_2(PR_3)_3$ complexes with $PR_3 = PMe_2Ph$. It was suggested [278] that these complexes existed as two isomers with noticeably different Mo=O bond lengths of 1.676 and ~1.80 Å,

O
1.676 || PR₃
R₃P —— Mo —— Cl
R₃P | 2.551
Cl

O
1.80 || PR₃
R₃P —— Mo —— Cl
R₃P |
Cl

(more recently [279], the existence of these isomers was questioned).

Such isomers were considered in [280] in terms of the theory of the pseudo-Jahn–Teller effect. Similar treatment can be applied to $[Os(NO)(NH_3)_4Hal]^{2+}$-type complexes on the assumption that the corresponding homoligand ML_6 complex of O_h symmetry is characterized by a strong pseudo-Jahn–Teller effect involving the Q_7, Q_8, and Q_9 modes (Fig. 7). The $\mathcal{E}(Q_9)$ potential energy curve of such a complex has two equivalent minima (Section 2.2), but the $ML_6 \longrightarrow$ *trans*-ML_4XY

substitution taking place on the z axis makes the minima nonequivalent or even removes one of them depending on the ligands X and Y:

As a result, an ML_4XY heteroligand complex can exist as either two distortion isomers with XML angles wider or narrower than 90° or one isomer. The $[Os(NO)(NH_3)_4Hal]^{2+}$ complexes in $[Os(NO)(NH_3)_4Hal]Cl_2$ (Hal = Cl, Br, and I) crystals can then be classified as isomers A, whereas the $[Os(NO)(NH_3)_4Br]^{2+}$ complex in $[Os(NO)(NH_3)_4Br]Br_2$ having a different geometry, as isomer B [275].

6.3. Stability of *cis* and *trans* Isomers for Transition Metal Complexes with Multiply Bonded Substituents

1. In this section, our concern will be the relative stability of geometric isomers of transition metal complexes with multiply bonded substituents (nontransition element complexes will not be considered because of insufficiency of relevant experimental data). Most simply, the problem is solved for d^0 complexes. As has been shown, *trans*-influence of a multiply bonded ligand in these complexes is explained by mixing of the even t_{2g}^* LUMO with π MOs having lower energies, among which the predominant contribution is made by the odd $1t_{1u}$, $2t_{1u}$, and t_{2u} MOs, which determines the character of mutual ligand effects. Following the same line of reasoning as in Section 3.8, we arrive at the conclusion that predominant stabilization should be expected for the isomers of ML_4X_2 and ML_3X_3 that are characterized by substitution operators with larger odd components. According to Table 3, such isomers will be *cis*-ML_4X_2 and *cis*-ML_3X_3, because for their *trans* counterparts, the odd component of the substitution operators is either comparatively small (for *trans*-ML_3X_3) or zero (for *trans*-ML_4X_2).

The available experimental data substantiate this conclusion. The structures of several dozens of transition metal d^0 complexes with two multiply bonded ligands (O atoms) were studied by the IR spectroscopy, NMR, and X-ray structure analysis methods [251a]; none of these structures was at variance with theoretical predictions, according to which *cis* isomers should predominantly be stabilized.

2. Applying the theory to d^1 and d^2 complexes, consider the system of t_{2g}^* electrons in the corresponding ML_6 homoligand complexes with

an ideal octahedral geometry. First, for one electron on three t_{2g}^* MOs, we have

$$\begin{aligned}\psi_{xy}^* &= l_\tau d_{xy} - (c_\tau/2)(\pi_{3y} - \pi_{4y} + \pi_{5x} - \pi_{6x}),\\ \psi_{xz}^* &= l_\tau d_{xz} - (c_\tau/2)(\pi_{1x} - \pi_{2x} + \pi_{3z} - \pi_{4z}),\\ \psi_{yz}^* &= l_\tau d_{yz} - (c_\tau/2)(\pi_{1y} - \pi_{2y} + \pi_{5z} - \pi_{6z}).\end{aligned} \tag{6.22}$$

The ground state of the d^1 ML_6 complex is therefore orbitally degenerate, and determining the adiabatic potential energy surface of an $ML_{6-k}X_k$ complex therefore requires combined treatment in terms of Jahn–Teller and substitution effects [281].

For this purpose, let us calculate the dependence of the energy of MO (6.22) on deformation and substitution. According to Table 3, the H_S substitution operator for $ML_{6-k}X_k$ includes contributions of the A_{1g}, E_g, and T_{1u} symmetry types, of which in the basis set of even MOs (6.22), only the E_g component is active. It is easy to see that the matrix of the H_S operator in the basis set of the t_{2g}^* MOs is diagonal; E_g-type Q_2 and Q_3 deformations also do not mix different t_{2g}^* MOs (we only consider deformations affecting bond lengths). As a result, the energies of the MOs under consideration in an $ML_{6-k}X_k$ complex with respect to the t_{2g}^* level in ML_6 will be [14, 15]

$$\begin{aligned}\varepsilon_{xy} &= -AQ_3 + S_{xy},\\ \varepsilon_{xz} &= (1/2)AQ_3 + (\sqrt{3}/2)AQ_2 + S_{xz,xz},\\ \varepsilon_{yz} &= (1/2)AQ_3 - (\sqrt{3}/2)AQ_2 + S_{yz,yz},\end{aligned} \tag{6.23}$$

where A is the orbital vibronic coupling constant

$$A = -\langle\psi_{xy}|(\partial H/\partial Q_3)_0|\psi_{xy}\rangle = (c_\tau l_\tau/2\sqrt{3})(\partial\beta(\pi, d_\pi)/\partial R)_0,$$

which, according to the equation just given, is positive.

The signs of the $\Delta\alpha_\pi$ and $\Delta\beta_\pi$ parameters describing the substitution of a multiply bonded ligand for a σ-bonded one were discussed in Section 6.1. It follows from the results obtained in that section that the $S_{jj} = \langle\psi_j|H_S|\psi_j\rangle$ matrix elements are also positive. Their values for singly and doubly substituted complexes are given in Table 23, where $S = (c_\tau/2)^2\Delta\alpha_\pi - (c_\tau/2)l_\tau\Delta\beta(\pi, d_\pi)$. It follows from (6.23) and positive S_{jj} values that the replacement of σ-bonded ligands by multiply bonded ones shifts all or some of MOs (6.22) to the higher energies and results in partial or complete removal of their degeneracy. The dependences of MO energies on interatomic distances remain the same as in ML_6, and the adiabatic potential energy surface of $ML_{6-k}X_k$ complexes is described by the equation

$$\mathcal{E} = (K_E/2)(Q_2^2 + Q_3^2) + \sum_i n_i\varepsilon_i,$$

Table 23. H_S operator matrix elements in the basis set of t_{2g}^* MOs for quasi-octahedral complexes

Complex	Substituted ligands	$S_{xy,xy}$	$S_{xz,xz}$	$S_{yz,yz}$
ML_5X	1	0	S	S
trans-ML_4X_2	1, 2	0	$2S$	$2S$
cis-ML_4X_2	3, 5	$2S$	S	S

Note: Numbering of ligands is given in Fig. 4.

where n_i are the populations of levels (6.23).

Compare the total energies of *cis*- and *trans*-ML_4X_2 d^1 complexes. It follows from Table 23 that in *trans* complexes with substituents situated on the z axis, the unpaired d electron is localized on the ψ_{xy}^* MO ($n_{xy} = 1$, $n_{xz} = n_{yz} = 0$), whereas in *cis* complexes with substituents on the x and y axes, this electron occurs on either the ψ_{xz}^* or the ψ_{yz}^* MO. We see that at the point $Q_2 = Q_3 = 0$, the *trans* isomer is stabler than the *cis* one, and the $\mathcal{E}_{c \to t}$ isomerization energy equals the S value given above.

Clearly, this conclusion remains valid if complete calculations are made, which implies substituting the populations of MOs into the equation for the adiabatic potential energy surface to determine the energies of the isomers at their equilibrium geometries rather than at the $Q_2 = Q_3 = 0$ point. For fairly large S values, the conclusion of stabilization of *trans* isomers can be extended to diamagnetic d^2 complexes, for which population values n_i are two times larger than for d^1 and d^2 complexes. In a similar way, it can be shown that ML_3X_3 d^1 complexes with three multiply bonded ligands X are also characterized by predominant stabilization of their *trans* isomers.

3. The conclusion of stabilization of *trans* isomers of d^2 complexes is in agreement with experimental data. It is noted in monographs [248, 86] and reviews [250, 251a] that all available X-ray structural and spectral data evidence that ML_4O_2 metal dioxo complexes such as anions in the molybdenum(IV) $K_4[MoO_2(CN)_4] \cdot 6H_2O$ and $NaK_3[MoO_2(CN)_4] \cdot 6H_2O$ compounds, have *trans* configurations. According to the spectral data, the *trans* arrangement of oxygen atoms is also characteristic of $[ReO_2L_4]^+$ dioxotetramine complexes. For L_2 = ethylenediamine and L = pyridine, the spectral data were substantiated by the results of X-ray structural analyses. The O=Os=O group in $K_2[OsO_2Cl_4]$ and

$K_2[OsO_2(OH)_4]$ hexavalent osmium compounds is also linear. Unfortunately, theoretical predictions on stabilization of *trans* isomers of metal d^1 complexes are difficult to compare with experiment because of insufficiency of structural data on mononuclear d^1 complexes with two multiply bonded ligands.

6.4. Mutual Influence of Ligands and Isomers in Actinide Complexes

1. Curiously, although mutual influence of ligands in transition metal and nontransition element complexes has long been the focus of interest of researchers, studies of similar effects in actinide compounds were initiated as late as in the 1980s [71, 282], although the concept of mutual influence of ligands might be applied to interpret the properties of, say, uranyl complexes much earlier.

A theory of mutual influence of ligands in actinide complexes was for the first time suggested in [71] to describe quasi-octahedral ML_5X and ML_4X_2 complexes, where M is an actinide, L is a halogen atom, and X is a multiply bonded ligand (oxygen). This theory was in essence very similar to that described in Section 6.1. As usual, our starting point will be the scheme and the atomic compositions of MOs of the corresponding homoligand ML_6 complexes (reliable data exist for actinide hexafluorides). According to the X_α calculation results [283–287], which are in agreement with X-ray photoelectron spectroscopy data, ligand substitution effects in uranium, neptunium, and plutonium hexafluorides should be described with the use of the $2t_{1u}$ and t^*_{2u} ML_6 MOs as the HOMO and LUMO, respectively (Fig. 34). The a_{2u} MO, which has a lower energy than the t^*_{2u} one, is a purely atomic orbital in the basis set of the valence s, p, d, and f AOs of actinides and the s and p AOs of ligands; it coincides with the f_{xyz} AO of atom M and, therefore, makes no contribution to mutual influence of ligands. (We use a nonrelativistic approach; applying relativistic schemes of MOs and double symmetry groups would require reformulating selection rules (2.25) and (2.43a) for MOs, vibrational modes, and substitution operators.)

It follows that in the basis set constructed of the $5f$, $6d$, $7s$, and $7p$ metal AOs and the σ and π ligand AOs, the HOMO and LUMO have the form

$$\begin{aligned}\text{HOMO:}\quad \psi_{2t_{1u},x} &= c_1\varphi_x + (l_1/2)(\sigma_3 - \sigma_4)\\ &\quad + (l_2/2)(\pi_{1x} + \pi_{2x} + \pi_{5x} + \pi_{6x}),\\ \text{LUMO:}\quad \psi^*_{t_{2u},x} &= l_\theta f_{x(z^2-y^2)} - (c_\theta/2)(\pi_{1x} + \pi_{2x} - \pi_{5x} - \pi_{6x})\end{aligned} \tag{6.24}$$

(the arrangement of ligands is shown in Fig. 4). In (6.24), only one MO is written for each of the T_{1u} and T_{2u} representations. The t^*_{2u}

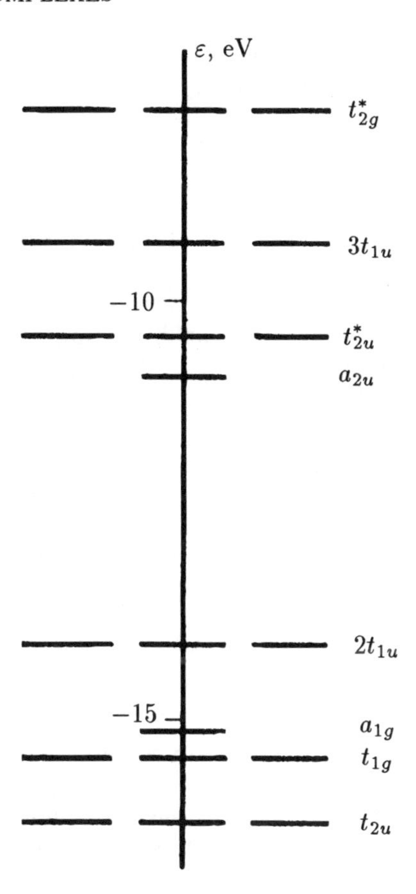

FIG. 34. SCHEME OF MOs FOR UF_6 ACCORDING TO THE NONRELATIVISTIC DVM-X_α CALCULATIONS [264]. SEVERAL HIGHER OCCUPIED AND LOWER UNOCCUPIED MOs ARE SHOWN; THE $2t_{1u}$ MO IS THE HOMO.

MO contains metal f and ligand π AOs (they are written in the coordinate system fixed on M). The normalized combination of actinide f and p AOs is denoted by φ_x. The c_θ and l_θ coefficients in (6.24) obey the inequality $c_\theta < l_\theta$, because ligands (halogen atoms) are more electronegative than M (according to the calculations performed in [286], $c_\theta^2 \lesssim l_\theta^2/4$, and $c_1^2 \approx l_1^2 \approx l_2^2 \approx 1/3$).

Consider the replacement of σ-bonded ligand L in position 1 of an ML_6 octahedral complex (see Fig. 4) by multiply bonded ligand X. Using the same form of the H_S operator as in Section 6.1 and selection rules (2.25), we find that the Q_3 mode only contributes to mutual influence of ligands, and equation (3.8) with $Q_\nu^f = Q_3^f$ is satisfied. It is easy to show that in this equation,

$$\begin{aligned} S &= -(c_\theta l_2/4)\Delta\alpha_\pi + (l_\theta l_2/2)\Delta\beta_\pi, \\ A &= (\sqrt{3}\, l_\theta l_2/2)(\partial\beta(\pi, d_\pi)/\partial R)_0, \end{aligned} \tag{6.25}$$

where $\Delta\alpha_\pi$ and $\Delta\beta_\pi$ are the same values as previously, and $\beta_\pi = \beta(\pi, f_\pi)$.

As mentioned, the replacement of a σ-bonded ligand by a multiply bonded one is described by the inequalities $\Delta\alpha_\pi > 0$ and $\Delta\beta_\pi < 0$. It then follows from (6.25) that $Q_3^f < 0$. That is, the $ML_6 \longrightarrow ML_5X$ replacement of σ-bonded L by multiply bonded X such as $O\equiv$ results in octahedral complex contraction along the X$-$M$-$L axis and expansion in the equatorial plane. It follows that mutual influence of ligands in actinide complexes is not similar to *trans*-influence in transition metal complexes; rather, the corresponding effects resemble *cis*-influence in 12-electronic nontransition element compounds. This similarity stems from the odd character of frontier MOs. The parities of the HOMO and LUMO coincide in actinide complexes, as they do in 12-electronic nontransition element complexes.

2. These theoretical predictions are in agreement with the experimental data on mutual influence of ligands in actinide complexes listed in Table 24. In the $UOCl_5^-$ and $PaOCl_5^{2-}$ uranium and protactinium complexes studied by X-ray diffraction, the axial $M-L_{trans}$ bonds are shorter than the equatorial $M-L_{cis}$ ones. A comparison of the UCl_6 and $UOCl_5^-$ complexes shows that shortening of the axial and elongation of the equatorial M$-$L bonds is both absolute and relative, because the $U-Cl_{trans}$ bonds are shorter and the $U-Cl_{cis}$ ones longer than the bonds in the isoelectronic UCl_6 molecule. The introduction of a second multiply bonded substituent positioned *trans* with respect to the first one increases the steric effect of substitution approximately twofold, as expected based on the superposition rule for the total substitution effect. This means that the well documented stability of the uranyl (more generally, actinyl) $O\equiv M\equiv O$ group can be considered a consequence of mutual influence of ligands in actinide complexes.

It should, however, be borne in mind that the analysis made above was based on the assumption that the schemes of MOs of all hexahalide actinide complexes are similar to the scheme of MOs of their hexafluorides. The calculations made in [288a] for UCl_6 and in [288b] for the PaL_6^{2-} and UL_6^- anions (L = Cl, Br, and I), however, show that the scheme of the MOs of ML_6 (L = Cl, Br, and I) actinide hexahalides may be close to that represented in Fig. 35. This scheme requires a more detailed analysis similar to that performed in Section 6.1.

Table 24. M—O and M—Hal bond lengths (Å) in uranium and protactinium compounds

Compound	M—O	M—L_{ax}	M—L_{eq}	Refs.
UCl_6	—	2.47	2.47	[69]
$[PPh_4][UOCl_5]$	1.76	2.43	2.53	[69]
$[NEt_4]_2[PaOCl_5]$	1.74	2.42	2.64	[70]
$[NEt_4]_2[UO_2Cl_4]$	1.75	—	2.67	[69]

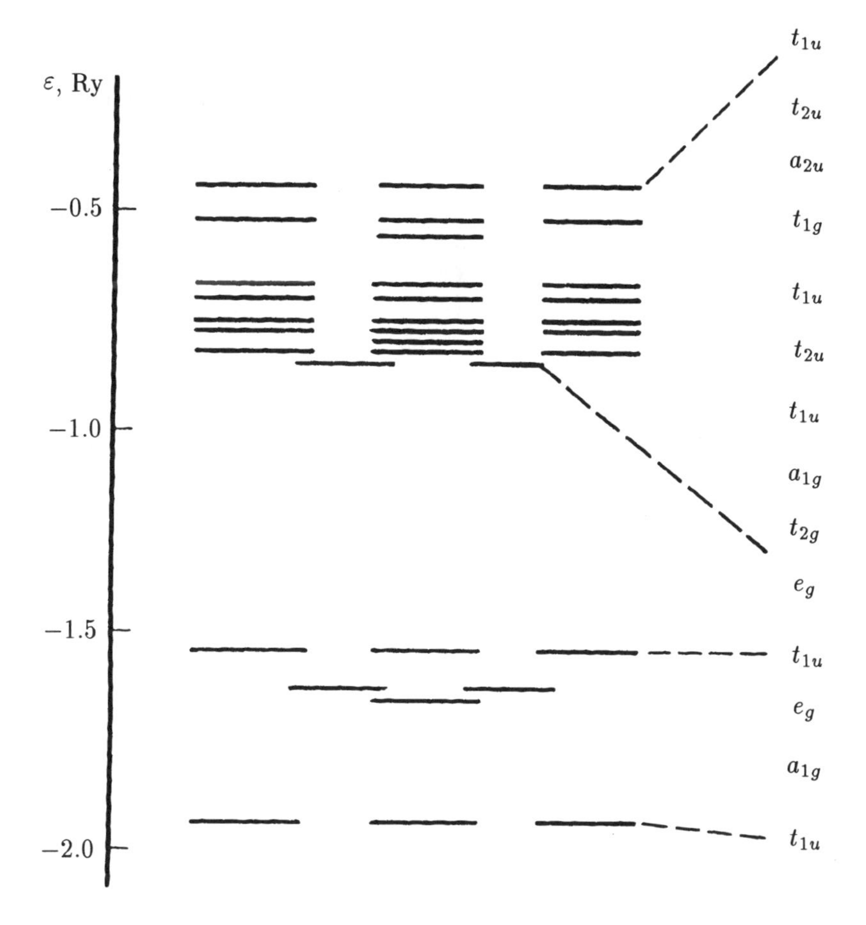

FIG. 35. SCHEME OF MOs FOR UCl_6 ACCORDING TO THE NONRELATIVISTIC SW-X_α CALCULATIONS [266]; THE t_{1g} MO IS THE HOMO.

A comparison of the schemes given in Figs. 35 and 33a shows them to be similar so far as the positions of occupied MOs are concerned, but their LUMOs (t^*_{2u} and t^*_{2g}, respectively) have different parities. For this reason, mutual influence of ligands in actinide complexes is dominated by the Q^f_3 even mode, rather than the Q^f_{12} odd one as in transition metal complexes. A more detailed analysis of mutual influence of ligands in actinide complexes, however, requires the use of data on the atomic compositions of MOs not given in [288a, 288b].

4. To conclude, consider the relative stability of isomers of actinide complexes. Mutual influence of ligands in these complexes is determined by MOs of the same parity, and stability of isomers therefore depends on the even component of the substitution operator. (Recall that in transition metal complexes, the determining role is played by the odd component.) For this reason, *trans* isomers of actinide complexes with two multiply bonded ligands should be more stable than *cis* ones, which is in agreement with the linear geometry of the actinyl group.

CHAPTER 7

ISOMERS OF HETEROLIGAND BIPYRAMIDAL MOLECULES

In the preceding chapters, we discussed heteroligand molecules and complexes that can be considered derivatives of homoligand molecular systems ML_n with ligands equivalent by symmetry. For closed-shell ML_n systems, treatment of the problem was based on second-order perturbation theory.

In this chapter, we discuss systems such as heteroligand trigonal–bipyramidal molecules, which are derivatives of homoligand (ML_n) closed-shell molecules containing ligands nonequivalent by symmetry. To analyze the properties of such systems, it suffices to apply first-order perturbation theory.

The presence of ligands nonequivalent by symmetry slightly changes the character of problems to be solved. For instance, a discussion of mutual influence of ligands in heteroligand derivatives of systems with ligands nonequivalent by symmetry is impeded by uncertainty inherent in the very formulation of the problem. One of the principal tasks of the theory of mutual influence of ligands in heteroligand $ML_{n-k}XY...Z$ derivatives of homoligand ML_n systems with equivalent ligands is an analysis of substituent effects on M—L bonds (e. g, *cis* and *trans* bonds) equivalent in the initial ML_n system but differently situated with respect to substituents X, Y, ..., Z. However, in a homoligand ML_n system with nonequivalent ligands, there are M—L bonds nonequivalent even prior to substitution.

At the same time, the problem of the relative stability of isomers of such heteroligand systems does not involve uncertainties of this kind, and in this chapter, we will confine the discussion to the relative stability of isomers. It should be borne in mind that isomerism of heteroligand systems with ligands nonequivalent by symmetry differs from isomerism of systems considered in the preceding chapters. In complexes with ligands equivalent by symmetry, isomers appear when two or more ligands L in ML_n undergo substitution, whereas even monosubstituted systems with nonequivalent atoms can form isomers.

7.1. Stability of Isomers in First-Order Perturbation Theory

1. First consider the question why it is necessary to use second-order perturbation theory in analyzing derivatives of homoligand closed-shell molecular systems with equivalent ligands, whereas derivatives of systems with nonequivalent ligands can be treated in terms of first-order perturbation theory. To answer this question, let us turn to (2.47) and (2.50), where substitution operator $\mathcal{H}_S$ matrix elements appear in both first- and second-order perturbation theory (S_0 and two last terms in (2.50), respectively).

Equation (2.47) shows that S_0 disappears from minimum conditions for the potential energy surface of a heteroligand system $(\partial E/\partial Q_\nu)_0 = 0)$, because S_0 is independent of Q_ν. For this reason, first-order perturbation terms are unimportant for solving problems of static mutual influence of ligands, that is, problems of the effects of substitution on molecular geometry. The S_0 value, however, appears in the equation for the total energy of a heteroligand system and determines this energy in first-order perturbation theory. At first sight, this means that certain problems such as relative stability of isomers can be studied at the level of first-order perturbation theory. Symmetry considerations show that this is not always so. We saw in Section 3.4 that first-order contribution S_0 to the energy of a heteroligand derivative of a molecule with ligands equivalent by symmetry was structurally insensitive. It follows that this term cannot be used to distinguish between isomers. This conclusion, however, does not apply to derivatives of molecules having lower symmetry and containing nonequivalent ligands.

2. Following the same line of reasoning as in the end of Section 2.4.5, let us divide ligands of an ML_n molecule or complex into "transitive sets" such that each ligand of each set can be transformed into any ligand of the same set (but never into a ligand of another set) under symmetry transformations of point symmetry group G of the molecule. For instance, in octahedral ML_6 complexes, all ligands make up a single transitive set, whereas there are two such sets in trigonal–bipyramidal ML_5 complexes, one including axial and the other equatorial ligands (Fig. 36).

Consider the $\mathcal{H}_S$ substitution operator assumed to satisfy (2.33), that is, the requirement of additivity with respect to substituents. Each partial substitution operator $\mathcal{H}^l_S$ will be separated into operators $\mathcal{H}^{l,\gamma}_\Gamma$ symmetrized according to the group G irreducible representations (Γ) and representation rows (γ). It is shown in Section 2.4.5 that for all ligands of the same transitive set, the a^l_A coefficients of the totally symmetric component of $\mathcal{H}^l_S$ coincide. In other words, we then have only one coefficient a_A independent of the number labeling the ligand. But

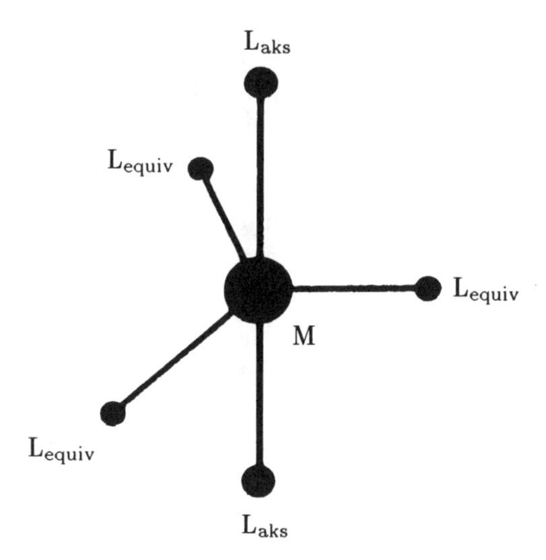

FIG. 36. TRIGONAL–BIPYRAMIDAL ML_5 MOLECULE.

generally, a_A coefficients do not coincide for ligands from different transitive sets. For instance, for trigonal-bipyramidal molecules ML_5, we have two independent coefficients a_A^{ax} and a_A^{eq} for axial and equatorial ligands, respectively.

Using (2.48) for S_0, we obtain

$$S_0 = \sum_t a_A^t \sum_{l_t} \langle \Psi_0 | \mathcal{H}_S^{l_t} | \Psi_0 \rangle, \tag{7.1}$$

or, which is the same,

$$S_0 = \sum_t n_t a_A^t \langle \Psi_0 | \mathcal{H}_S^{l_t} | \Psi_0 \rangle. \tag{7.2}$$

Here, the summation over t is over different transitive sets of ligands, and the sum over l_t is taken over ligands within each tth transitive set; n_t is the number of substituted atoms in the tth transitive set; and $\mathcal{H}_S^{l_t}$ is the operator corresponding to the substitution of one (l_tth) atom in the tth set.

It follows from (7.1) and (7.2) that in the presence of two or more transitive sets, S_0 depends on not only the number but also the mutual arrangement of substituents. To put it more clearly, we must write S_0 in terms of MOs. Using the same approach as in Section 2.4 and applying (2.48), we obtain

$$S_0 = \sum_l \sum_i P_{li} \Delta \alpha_{li}, \tag{7.3}$$

where P_{li} is the population of the ith AO of the lth ligand in the initial homoligand molecule, and $\Delta\alpha_{li}$ is the difference of the Coulomb integrals between the AOs of the substituent and ligand L occupying position l in the homoligand ML_n molecule.

7.2. Isomers of 10-Electron Trigonal–Bipyramidal Molecules

We now turn to the relative stability of isomers of derivatives of 10-electron σ-bonded ML_5 molecules having the geometry of a trigonal bipyramid [289, 290], Fig. 36. As mentioned, trigonal–bipyramidal molecules contain nonequivalent L_{ax} and L_{eq} ligands in physically different axial and equatorial positions. For this reason, populations P_i of σ orbitals are different for L_{ax} and L_{eq}. For instance, according to the calculation data [291–293], the electronic populations of the σ AOs of axial ligands in PH_5 and PF_5 are larger than those of equatorial atoms by ~0.2 e. Applying (7.3) to estimate S_0 yields

$$S_0 = P_{eq}\sum_i^{eq}\Delta\alpha_i + P_{ax}\sum_i^{ax}\Delta\alpha_i. \tag{7.4}$$

Taking into account (2.57), we arrive at the conclusion that, in first-order perturbation theory, (7.4) determines the total energy of the heteroligand molecule under consideration at the point corresponding to the potential energy surface minimum,

$$\mathcal{E}(ML_{5-k}XY\ldots Z) \approx S_0. \tag{7.5}$$

3. First, consider an ML_4X heteroligand molecule, which can exist as two isomers, one with axial and the other with equatorial ligand X. According to (7.4) and (7.5), the isomerization energy for ML_4X will be

$$\mathcal{E}_{ax\to eq} = \mathcal{E}(ML_4X_{ax}) - \mathcal{E}(ML_4X_{eq}) = (P_{ax} - P_{eq})\Delta\alpha. \tag{7.6}$$

It follows that the relative stability of the eq-ML_4X and ax-ML_4X isomers is determined by the $P_{ax} - P_{eq}$ difference of the populations of ligand σ AOs in the ML_5 homoligand molecule and the nature of ligands L and X, that is, the $\Delta\alpha = \alpha(X) - \alpha(L)$ difference.

According to the calculations performed in [291–293], $P_{ax} - P_{eq} > 0$. On the other hand, $\Delta\alpha$ is positive if ligand X is more electropositive than L and negative otherwise. As a result, ligands more electropositive than L occupy equatorial positions in 10-electron trigonal–bipyramidal

σ-bonded ML_4X molecules, and ligands more electronegative than L occupy axial positions. It follows from (7.4) that this conclusion is valid for not only singly substituted ML_4X molecules but also systems with several identical or different substituents. Note that these rules resemble those governing electro- and nucleophilic substitution in organic chemistry (the site at which substitution occurs is determined by the π electronic densities on carbon atoms) and predict the same arrangement of substituents in bipyramidal molecules as the Gillespie–Nyholm model [102].

4. A comparison with the available experimental data shows that equations (7.4) and (7.5) correctly describe the structure of trigonal–bipyramidal molecules. According to [102b], all known ML_5 molecules formed by Group V elements except $Sb(C_6H_5)_5$ have trigonal–bipyramidal structures, and in almost all instances of $ML_{5-k}X_k$ derivatives, the more electronegative ligands are in axial positions, whereas the more electropositive ones occur in the equatorial plane. For instance, for PF_4H and PF_4Cl molecules as examples of ML_4X systems with monoatomic ligands, the more electropositive substituents, H and Cl, occupy equatorial positions [102, 294]. The PF_4CH_3 methyl derivative was found [102] to have a similar structure, whereas the axial position of a more electronegative substituent (F) was established for PCl_4F [102]:

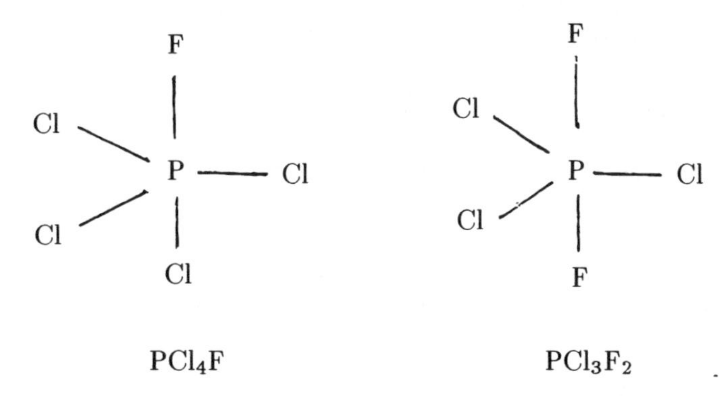

The conclusion that more electropositive substituents replace equatorial ligands remains valid for ML_4E molecules with lone electron pairs, because the transition from ML_5 to ML_4E can be considered the limiting case of electropositive substitution (such a treatment, however, goes beyond the scope of perturbation theory applicability). Of two ML_4E structures having C_{3v} and C_{2v} symmetries, the low-symmetry structure with the lone pair in the equatorial position should be the more stable. This prediction is in agreement with the structural data on SF_4, SeF_4, $TeCl_4$, BrF_4^+, IF_4^+, and ClF_4^+ [102, 293, 294].

Trigonal–bipyramidal molecules with two substituents X can exist as three different isomers. Doubly substituted compounds may have D_{3h} symmetry, when both substituents occupy axial positions, C_s symmetry, when one substituent occupies an equatorial position and the other, an axial one, and C_{2v} symmetry, when both substituents occur in the equatorial plane. Electropositive substitution should favor stabilization of the C_{2v} form. Such a structure was observed in PF_3H_2, PF_3Cl_2, and PF_3Br_2 [102, 294] and the $PF_3(CH_3)_2$ dimethyl derivative [102]. Conversely, two more electronegative substituents should stabilize the D_{3h} form, as in PCl_3F_2.

Of three different structures of ML_3E_2 molecules with two lone pairs (triangular, pyramidal, and T-shaped), the T-shaped one should be more stable than the other two. This explains the observed configurations of ClF_3, BrF_3, and XeF_3^+ [102, 294]. The SF_2H_2E compound is an example of a system containing different monoatomic ligands and a lone pair [102, 294]. The F—S—F angle equals 167°, which is fairly close to 180°, whereas the H—S—H angle approximates to 120°, and the molecule can therefore be assigned the structure with the equatorial arrangement of the lone pair and the less electronegative substituents (hydrogen atoms) and the axial arrangement of the more electronegative ligands (fluorine atoms). A similar structure with the equatorial arrangement of two less electronegative ligands and the lone electron pair is characteristic of ML_2X_2E with M = Sc and Te, L = Cl and Br, and X = Me, Ph, and MeC_6H_4. Lastly, an example of ML_2XE_2 systems is ICl_2Ph, where the I—C bond is perpendicular to the Cl—I—Cl axis [294].

The analysis performed above and equation (7.4) are valid not only for 10-electron σ-bonded trigonal–bipyramidal molecules but also σ-bonded pentagonal–bipyramidal systems [295]. However for such systems, a comparison of theoretical predictions with experiment is impeded by rarity of heteroligand σ-bonded molecules of this type.

CHAPTER 8

ORDER–DISORDER FERROELECTRICS WITH INTERMOLECULAR H-BONDS

Fairly often, theoretical ideas and methods initially developed for treatment of free molecules are afterward applied to solids. The principal concepts of the theory of the Jahn–Teller and pseudo-Jahn–Teller effects were transferred to crystals, which led to the development of the theory of the cooperative Jahn–Teller effect and the vibronic theory of ferroelectricity [14, 15, 296]. Similarly, the ideas of the theory of heteroligand systems described in the preceding chapters can be applied to solids. In this chapter, we will, very briefly and without going into details, consider extended systems such as ferroelectrics and related materials comprising molecules and complexes linked with each other by intermolecular H-bonds [297].

8.1. Rearrangement of H-Bond Protons as Cooperative Substitution of Ligands

1. Typical representatives of H-bonded extended systems are three-dimensional ferroelectrics of the KH_2PO_4 (KDP) family having the composition $M^IH_2MO_4$ and their deuterated analogues ($M^I = K$, Rb, and Cs and $M = P$ and As) and a "two-dimensional" ferroelectric, a layer in the layered structure of $H_2C_4O_4$ (H_2SQ) and $D_2C_4O_4$ squaric acid crystals [298, 299] (Figs. 37 and 38). To be more specific, the further discussion will be confined to these ferroelectrics, although the approach that we use can also be applied to other systems. The systems of the two types specified above are similar in that they are built of molecular structural units, *viz.*, tetrahedral phosphate or arsenate $[MO_4]$ groups or $[C_4O_4]$ groups (for shortness, below referred to as molecules) linked into a continuous network by strong H-bonds. Accordingly, each proton can occupy two energetically equivalent positions along the H-bond that it forms, O—H...O or O...H—O. To put it differently, each oxygen can be in one of two states in the crystal, which can be denoted as O(—H) and O(...H) to indicate bonding with H by either a covalent or an H-bond. Rearrangement of protons along

H-bonds results in the transition of some or even all crystal oxygen atoms from one state to the other. This transition can formally be treated as ligand substitution in structural units of the crystal. It is important that this substitution is a cooperative process, because the O(−H)→O(...H) transition in one molecule automatically causes the reverse O(...H)→O(−H) transition in the neighboring molecule [300, 301].

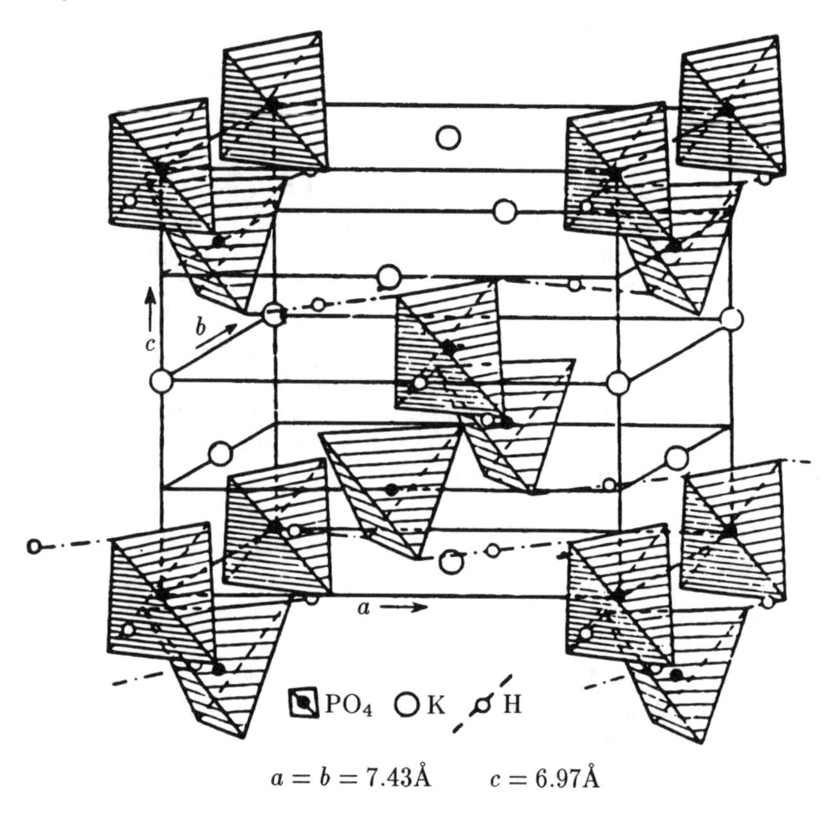

$a = b = 7.43$Å $c = 6.97$Å

FIG. 37. CRYSTAL STRUCTURE OF KH_2PO_4.

2. Charges on H-bond protons involved in the rearrangement can be considered constant, because protons shift between equivalent positions along H-bonds (taking into account proton-proton correlations may make these positions not exactly equivalent). Other atomic charges, especially oxygen charges, however, change because of electron density transfer between oxygen atoms of neighboring molecules mediated by H atoms and because of electron density transfer between O atoms of the same molecule through M (in KDP) or C (in H_2SQ) atoms. A

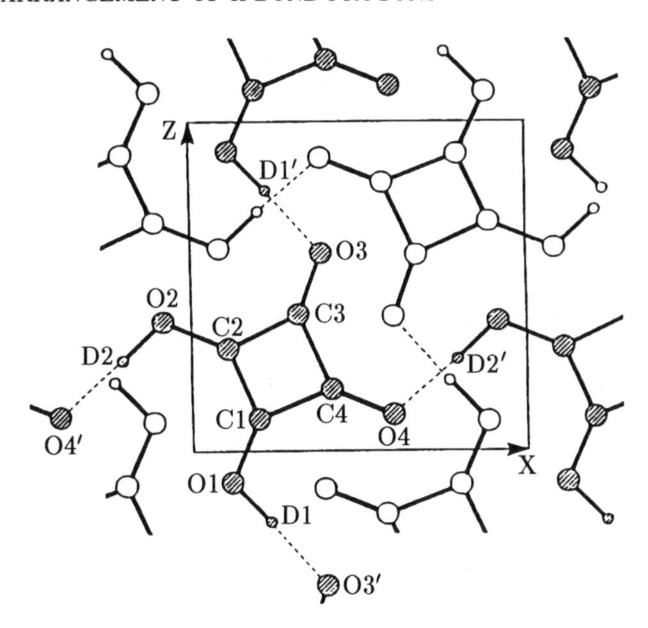

FIG. 38. STRUCTURE OF DEUTERATED SQUARIC ACID $D_2C_4O_4$ (LOW-TEMPERATURE PHASE) [308].

detailed description of redistribution of charges for various types of ordering of H-bond protons was obtained in quantum-chemical calculations of various clusters simulating KDP, H_2SQ, and other H-bonded systems [302–304].

It was found that the character of transfer and the amount of transferred charge are different for inorganic crystals built of σ-bonded complexes (KDP) and organic materials comprising molecules with π bonds (H_2SQ). However in any event, O(−H) atoms play the role of intermolecular electron donors and intramolecular electron acceptors. Conversely, O(...H) atoms are intermolecular electron acceptors and intramolecular electron donors (to avoid confusion, it should be stressed once more that we speak of charge redistribution between oxygen atoms induced by redistribution of protons rather than mere charge transfer between oxygens and the other atoms in the lattice). The total electronic population of O(...H) atoms is always larger than that of O(−H) because of the presence of the intermolecular transfer channel. This is contrary to the situation considered standard in coordination chemistry, when the population of the donor atom in an isolated complex decreases and that of the acceptor increases. Note that the calculations [305, 306] reproduce macroscopic spontaneous polarization P_s

of crystals in the ferroelectric (FE) phase with reasonable accuracy and show that P_s, at least in the KDP family, is related to electronic polarization of molecular structural units and can be calculated from effective atomic charges.

8.2. Molecular Geometry in the Ferroelectric Phase

1. Spontaneous polarization of H-bonded ferroelectrics in the FE phase is due to ordering of H-bond protons accompanied by ordering of O(...H) and O(—H) atoms in the crystal and, therefore, in each molecule. In the F phase of KDP, O(...H) atoms occupy two "upper" (or two "lower") vertices of PO_4 tetrahedra, whereas O(—H) atoms occupy two other vertices. Similarly, in all molecules of a layer in the low-temperature phase of H_2SQ, O(...H) atoms occur on one side of the C_4O_4 square, whereas O(—H) atoms, on the opposite side. Such an arrangement of oxygen atoms causes the appearance of steric strain, which has been studied by diffraction methods [307, 308] and can be given a simple explanation in terms of the theory of heteroligand systems taking into account intramolecular donor-acceptor behavior of the O(...H) and O(—H) oxygen atoms.

2. First consider an MO_4 tetrahedron in a crystal of the KDP family. Applying the MO LCAO approach, we can describe its electronic structure in terms of four bonding and four antibonding σ MOs. These MOs include the valence ns and np AOs of the central atom ($n = 3$ for M = P and $n = 4$ for M = As) and four oxygen σ AOs. The other valence AOs of each oxygen atom are either occupied by lone electron pairs, which virtually do not participate in M—O bonding, or are involved in O—H...O bonds.

The stereochemistry of this system can conveniently be analyzed with the use of a hypothetical "homoligand" $M[O(\text{--}H)]_4$ tetrahedron as the parent structure. In this tetrahedron, all oxygen atoms are in a state intermediate between O(—H) and O(...H) (this state models the averaged state of oxygen atoms in the paraelectric phase). Equally effectively, we might choose the state of an MO_4 tetrahedron in the hypothetical "protophase", in which all protons occur in the middle of H-bonds. Because of the geometry of the arrangement of H-bonds around such an MO_4 tetrahedron, its T_d symmetry is lowered to S_4, and the qualitative scheme of MOs has form I:

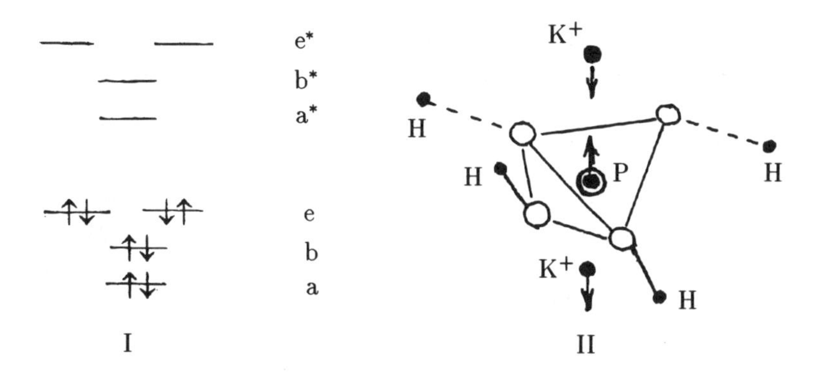

(the b–e and b^*–e^* splittings are exaggerated, in reality, they are small).

When the paraelectric phase transforms into the FE phase, each M[O(- -H)]$_4$ tetrahedron transforms into the M[O(—H)]$_2$[O(...H)]$_2$ one, that is, two ligands in its upper vertices are replaced by the O(...H) donor ligands, and ligands in two other vertices, by the O(—H) acceptor ligands (or vice versa). The character of the accompanying steric distortions (see scheme II above) is described by the theory of mutual influence of ligands in quasi-tetrahedral molecular systems (Section 4.7).

Let us first assume that in the protophase, the MO_4 tetrahedron has T_d symmetry. Its deformation will then involve the Q_6 and Q_9 modes (Fig. 20), and the magnitudes and signs of atomic displacements will be determined by (4.28), (4.29) and (4.30), (4.31) (it should be taken into account that two pairs of ligands undergo replacement). The resulting deformation has the same form as for a free ML_2X_2 quasi-tetrahedral molecule of the SiH_2F_2 or $PH_2F_2^+$ type (Tables 14 and 15). This deformation can be described as a displacement of M from the center of the tetrahedron in the direction from O(—H) toward O(...H) atoms.

It should be noted that the formulas applied to obtain this result were derived in the approximation of frontier MOs based on the scheme of σ MOs of a regular ML_4 tetrahedron (Fig. 18). However, distortions of the same type are obtained in the approximation of frontier MOs if it is taken into account that in the protophase, the MO_4 tetrahedron has S_4 symmetry, and its MOs are arranged as shown in Scheme I above.

According to the diffraction data [307], distortions of MO_4 tetrahedra exactly like those shown in Scheme II above accompany the paraelectric phase→FE phase transition of KDP family crystals. Simultaneously, outer-sphere M^{I+} cations shift in the direction opposite to phosphorus atom shifts, which conforms with an increase in the electronic population of the O(...H) atoms and a decrease in the electronic population of the O(—H) atoms in comparison with O(--H).

3. An analysis of molecular distortions in organic materials such as H_2SQ at first sight involves considerable difficulties, because the role of central atoms M in H_2SQ-type molecules is played by atomic groups (C_4). The simplest method for reducing such situations to that considered in the theory of mutual influence of ligands is treatment of atomic groups as pseudoatoms with the MOs of C_4 as pseudoatom AOs [301, 309]. In this model, the π MOs of the $C_4[O(\text{--}H)]_4$ parent structural unit treated as a square-planar complex of the C_4 pseudoatom with four oxygen ligands have form III:

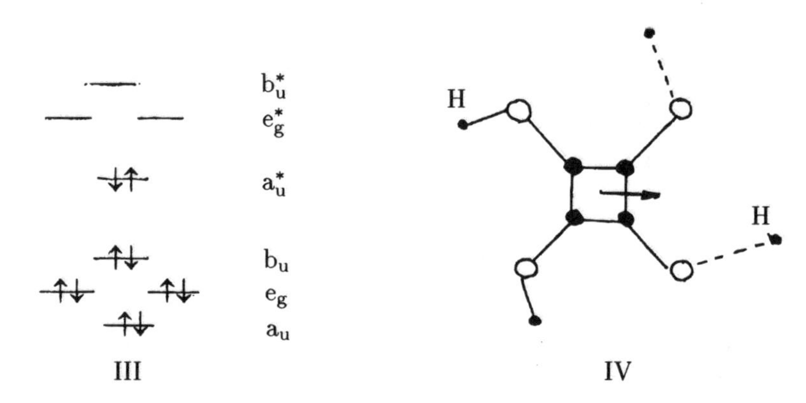

Steric distortions of C_4O_4 units that accompany the transition of the H_2SQ layer from the disordered paraelectric state to the state with ferroelectric ordering (Fig. 38) can be treated as resulting from the transformation of the $C_4[O(\text{--}H)]_4$ homoligand form into the *cis*-$C_4[O(-H)]_2[O(...H)]_2$ form. This transition can be analyzed in terms of the theory of mutual influence of ligands in quasi-square-planar non-transition element complexes (Section 4.2). The a_u^* and e_g^* frontier π MOs (see Scheme III above) play the role of the a_g^* and e_u^* frontier σ MOs in Te(II) complexes (Fig. 15). Accordingly, the resulting distortion (Scheme IV) of the geometry of C_4O_4 can be described as a shift of the C_4 pseudoatom from the O(—H) atoms toward the O(...H) ones. This prediction is also in agreement with the diffraction data [308].

8.3. Ising Model Interaction Parameters and Slater Parameters

1. In the preceding section, we applied (2.56) to molecules in H-bonded crystals. Here, we consider equation (2.57), which allows the total energy of an H-bonded ferroelectric to be written as depending on the distribution of protons. As is customary [298], the position of a pro-

ton on the ith H-bond will be characterized by pseudospin $\sigma_i^{(z)}$ taking on values ± 1 depending on whether the proton is in the O$-$H...O or O...H$-$O position. Any distribution of protons can then be described by the distribution of pseudospins or, which is the equivalent, the distribution of O($-$H) and O(...H) oxygen atoms in the whole lattice and each molecule.

If proton-proton correlations are ignored, the energy of a proton is independent of its position along the H-bond, and the energy of the crystal as a whole can be written as the sum of the energies of its nonhydrogen structural units. The distributions of oxygen atoms of two kinds in these units are, however, not independent but obey the cooperative rule formulated above, that is, the presence of an O atom of a certain kind in some molecule automatically implies the presence of an O atom of the other kind in the neighboring one.

The summation of (2.57) over the lattice taking into account cooperative relations between the S_{nm} matrix elements of operators H_S of neighboring molecules gives the energy of the crystal in the form [300, 301]

$$E = C - 1/2 \sum_{i,j} J_{ij} \sigma_i^{(z)} \sigma_j^{(z)}, \qquad (8.1)$$

where C is a constant; (8.1) coincides with the equation for the energy in the Ising model extensively used in statistical thermodynamics of systems of spins and pseudospins [310]. Importantly, equation (8.1) is not postulated but follows from (2.57), which allows Ising model parameters J_{ij} to be written in terms of the electronic structure of molecules in the crystal. We will not write the corresponding expressions for J_{ij} (see [300, 301]). Instead, consider the Slater parameters $e_0 = 4J_{||} + 4J_{\perp}$ and $w = 2J_{||} + 4J_{\perp}$ for the KDP family and $e_0 = 4J_t + 4J_c$ and $w = 2J_t$ for the H_2SQ layer, which are closely related to Ising parameters ($J_{||}$ and $J_{\perp}$ or J_t and J_c are two independent Ising parameters describing nearest-neighbor pseudospin interactions in systems with H-bonds).

The Slater parameters have a clear physical meaning and characterize the isomerization energy of separate molecules (e_0)

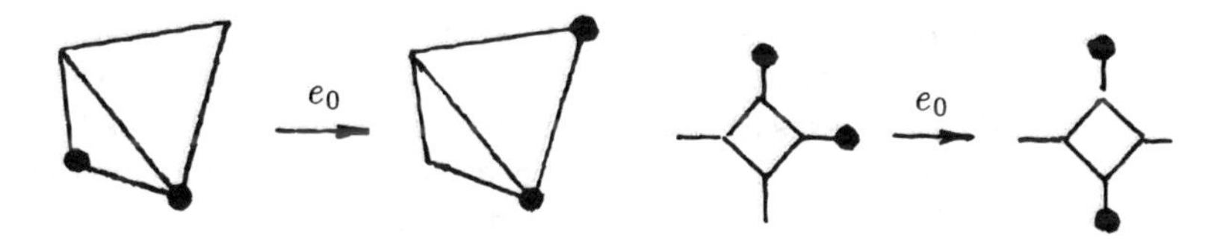

and pairs of molecules (w)

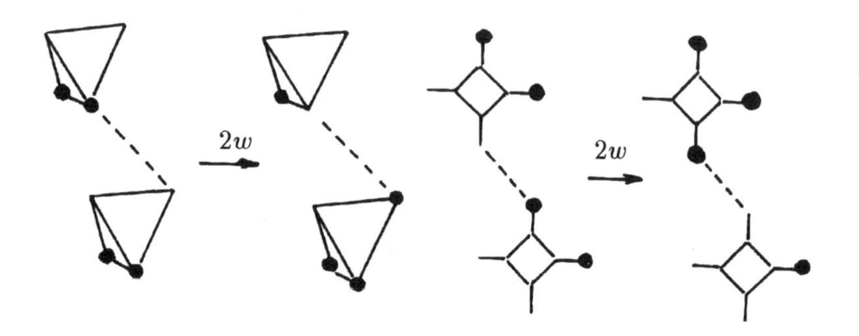

Solid circles and unoccupied vertices denote O(—H) and O(...H) atoms, respectively.

2. Equations for e_0 and w can be written in terms of Ising parameters or obtained directly from (2.39) and (2.57) on the assumption that the initial structural units are $M[O(\text{--}H)]_4$ tetrahedra or $C_4[O(\text{--}H)]_4$ structural units. For example, for KDP and its analogues, we have

$$\begin{aligned} e_0 &\approx 2(c_a^2 l_t^2 \Delta\alpha_\sigma^2/\omega_\perp)[(1 + 4A_\perp^2/K_\perp\omega_\perp) \\ &\qquad - (1 + 4A_{||}^2/K_{||}\omega_{||})(\omega_\perp/\omega_{||})], \\ w &\approx (c_a^2 l_t^2 \Delta\alpha_\sigma^2/\omega_\perp)[3(1 + 4A_\perp^2/K_\perp\omega_\perp) \\ &\qquad - 2(1 + 4A_{||}^2/K_{||}\omega_{||})(\omega_\perp/\omega_{||})]. \end{aligned} \tag{8.2}$$

Here, K, A, and ω with various indices are the force constants, orbital vibronic coupling constants, and energy gaps present in (2.57) and corresponding to the approximation of the b, e, and a^* frontier σ MOs (Scheme I above) for the $M[O(\text{--}H)]_4$ complex. The $c_a^2 l_t^2 \Delta\alpha_\sigma^2$ product in (8.2) appears in calculating the matrix element of substitution operator (2.39) describing the $M[O(\text{--}H)]_4 \longrightarrow M[O(-H)]_2[O(...H)]_2$ transition. The Slater parameters for the H_2SQ layer have the form similar to (8.2).

We have shown in the preceding chapters that for isolated molecules and complexes, explicit equations such as (2.56) and (2.57) can be used to qualitatively explain and predict diverse phenomena related to mutual influence of ligands and the relative stability and transformations of heteroligand complexes as depending on their chemical composition. Similarly, relations of type (8.2) can be used to explain and predict chemical trends in the behavior of H-bonded ferroelectrics and related systems [311].

For instance, it is common knowledge that in deuterated systems, O...D bonds are longer and O—D ones shorter than the O...H and O—H bonds, respectively, that is, deuteration increases $|\Delta\alpha|$. It can be shown that an increase in $|\Delta\alpha|$ in turn increases the critical temperature (T_c) of the paraelectric phase→FE phase transition of deuterated KDP-type crystals, which agrees with experiment. For reasons of space, we shall not discuss other chemical trends [311].

Numerical estimates of e_0 and w Slater parameters for KDP obtained by applying (8.2) (the necessary values were calculated by quantum-chemical methods [300, 301]) are in reasonable agreement with those found from the thermodynamic data [312, 313]. Acceptable estimates of Slater parameters for the H_2SQ layer can be obtained in a similar way, although the approximation of frontier MOs is then insufficient, and all π MOs of the C_4O_4 structural unit should be taken into account. This lends support to the claim that the ideas of the theory of heteroligand systems can be used to describe H-bonded ferroelectrics and, possibly, other materials.

Note also that this approach is of interest for understanding the molecular mechanism of structural phase transitions in H-bonded systems, because this mechanism is explicitly related to reorganization of the electronic and geometric structure of the nonhydrogen framework of crystals.

CONCLUDING REMARKS

In conclusion, we will consider some special problems not covered in this book or still unsolved. Several points should also be mentioned to throw light on the general meaning and possible significance of the approach that we advocate.

The book contains many examples that show what conclusions can be drawn if we know how the replacement of one or several atoms in a molecule changes the shape of its adiabatic potential energy surface. We considered a broad class of molecular systems, namely, free molecules and complexes comprising a central atom (M) and n peripheral atoms or atomic groups attached to it. The replacement of atoms was understood to be the replacement of ligands, $ML_n \longrightarrow ML_{n-k}XY...Z$, and by the adiabatic potential energy surface, we meant its lower sheet corresponding to the ground electronic state of the system. In this way, we studied the mutual influence of ligands in molecules and complexes of various types and analyzed the relation between the energy and geometry of isomers of molecular systems; specifically, we discussed geometric (*cis* and *trans*) and distortion isomers.

Even within the scope of these problems, new results can be obtained through introducing certain methodological improvements. For instance, in Section 6.4, the mutual influence of ligands and stability of *cis* and *trans* isomers of actinide complexes are treated in terms of nonrelativistic MOs. Of course, with heavy element compounds, using relativistic or quasi-relativistic wave functions would be an improvement [11]. Such an analysis would be of methodological interest and, probably, useful for explaining the special features of the structure of heavy element compounds.

Another problem is extension of the theory of heteroligand systems to isomeric forms different than geometric and distortion isomers. Transition metal heteroligand complexes are known to exist in low- and high-spin forms (spin isomerism) or dia- and paramagnetic forms (complexes with an even number of electrons). The relation between spin isomerism and the heteroligand character of complexes was considered in the Russian version of this book [315] for the example of six-coordinate Ni(II) and Fe(III) complexes. These sections were excluded from the revised English version from size considerations. The results obtained in [315] allow us to hope that a more detailed elucidation of the relation between spin isomerism and the heteroligand character of complexes can provide a deeper insight into the problem of spin isomers and transitions between them.

An important and interesting domain is studies of photoisomerization of heteroligand molecular systems through analyzing their adiabatic potential energy surfaces in electronically excited states. This problem was touched upon in [315] for the example of $Pt^{II}L_2X_2$ four-coordinate and some d^6 six-coordinate complexes. Of course, several results obtained for particular systems do not exhaust the possibilities of applying the vibronic theory of heteroligand systems to electronically excited states of molecules and complexes.

Clearly, problems of mutual influence of ligands and stability of various isomers arise not only with mononuclear $ML_{n-k}XY...Z$ complexes. They are also discussed for binuclear complexes (of the most recent works, see [316, 317]). Generalization of the theory to binuclear and polynuclear complexes is at present a topical problem.

The class of molecular systems with a central polyatomic "nucleus" and ligands (addends) attached to this nucleus is fairly broad; it includes such exotic species as fullerene derivatives. The problem of relative stabilities of their various isomers was considered in several works, for instance, in [318, 319], where the relative stabilities of the isomers of $C_{60}X_k$, $C_{70}X_k$, and other substituted fullerenes were studied; here, X are various simple (H, F) or complex addends including coordination addends such as $>Pt(PH_3)_2$. The analysis performed in [318, 319] was based on the assumption that the relaxation energy of a substituted molecule can be separated into the electronic and geometric contributions, see Section 2.4. (Electronic relaxation was treated in terms of the simple Hückel method, and geometric relaxation, by the method of molecular mechanics.) Works [318, 319] raised the question of applying the theory of heteroligand systems to fullerene derivatives. The use of this theory as a tool for studying such unusual objects as derivatives of carbon and similar complex clusters is doubtless of interest, and new developments in this direction are to be expected.

One more problem is transition to molecules and molecular ions built of more or less equivalent atoms, of which any can be considered central. The problem was studied in [315] for the example of linear and angular triatomic XYX and XYZ molecules, which were treated as heteroatomic derivatives of isoelectronic $(AAA)^{q\pm}$ precursors with a linear geometry.

This simple approach based on first-order perturbation theory with respect to the $\mathcal{H}_S$ operator was used to characterize the type of the most stable isomer (YXZ, XYZ, or XZY) for fifteen triatomic molecules comprising first and second row nontransition elements. It is likely that this approach can be applied to other similar molecules including those more complex than triatomic.

Lastly, we must consider the frequent situation of extended structures such as polymer chains, layers, or three-dimensional crystals, in which separate "centers" are integrated into one system by cooperative interactions between monomers. We already encountered such a situation in Section 5.5, where distortion isomerism of chain $Cu(NH_3)_2Hal_2$ complexes was analyzed, however, without taking into account cooperative interactions between Cu chromophores in chains and three-dimensional crystals. Another example is provided by various silicates comprising $[SiO_4]$ tetrahedra, which can be considered heteroligand because of the presence of bridge and terminal oxygen atoms. (It was shown in [320] that certain stereochemical and spectral characteristics of silicates can be explained in terms of the theory of heteroligand systems within the framework of the one-center approach.) There also exist solid-state heteroligand coordination compounds of Fe(II) and Fe(III) experiencing spin-state transitions.

In all these and similar situations, the problem is to pass from the simplified one-center analysis to a more complete treatment explicitly including intercenter interactions. This field of research can be characterized as the study of cooperative effects in the theory of heteroligand systems or "heteroligand" features in the vibronic theory of cooperative effects. It is easy to see that we here encounter many new problems that should be studied and solved.

This list of insufficiently studied and unsolved problems of the theory of heteroligand systems is certainly incomplete. Note, however, that all of them are of a fairly special character. The scope of the approach described in this book, for which heteroligand species are but an illustration, is much broader. Its essence is joint consideration of substitution and vibronic coupling operators. This approach is therefore applicable to all problems in which the standard PMO theory should be augmented by analysis of spatial distortions induced by replacement of atoms. It can be equally useful in vibronic coupling theory, when vibronic effects experience perturbation caused by a change in the atomic composition of a system. Most importantly, the suggested approach combines perturbation molecular orbital (PMO) theory and the theory of vibronic interactions, which are two major branches of quantum chemistry based on perturbation theory.

REFERENCES

1. L.D.Landau and E.M.Lifshitz, *Quantum Mechanics. Nonrelativistic Theory*, Oxford: Pergamon, 1977.
2. L.I.Shiff, *Quantum Mechanics*, New York: McGraw-Hill, 1968.
3. R.Messiah, *Quantum Mechanics*, Amsterdam: North-Holland, 1961.
4. P.W.Atkins, *Molecular Quantum Mechanics*, Oxford: Clarendon Press, 1983.
5. R.McWeeny, *Methods of Molecular Quantum Mechanics*, London: Academic Press, 1989.
6. R.L.Flurry, *Quantum Chemistry. An Introduction*, New Jersey: Prentice-Hall, 1983.
7. J.C.Slater, *Quantum Chemistry of Molecules and Solids, vol. 1: Electronic Structure of Molecules*, New York: McGraw-Hill, 1963.
8. J.C.Slater, *Quantum Chemistry of Molecules and Solids, vol. 4: The Self-Consistent Field for Molecules and Solids*, New York: McGraw-Hill, 1974.
9. (a) J.N.Murrell, S.F.A.Kettle, and J.M.Tedder, *The Chemical Bond*, Chichester: Wiley, 1978.
   (b) R.McWeeny, *Coulson's Valence*, Oxford: University Press, 1979.
10. J.K.Burdett, *Molecular Shapes: Theoretical Models of Inorganic Stereochemistry*, New York: Wiley, 1980.
11. (a) I.B.Bersuker, *Elektronnoe stroenie i svoistva koordinatsionnykh soedinenii (The Electronic Structure and Properties of Coordination Compounds)*, Leningrad, 1986.
    (b) I.B.Bersuker, *Electronic Structure and Properties of Transition Metal Compounds. Introduction to the Theory*, New York: Wiley, 1996.
12. M.J.S.Dewar, *The Molecular Orbital Theory of Organic Chemistry*, New York: McGraw-Hill, 1969.
13. W.H.Flygare, *Molecular Structure and Dynamics*, Englewood Cliffs, New Jersey: Prentice-Hall, 1978.
14. I.B.Bersuker and V.Z.Polinger, *Vibronic Interactions in Molecules and Crystals*, Berlin: Springer, 1989.
15. I.B.Bersuker, *The Jahn–Teller Effect and Vibronic Interactions in Modern Chemistry*, New York: Plenum Press, 1984.
16. A.Hinchliffe, *Modelling Molecular Structure*, Chichester: Wiley, 1996.
17. J.Sadlej, *Semi-Empirical Methods of Quantum Chemistry CNDO, INDO, NDDO*, Warszawa: PWN, Chichester: Ellis Harwood, 1985.
18. (a) W.J.Hehre, L.Radom, P.R.Schleyer, and J.A.Pople, *Ab Initio Molecular Orbital Theory*, New York: Wiley, 1986.

(b) C.E.Dykstra, *Ab Initio Calculation of the Structures and Properties of Molecules*, Amsterdam: Elsevier, 1988.

19. (a) J.M.Seminario and P.Politzer (eds.), *Modern Density Functional Theory. A Tool for Chemistry*, Amsterdam: Elsevier, 1995.
    (b) J.K.Lobanovski and J.W.Andzelm (eds.), *Density Functional Methods in Chemistry*, New York: Springer, 1991.
20. I.B.Bersuker, A.S.Dimoglo, and A.A.Levin, *Sovremennye problemy kvantovoi khimii. Sbornik (Modern Problems in Quantum Chemistry: Collected Papers)* (M.G.Veselov, ed.), Leningrad: Nauka, 1986, pp. 78–122.
21. A.Veillard, *Quantum Chemistry: The Challenge of Transition Metals and Coordination Chemistry*, Dordrecht: Reidel, 1985.
22. R.M.Hochstrasser, *Molecular Aspects of Symmetry*, New York: Benjamin, 1966.
23. R.L.Flurry, *Symmetry Groups. Theory and Application*, Englewood Cliffs, New Jersey: Prentice-Hall, 1980.
24. V.Heine, *Group Theory in Quantum Mechanics*, London: Pergamon Press, 1960.
25. S.F.A.Kettle, *Symmetry and Structure*, New York: Wiley, 1995.
26. (a) J.M.Tedder and A.Nechvatal, *Pictorial Orbital Theory*, London: Pitman, 1985.
    (b) J.G.Verkade, *A Pictorial Approach to Molecular Bonding*, New York: Springer, 1986.
27. D.R.Bates and T.R.Carson, *Proc. Roy. Soc.*, **A234** (1956), 207–218.
28. G.A.Korn and T.M.Korn, *Mathematical Handbook for Scientists and Engineers*, New York: McGraw-Hill, 1968.
29. H.B.Gray, *Electrons and Chemical Bonding*, New York: Benjamin, 1965.
30. E.Clementi and M.L.Roetti, *At. Data Nucl. Data,* **14** (1974), 977–978.
31. F.Herman and S.Sckillman, *Atomic Structure Calculations*, New York: Prentice-Hall, 1963.
32. R.G.Pearson, *Symmetry Rules in Chemical Reactions*, New York: Wiley, 1976.
33. R.G.Pearson, *J. Am. Chem. Soc.,* **91** (1969), no. 18, 4947–4954.
34. S.Reims, *Many-Electron Theory*, Amsterdam: North-Holland, 1972.
35. A.A.Levin, A.N.Klyagina, S.P.Dolin, and P.N.D'yachkov, *Khimicheskaya svyaz' i stroenie molekul (Chemical Bond and the Structure of Molecules)* (V.I.Nefedov, ed.), Moscow: Nauka, 1984, pp. 67–82.
36. A.A.Levin, *Zh. Strukt. Khim.,* **29** (1988), no. 6, 9–18.
37. P.-O.Lowdin, *J. Chem. Phys.,* **49** (1951), no. 11, 1396–1401.
38. K.Nakamoto, *Infrared and Raman Spectra of Inorganic and Coordination Compounds*, New York: Wiley, 1986.
39. E.B.Wilson, J.C.Decius, and P.C.Cross, *Molecular Vibrations*, New York: McGraw-Hill, 1955.
40. I.B.Bersuker, N.N.Gorinochoi, and W.Z.Polinger, *Theor. Chim. Acta,* **66** (1984), no. 3/4, 161–172.
41. I.B.Bersuker, B.G.Vekhter, and A.A.Levin, *Sovremennye problemy kvantovoi khimii (Modern Problems in Quantum Chemistry)* (M.G.Veselov, ed.), Leningrad: Nauka, 1986, pp. 212–252.
42. C.A.Coulson and H.C.Longuet-Higgins, *Proc. Roy. Soc. London,* **191** (1947), 39–60.
43. M.J.S.Dewar and H.C.Longuet-Higgins, *Ibid.,* **214** (1952), 482–493.

44. M.J.S.Dewar and R.C.Dougherty, *The PMO Theory of Organic Chemistry*, New York: Plenum Press, 1975.
45. V.I.Baranovskii and O.V.Sizova, *Teor. Eksp. Khim.*, **10** (1974), no. 4, 678–681.
46. N.A.Popov, *Koord. Khim.*, **1** (1975), no. 6, 732–739.
47. N.A.Popov, *Ibid.*, **2** (1976), no. 9, 1155–1163.
48. A.A.Levin, A.N.Klyagina, and S.P.Dolin, *Zh. Neorg. Khim.*, **24** (1979), no. 9, 2307–3216.
49. A.A.Levin, A.N.Klyagina, and S.P.Dolin, *Koord. Khim.*, **4** (1978), no. 3, 354–360.
50. A.A.Levin and S.P.Dolin, *Ibid.*, **4** (1978), no. 12, 1812–1817.
51. P.N.D'yachkov, *Dokl. Akad. Nauk SSSR,* **254** (1980), no. 6, 1481–1485.
52. P.N.D'yachkov and A.A.Levin, *Zh. Neorg. Khim.*, **26** (1981), no. 10, 2595–2602.
53. (a) C.J.Balhausen, *Introduction to Ligand Field Theory*, New York: McGraw-Hill, 1955.
    (b) J.S.Griffith, *The Theory of Transition Metal Ions*, London, 1961.
54. P.N.D'yachkov and A.A.Levin, *Vibronnaya teoriya otnositel'noi stabil'nosti izomerov v neorganicheskikh molekulakh i kompleksakh (Vibronic Theory of Relative Isomer Stabilities of Inorganic Molecules and Complexes) (Itogi nauki i tekhniki, Ser. Stroenie molekul i khimicheskaya svyaz', vol. 11)*, Moscow: VINITI, 1987.
55. P.N.D'yachkov, *Dokl. Akad. Nauk SSSR,* **254** (1980), no. 4, 920–922.
56. P.N.D'yachkov and A.A.Levin, *Zh. Neorg. Khim.*, **26** (1981), no. 10, 2763–2774.
57. Yu.N.Kukushkin and R.I.Bobokhodzhaev, *Zakonomernost' transvliyaniya I.I.Chernyaeva (I.I. Chernyaev's Law of Trans-Influence)*, Moscow: Nauka, 1977.
58. I.I.Chernyaev, *Izv. Sekt. Platiny Inst. Obshch. Neorg. Khim. Akad. Nauk SSSR,* (1926), no. 4, 243–275.
59. I.I.Chernyaev, *Ibid.*, (1927), no. 5, 118–158.
60. G.B.Bokii, *Kristallografiya,* **2** (1957), no. 3, 400–407.
61. T.G.Appleton, H.C.Clark, and L.E.Manzer, *Coord. Chem. Rev.*, **10** (1973), no. 3, 335–422.
62. E.M.Schustorovich, M.A.Porai-Koshits, and Yu.A.Buslaev, *Ibid.*, **17** (1975), no. 1, 1–95.
63. (a) M.A.Porai-Koshits, *Koord. Khim.*, **4** (1978), no. 6, 842–866.
    (b) A.Weisman, M.Gozin, H.B.Kraatz, and D.Milstein, *Inorg. Chem.*, **35** (1996), no. 7, 1792–1797.
64. O.Foss, *Pure Appl. Chem.*, **24** (1970), no. 1, 31–48.
65. C.W.Hobbs and R.S.Tobias, *Inorg. Chem.*, **9** (1970), no. 5, 1037–1044.
66. M.A.Porai-Koshits and V.N.Shchurkina, *Koord. Khim.*, **2** (1976), no. 1, 99–102.
67. C.J.Marsden and L.S.Bartell, *Inorg. Chem.*, **15** (1976), no. 12, 3004–3008.
68. L.S.Bartell, F.B.Clippard, and E.J.Jacob, *Ibid.*, **15** (1976), no. 12, 3009.
69. J.F. de Wet and J.G.Praez, *J. Chem. Soc., Dalton Trans.*, (1978), no. 6, 593–597.
70. D.Brown, G.T.Reinolds, and P.T.Moesely, *Ibid.*, (1972), no. 7, 857–959.
71. A.A.Levin, A.N.Klyagina, and G.L.Gutsev, *Dokl. Akad. Nauk SSSR,* **250** (1980), no. 3, 638–640.

72. A.Pidcock, R.E.Richards, and L.Venanzi, *J. Chem. Soc., A,* (1966), 1707–1710.
73. (a) I.I.Chernyaev, *Izv. Inst. Izuch. Platiny Akad. Nauk SSSR,* (1927), no. 5, 102–118.
(b) I.I.Chernyaev, *Izv. Sekt. Platiny Inst. Obshch. Neorg. Khim. Akad. Nauk SSSR,* (1945), no. 18, 8–18.
(c) I.I.Chernyaev and A.D.Gel'man, *Izv. Sekt. Platiny Inst. Obshch. Neorg. Khim. Akad. Nauk SSSR,* (1937), no. 14, 77–122.
(d) A.A.Grinberg, *Izv. Inst. Izuch. Platiny Akad. Nauk SSSR,* (1932), no. 10, 47–64.
(e) B.V.Nekrasov, *Kurs obshchei khimii (Course in General Chemistry),* Moscow: ONTI, 1935.
(f) A.A.Grinberg, *Izv. Inst. Izuch. Platiny Akad. Nauk SSSR,* (1928), no. 6, 122–177.
74. (a) Ya.K.Syrkin, *Izv. Akad. Nauk SSSR, Otd. Khim. Nauk,* (1948), no. 1, 69–82.
(b) A.A.Grinberg and Yu.N.Kukushkin, *Zh. Neorg. Khim.,* **3** (1958), no. 8, 1810–1817.
(c) H.M.E.Cardwell and M.A.Phil, *Chem. Ind.,* (1955), no. 16, 422–424.
(d) J.Chatt, L.A.Duncanson, and L.M.Venanzi, *J. Chem. Soc.,* (1955), 4456–4460.
(e) J.Chatt, L.A.Duncanson, and L.M.Venanzi, *Chem. Ind.,* (1955), no. 26, 749.
(f) L.E.Orgel, *J. Inorg. Nucl. Chem.,* **2** (1956), no. 3, 137–140.
(g) F.Basolo and R.G.Pearson, *Progr. Inorg. Chem.,* **4**, 381–453.
(h) F.Basolo and R.G.Pearson, *Mechanisms of Inorganic Reactions,* New York: Wiley, 1969.
(i) C.H.Langford and H.B.Gray, *Ligand Substitution Processes,* New York: W.A.Benjamin, 1965.
75. (a) E.M.Shustorovich and Yu.A.Buslaev, *Zh. Strukt. Khim.,* **13** (1972), no. 1, 111–121.
(b) E.M.Shustorovich, M.A.Porai-Koshits, T.S.Khodasheva, and Yu.A.Buslaev, *Ibid.,* **14** (1973), no. 4, 706–716.
(c) E.M.Shustorovich, *Ibid.,* **15** (1974), no. 1, 123–127.
76. V.I.Nefedov and Yu.A.Buslaev, *Zh. Neorg. Khim.,* **18** (1973), no. 11, 3163–3170.
77. V.I.Nefedov and M.M.Gofman, *Vzaimnoe vliyanie ligandov v neorganicheskikh soedineniyakh (Mutual Influence of Ligands in Inorganic Compounds),* Moscow: VINITI, 1978.
78. R.S.Tobias, *Inorg. Chem.,* **9** (1970), no. 5, 1296–1298.
79. I.B.Bersuker, *Kinet. Katal.,* **18** (1977), no. 5, 1268–1282.
80. (a) T.G.Appleton, *Coord. Chem. Revs.,* **166** (1997), 313–359.
(b) J.Chatt and B.T.Heaton, *J. Chem. Soc. A,* (1968), no. 11, 2745–2757.
(c) R.G.Pearson, *Inorg. Chem.,* **12** (1973), no. 3, 712–713.
81. (a) M.-D.Su, *Mol. Phys.,* **80** (1993), no. 5, 1223–1251.
81. (b) M.-D.Su, *Ibid.,* **82** (1994), no. 3, 567–595.
82. (a) S.S.Zumdahl and R.S.Drago, *J. Am. Chem. Soc.,* **90** (1968), no. 24, 6669–6675.
(b) J.K.Burdett, *Inorg. Chem.,* **15** (1976), no. 1, 212–219.

(c) D.J.A.De Ridder, E.Zangrando, and H.B.Burgi, *J. Mol. Struct.,* **374** (1996), 63–83.

(d) P.D.Lyne and D.M.P.Mingos, *J. Organomet. Chem.,* **478** (1994), no. 1–2, 141–151.

(e) L.M.Hansen, P.N.Kumar, and D.S.Marinek, *Inorg. Chem.,* **33** (1994), no. 4, 728–735.

(f) S.Sakaki, M.Ogawa, Y.Musashi, and T.Arai, *J. Am. Chem. Soc.,* **116** (1994), no. 16, 7258–7265.

(g) J.S.Craw, G.B.Bacskay, and N.S.Hush, *J. Am. Chem. Soc.,* **116** (1994), no. 13, 5973–5948.

(h) K.Kurzak and A.Kolkowicz, *Pol. J. Chem.,* **68** (1994), no. 9, 1519–1528.

(i) S.K.Kang, J.S.Song, J.H.Moon, and S.S.Yun, *Bull. Korean Chem. Soc.,* **17** (1996), no. 12, 1149–1153.

(j) R.J.Deeth, *J. Chem. Soc., Dalton Trans.,* (1977), no. 22, 4203–4207.

(k) C.Bianchini, D.Masi, M.Peruzzini, *et al., Inorg. Chem.,* **36** (1997), no. 6, 1061–1069.

(l) G.Barea, A.Lledos, F.Maseras, and Y.Jean, *Ibid.,* **37** (1998), no. 13, 3321–3325.

83. (a) R.Mason and A.D.Towel, *J. Chem. Soc. A,* (1970), no. 9, 1601–1613.

(b) C.Bruhn and W.Preetz, *Acta Crystallogr., Sect. C: Cryst. Struct. Commun.,* **50** (1994), 1687–1690.

(c) A.J.Canty, S.D.Fritsche, H.Jin, *et al., J. Organomet. Chem.,* **510** (1996), no. 1–2, 281–286.

(d) V.Kettmann, M.Dunaj-Jurčo, D.Steinborn, and M.Ludwig, *Acta Crystallogr., Sect. C: Cryst. Struct. Commun.,* **50** (1994), 1239–1241.

(e) S.Geremia, R.Dreos, L.Randaccio, *et al., Inorg. Chim. Acta,* **216** (1994), no. 1–2, 125–129.

(f) J.S.Sommers, J.L.Petrsen, and A.M.Stolzenberg, *J. Am. Chem. Soc.,* **116** (1994), no. 16, 7189–7195.

84. N.W.Alcock, *Adv. Inorg. Chem. Radiochem.,* **15** (1972), 1–58.
85. O.A.Reutov, L.A.Aslanov, and V.S.Petrosyan, *Zh. Strukt. Khim.,* **29** (1988), no. 6, 112–118.
86. Yu.A.Buslaev and E.G.Il'in, *Mezhligandnye vzaimodeistviya i stereokhimiya oktaedricheskikh raznoligandnykh kompleksov (Interligand Interactions and Stereochemistry of Octahedral Complexes) (Itogi nauki i tekhniki, Ser. Neorgan. Khimiya),* Moscow: VINITI, 1974.
87. W.Strorzer, *Chem. Ber.,* **116** (1983), no. 2, 367–374.
88. W.S.Sheldrick, *J. Chem. Soc., Dalton Trans.,* (1974), no. 8, 1402–1406.
89. S.Bellard, A.V.Rivera, and W.S.Sheldrick, *Acta Crystallogr., Sect. B: Struct. Sci.,* **34** (1978), no. 5, 1934–1935.
90. J.W.Moore, H.M.Beird, and H.B.Miller, *J. Am. Chem. Soc.,* **90** (1968), no. 5, 1358–1359.
91. E.G.Zaitseva, *Synthesis and Structure of Octahedral Phenyl Halide Complexes of Sb(V) Compounds: Cand. Sci. (Chem.) Dissertation,* Moscow, 1988.
92. E.G.Zaitseva, S.V.Medvedev, and L.A.Aslanov, *Zh. Strukt. Khim.,* **31** (1990), no. 1, 104–109.
93. E.G.Zaitseva, S.V.Medvedev, and L.A.Aslanov, *Ibid.,* **31** (1990), no. 1, 110–116.
94. E.G.Zaitseva, S.V.Medvedev, and L.A.Aslanov, *Ibid.,* **31** (1990), no. 2, 92–98.

95. E.G.Zaitseva, S.V.Medvedev, and L.A.Aslanov, *Ibid.*, **31** (1990), no. 5, 133–138.
96. E.G.Zaitseva, S.V.Medvedev, and L.A.Aslanov, *Metalloorg. Khim.*, **1** (1988), no. 6, 1360–1364.
97. L.S.Bartell, S.Down, and C.J.Marsden, *J. Mol. Struct.*, **75** (1981), no. 2, 271–282.
98. H.Oberhammer, K.Seppelt, and R.Mews, *Ibid.*, **101** (1983), no. 3/4, 325–331.
99. C.J.Marsden, D.Christen, and H.Oberhammer, *Ibid.*, **131** (1985), no. 3/4, 299–307.
100. A.A.Levin and S.P.Dolin, *Koord. Khim.*, **5** (1979), no. 3, 320–335.
101. H.Goldstein, *Classical Mechanics*, Reading: Addison-Wesley, 1950.
102. (a) R.J.Gillespie, *Molecular Geometry*, London: Van Nostrand, 1972.
(b) R.J.Gillespie and I.Hargittai, *The VSEPR Model of Molecular Geometry*, Boston: Allyn and Bacon, 1991.
(c) R.J.Gillespie and E.A.Robinson, *Angew. Chem. Int. Edit.*, **35** (1996), no. 5, 495–514.
103. K.Leary, D.H.Templeton, A.Zalkin, and N.Bartlett, *Inorg. Chem.*, **12** (1973), no. 8, 1726–1730.
104. R.R.Ryan and D.T.Cramer, *Ibid.*, **11** (1972), no. 10, 2322–2324.
105. A.A.Levin, A.N.Klyagina, and S.P.Dolin, *Koord. Khim.*, **4** (1978), no. 4, 505–517.
106. B.Cohen and R.D.Peacock, *Adv. Fluor. Chem.*, **6** (1970), 343–385.
107. A.G.Legon, *J. Chem. Soc., Faraday Trans. 2,* (1973), no. 1, 29–35.
108. L.S.Bartell, *J. Chem. Educ.*, **45** (1968), no. 12, 754–767.
109. L.G.Allen, *Theor. Chim. Acta,* **24** (1972), no. 2/3, 117–131.
110. R.Boku, *J. Sci. Hiroshima Univ. A,* **45** (1981), no. 1, 123.
111. W.Schwarz and H.-J.Guder, *Z. Naturforsch., B: Chem. Sci.*, **33** (1978), no. 5, 485–488.
112. A.J.Edwards and G.R.Jones, *J. Chem. Soc. A,* (1969), no. 9, 1467–1470.
113. H.Krebs, *Z. Anorg. Allg. Chem.*, **397** (1973), no. 1, 1–15.
114. I.N.Marov and N.A.Kostromina, *EPR i YaMR v khimii koordinatsionnykh soedinenii (EPR and NMR in Chemistry of Coordination Compounds)*, Moscow: Nauka, 1979.
115. P.N.D'yachkov, *Koord. Khim.*, **8** (1982), no. 12, 1600–1604.
116. P.N.D'yachkov and V.O.Gel'mbol'dt, *Ibid.*, **10** (1984), no. 12, 1598–1606.
117. J.Burgess, J.W.Peacock, and R.D.Taylor, *J. Fluor. Chem.*, **3** (1973), no. 1, 55–59.
118. Yu.A.Buslaev, D.S.Dyer, and R.O.Ragsdale, *Inorg. Chem.*, **6** (1968), no. 12, 2208–2212.
119. R.O.Ragsdale and B.B.Stewart, *Ibid.*, **2** (1963), no. 5, 1002–1004.
120. R.S.Borden and R.N.Hammer, *Ibid.*, **9** (1971), no. 9, 2004–2009.
121. M.E.Ignatov, E.G.Il'in, V.G.Yagodin, and Yu.A.Buslaev, *Dokl. Akad. Nauk SSSR,* **243** (1978), no. 5, 1179–1181.
122. J.W.Downing and R.O.Ragsdale, *Inorg. Chem.*, **7** (1968), no. 8, 1675–1677.
123. D.S.Dyer and R.O.Ragsdale, *Ibid.*, **8** (1969), no. 5, 1116–1120.
124. P.A.W.Dean and B.J.Ferguson, *Can. J. Chem.*, **52** (1974), no. 4, 667–673.
125. Yu.A.Buslaev, Yu.V.Kokunov, V.D.Kopanev, and M.P.Gustyakova, *J. Inorg. Nucl. Chem.*, **36** (1974), no. 7, 1569–1574.

126. Yu.V.Kokunov, M.P.Gustyakova, and V.A.Bochkareva, *Koord. Khim.*, **7** (1981), no. 5, 725–730.
127. E.G.Il'in, M.E.Ignatov, and Yu.A.Buslaev, *Ibid.*, **3** (1977), no. 1, 46–50.
128. Yu.A.Buslaev, E.G.Ilyin, and M.E.Ignatov, *J. Fluor. Chem.*, **12** (1978), no. 5, 381–395.
129. L.B.Handy and F.E.Brinckman, *J. Chem. Soc., Chem. Commun.*, (1970), no. 4, 214–216.
130. L.B.Handy, C.Benham, F.E.Brinckman and R.B.Johnsen, *J. Fluor. Chem.*, **8** (1976), no. 1, 55–67.
131. A.M.Noble and J.M.Winfield, *J. Chem. Soc.*, (1970), no. 15, 2574–2578.
132. V.A.Shchipachev, S.V.Zemokov, S.V.Tkachev, and L.Ya.Al't, *Koord. Khim.*, **6** (1980), no. 12, 1877–1878.
133. M.E.Ignatov, E.G.Il'in, A.D.Granovskii, and Yu.A.Buslaev, *Ibid.*, **8** (1982), no. 10, 1368–1371.
134. Yu.A.Buslaev, G.A.Kirakosyan, and V.P.Tarasov, *Dokl. Akad. Nauk SSSR*, **264** (1982), no. 6, 1405–1408.
135. G.A.Kirakosyan, V.P.Tarasov, and Yu.A.Buslaev, *Abstracts of Papers, II Vsesoyuz. konf. "Spektroskopiya YaMR tyazhelykh yader elementoorganicheskikh soedinenii" (II All-Union Conf. on Heavy-Nucleus NMR Spectroscopy of Organoelement Compounds)*, Irkutsk, 1983.
136. (a) E.L.Muetterties (ed.), *Transition Metal Hydrides (The Hydrogen Ser., vol. 1)*, New York: M.Dekker, 1971.
(b) R.Bau (ed.), *Transition Metal Hydrides (Adv. in Chem. Ser., vol. 167)*, 1978.
(c) A.Dedieu (ed.), *Transition Metal Hydrides*, New York: VCH Corp., 1992.
137. S.J.S.Kerrison and P.J.Sadler, *J. Magn. Reson.*, **31** (1978), no. 2, 321–325.
138. V.P.Tarasov, V.I.Privalov, and Yu.A.Buslaev, *Mol. Phys.*, **35** (1978), no. 4, 1047–1056.
139. P.N.D'yachkov and V.O.Gel'mbol'dt, *Koord. Khim.*, **9** (1983), no. 4, 471–475.
140. R.F.Zahrobsky, *J. Am. Chem. Soc.*, **92** (1971), no. 14, 3313–3319.
141. V.O.Gel'mbol'dt, *Koord. Khim.*, **7** (1981), no. 12, 1814–1818.
142. V.O.Gel'mbol'dt, *Ibid.*, **8** (1982), no. 10, 1317–1318.
143. S.J.Ruzicka and A.E.Merbach, *Inorg. Chim. Acta*, **22** (19), no. 2, 191–200.
144. E.M.Shustorovich and Yu.A.Buslaev, *Koord. Khim.*, **1** (1975), no. 8, 1020–1030.
145. V.O.Gel'mbol'dt and P.N.D'yachkov, *Zh. Neorg. Khim.*, **34** (1989), no. 4, 840–843.
146. K.O.Christe, C.J.Schack, and E.C.Curtis, *Inorg. Chem.*, **15** (1976), no. 4, 843–848.
147. A.H.Cowley, P.J.Wisian, and M.Sanchez, *Ibid.*, **16** (1977), no. 6, 1451–1455.
148. W.Stadelmann, O.Stelzer, and R.Schmutzler, *J. Chem. Soc., Chem. Commun.*, (1971), no. 2, 1456–1458.
149. S.C.Peake, M.J.C.Hewson, and R.Schmutzler, *J. Chem. Soc. A*, (1970), no. 14, 2364–2367.
150. D.B.Denney, D.Z.Denney, and Y.F.Hsu, *J. Am. Chem. Soc.*, **95** (1973), no. 24, 8191–8192.
151. J.G.Malm, H.Selig, J.Jortner, and S.A.Rice, *Chem. Rev.*, **65** (1965), no. 2, 199–236.
152. K.O.Christe and D.Naumann, *Inorg. Chem.*, **12** (1973), no. 1, 59–62.

153. R.O.Ragsdale and B.B.Stewart, *Proc. Chem. Soc.,* (June 1964), 194.
154. C.E.Michelson and R.O.Ragsdale, *Inorg. Chem.,* **9** (1970), no. 12, 2718–2721.
155. R.Höfer and O.Glemser, *Angew. Chem.,* **9** (1973), no. 4, 465–468.
156. T.Abe and J.M.Shreeve, *Inorg. Nucl. Chem. Lett.,* **9** (1973), no. 4, 465–468.
157. G.W.Frazer and J.B.Millar, *J. Chem. Soc., Dalton Trans.,* (1974), no. 19, 2029–2031.
158. Yu.A.Buslaev, S.P.Petrosyants, and V.P.Tarasov, *Zh. Strukt. Khim.,* **10** (1969), no. 3, 411–416.
159. Yu.A.Buslaev, V.O.Gel'mbol'dt, S.P.Petrosyants, and A.A.Ennan, *Koord. Khim.,* **3** (1977), no. 4, 485–491.
160. V.O.Gel'mbol'dt, S.P.Petrosyants, A.A.Ennan, and Yu.A.Buslaev, *Dokl. Akad. Nauk SSSR,* **242** (1978), no. 6, 1343–1346.
161. P.A.W.Dean and D.F.Evans, *J. Chem. Soc. A,* (1968), no. 5, 1154–1166.
162. P.N.D'yachkov and A.A.Levin, *Zh. Neorg. Khim.,* **26** (1981), no. 9, 2291–2299.
163. V.I.Nefedov and V.I.Vovna, *Elektronnaya struktura khimicheskikh soedinenii (The Electronic Structure of Chemical Compounds),* Moscow: Nauka, 1987.
164. P.Biloen and R.Prins, *Chem. Phys. Lett.,* **16** (1972), no. 3, 611–613.
165. R.P.Messmer, L.V.Interrante, and K.H.Johnson, *J. Am. Chem. Soc.,* **96** (1974), no. 12, 3847–3854.
166. M.Barber, J.D.Clark, and A.Hinchliffe, *J. Mol. Struct.,* **57** (1979), no. 1, 305–307.
167. (a) M.A.Porai-Koshits and G.A.Kukina, *Uspekhi kristallokhimii kompleksnykh soedinenii (Advances in Crystal Chemistry of Coordination Compounds) (Itogi nauki i tekhniki, Ser. Kristallokhimiya),* Moscow: VINITI, 1974.

(b) V.Yu.Kukushkin, K.Lovqvist, B.Noren, A.Oskarsson, and L.I.Elding, *Inorg. Chim. Acta,* **219** (1994), no. 1–2, 155–160.
168. (a) G.W.Bushnell, A.Pidcock, and M.A.R.Smith, *J. Chem. Soc., Dalton Trans.,* (1975), no. 7, 572–575.

(b) S.Ganguli, J.T.Mague, and D.M.Roundhill, *Acta Crystallogr., Sect. C: Cryst. Struct. Commun.,* **50** (1994), 217–219.
169. R.Isenberg and J.A.Ibers, *Inorg. Chem.,* **4** (1965), no. 5, 773–778.
170. (a) M.R.Coller, C.Eaborn, and B.Ivanovich, *J. Chem. Soc., Chem. Commun.,* (1972), no. 10, 613–614.

(b) M.Hopp, A.Erxleben, I.Rombeck, and B.Lippert, *Inorg. Chem.,* **35** (1996), no. 5, 397–403.
171. P.B.Hitchcock, *Acta Crystallogr., Sect. B: Struct. Sci.,* **32** (1976), no. 7, 2014–2017.
172. (a) E.N.Yurchenko, T.S.Khodashova, M.A.Porai-Koshits, *et al.*, *Koord. Khim.,* **11** (1985), no. 3, 359–369.

(b) E.N.Yurchenko, T.S.Khodashova, M.A.Porai-Koshits, and V.P.Nikolaev, *Koord. Khim.,* **6** (1980), no. 8, 1290–1303.
173. (a) R.V.Parish, B.P.Howe, J.P.Wright, J.Mack, and R.G.Pritchard, *Inorg. Chem.,* **35** (1996), no. 6, 1659–1666.

(b) C.J.Levi, J.J.Vittal, and R.J.Puddephatt, *Organometallics,* **15** (1996), no. 8, 2108–2117.
174. I.I.Chernyaev, V.A.Sokolov, and V.A.Palkin, *Izv. Sekt. Platiny Inst. Obshch. Neorg. Khim. Akad. Nauk SSSR,* (1954), no. 28, 142–160.

175. Yu.N.Kukushkin, *Reaktsionnaya sposobnost' koordinatsionnykh soedinenii (Reactivity of Coordination Compounds) (Ser. Problemy Koord. Khim.)*, Moscow: Nauka, 1976.
176. (a) R.J.Puddephatt, *Chemistry of Gold*, Amsterdam: Elsevier, 1978.
(b) J.H.Nelson, W.L.Wilson, L.W.Cary, and N.W.Alcock, *Inorg. Chem.*, **35** (1996), no. 4, 883–892.
177. J.Chatt and R.G.Wilkins, *J. Chem. Soc.*, (1952), no. 11, 4300–4306.
178. A.W.Verstuyft and J.H.Nelson, *Inorg. Chem.*, **14** (1975), no. 7, 1501–1505.
179. D.A.Redfield and J.H.Nelson, *Ibid.*, **12** (1973), no. 1, 15–19.
180. L.M.Knight and J.H.Nelson, *Ibid.*, **16** (1977), no. 6, 1317–1321.
181. E.A.Allen, J.Del Gaudio, and W.Wilkinson, *Thermochim. Acta*, **11** (1975), no. 2, 197–203.
182. C.S.Scott and S.H.Mastin, *Ibid.*, **14** (1976), no. 1, 141–150.
183. Yu.N.Kukushkin, P.G.Antonov, and L.M.Mitronina, *Zh. Neorg. Khim.*, **23** (1978), no. 4, 1028–1035.
184. Yu.N.Kukushkin, G.N.Sedova, and L.M.Mitronina, *Ibid.*, **22** (1977), no. 10, 2785–2788.
185. R.Ronlet and G.Barbey, *Helv. Chim. Acta*, **56** (1973), no. 7, 2179.
186. E.A.Allen, N.P.Johnson, D.T.Rosenvear, and W.Wilkinson, *J. Chem. Soc., Chem. Commun.*, (1971), no. 3, 171–172.
187. Yu.N.Kukushkin, V.F.Budanova, G.N.Sedova, and V.G.Pogareva, *Zh. Neorg. Khim.*, **22** (1977), no. 5, 1305–1308.
188. Yu.N.Kukushkin, V.I.Vshivtsev, and L.V.Konovalov, *Ibid.*, **20** (1975), no. 7, 1938–1940.
189. (a) E.A.Andronov, Yu.N.Kukushkin, and T.M.Lukicheva, *Ibid.*, **21** (1976), no. 9, 2443–2447.
(b) V.A.Palkin, T.A.Kuzina, and N.N.Kuz'mina, *Ibid.*, **18** (1973), no. 3, 773–777.
190. J.Vicente, M.T.Chicote, M.C.Lagunas, P.G.Jones, and E.Bembenek, *Organometallics*, **13** (1994), no. 4, 1243–1250.
191. P.N.D'yachkov, *Zh. Neorg. Khim.*, **29** (1984), no. 1, 164–167.
192. M.Dub (ed.), *Organometallic Compounds, vol. 1: Compounds of Transition Metals*, Berlin: Springer, 1966.
193. J.D.Ruddic and B.L.Shaw, *J. Chem. Soc. A*, (1969), no. 18, 2801–2808.
194. (a) Yu.N.Kukushkin, V.F.Budanova, and G.N.Sedova, *Termicheskie prevrashcheniya koordinatsionnykh soedinenii (Thermal Transformations of Coordination Compounds)*, Leningrad: Izd. Leningr. Univ., 1981.
(b) L.A.Latif, C.Eaborn, A.P.Pidcock, and N.S.Weng, *J. Organomet. Chem.*, **476** (1994), no. 1–2, 217–221.
195. (a) P.S.Pregosin, *Coord. Chem. Rev.*, **44** (1982), no. 2, 247–291.
(b) W.Baratta, P.S.Pregosin, A.Bacchi, and G.Pelizzi, *Inorg. Chem.*, **33** (1994), no. 20, 4494–4501.
196. K.Kitaura, S.Obara, and K.Morokuma, *Chem. Phys. Lett.*, **77** (1981), no. 3, 452–454.
197. J.O.Noell and P.T.Hay, *Inorg. Chem.*, **21** (1982), no. 1, 14–20.
198. D.L.Packett, G.M.Jensen, and R.L.Cowen, *Ibid.*, **24** (1985), no. 22, 3578–3583.
199. P.N.D'yachkov and A.A.Levin, *Zh. Neorg. Khim.*, **28** (1983), no. 2, 275–278.
200. R.A.Vlasova, G.N.Sedova, and M.A.Kirrilova, *Koord. Khim.*, **12** (1986), no. 1, 118–121.

201. L.K.Shubochkin, L.D.Bol'shakova, and E.F.Shubochkina, *Ibid.*, **12** (1986), no. 3, 372–375.
202. P.N.D'yachkov and A.A.Levin, *Zh. Neorg. Khim.*, **27** (1982), no. 1, 242–245.
203. A.V.Ablov, P.A.D'yakon, and V.Ya.Ivanova, *Ibid.*, **10** (1965), no. 3, 628–630.
204. B.W.Delf, R.D.Gillard, and P.P.Brieb, *J. Chem. Soc., Dalton Trans.*, (1979), no. 8, 1301–1305.
205. M.Melnic, *Ibid.*, **47** (1983), no. 3, 239–261.
206. S.P.Dolin, Yu.P.Dikov, and V.I.Rekharskii, *Geokhimiya*, (1988), no. 7, 915–930.
207. R.M.Pitzer, *J. Chem. Phys.*, **46** (1967), no. 1, 48–71.
208. (a) A.E.Read and R.P. von Schleyer, *J. Am. Chem. Soc.*, **109** (1987), no. 24, 7362–7373.
(b) V.Jonas, G.Frenking, and M.T.Reetz, *J. Comput. Chem.*, **13** (1992), no. 8, 935–943.
(c) P.Briant and J.Green, *J. Am. Chem. Soc.*, **111** (1989), no. 24, 3434.
(d) V.Jonas, G.Frenking, and M.T.Reetz, *J. Comput. Chem.*, **13** (1992), no. 8, 919–934.
209. P.N.D'yachkov, I.S.Fedorova, and A.A.Levin, *Zh. Neorg. Khim.*, **28** (1983), no. 1, 135–140.
210. A.A.Levin, I.S.Fedorova, and P.N.D'yachkov, *Koord. Khim.*, **11** (1985), no. 6, 734–740.
211. (a) I.B.Bersuker, *Teor. Eksp. Khim.*, **1** (1965), no. 1, 5–10.
(b) I.B.Bersuker, *Coord. Chem. Rev.*, **14** (1975), no. 4, 357–412.
(c) B.J.Hathaway, *Struct. Bonding (Berlin)*, **57** (1984), 56.
(d) D.Reinen and M.Atanasov, *Magn. Res. Rev.*, **15** (1991), 167.
(e) R.Valiente, M.C.M.Delucas, and F.Rodrigues, *J. Phys.: Condens. Matter*, **6** (1994), no. 24, 4527–4540.
(f) I.B.Bersuker, *J. Coord. Chem.*, **34** (1995), no. 4, 289–338.
(g) C.J.Simmons, *New J. Chem.*, **17** (1993), no. 1–2, 77–95.
(h) D.Reinen and M.A.Hitchman, *Zs. Phys. Chem. - Int. J. Res. Phys. Chem. and Chem. Phys.*, **200** (1997), no. 1–2, 11–19.
(i) J.R.Naskar, S.Hati, and D.Datta, *Acta Crystallogr., Sect. B: Struct. Sci.*, **50** (1997), 885–894.
212. (a) C.Simmons, *J. Struct. Chem.*, **3** (1992), 25.
(b) J.Costa, R.Delgado, M.G.B.Drew, *et al.*, *J. Chem. Soc., Dalton Trans.*, (1998), no. 6, 1063–1071.
(c) C.J.Simmons, M.A.Hitchman, H.Stratemeier, and A.J.Schultz, *J. Am. Chem. Soc.*, **115** (1993), no. 24, 11304–11311.
(d) H.Stratemeier, B.Wagner, E.R.Krausz, *et al.*, *Inorg. Chem.*, **33** (1994), no. 11, 2320–2329.
(e) P.E.M.Wijnands, J.S.Wood, J.Reedijk, and W.J.A.Maaskant, *Ibid.*, **35** (1996), no. 5, 1214–1222.
(f) G.Golub, A.Lashaz, H.Cohen, *et al.*, *Inorg. Chim. Acta*, **255** (1997), no. 1, 111–115.
213. D.Attanasio, I.Collamati, and C.Ercobani, *J. Chem. Soc. A*, (1971), no. 15, 2516–2521.
214. F.A.Walker and H.Sigel, *Inorg. Chem.*, **11** (1972), no. 5, 1161–1164.
215. G.A.Sim and L.E.Sutton (eds.), *Molecular Structure by Diffraction Methods*, vol. 1–6, London: Chem. Soc., 1973–1978.

216. J.Wilkinson (ed.), *Comprehensive Coordination Chemistry: Properties and Application of Coordination Compounds*, vol. 1–7, Oxford: Pergamon Press, 1987.
217. (a) C.C.Su, Y.L.Lin, S.J.Liu, *et al.*, *Polyhedron,* **12** (1993), no. 22, 2687–2696.
(b) C.Tsiamis and N.Themeli, *Inorg. Chim. Acta,* **206** (1993), no. 1, 105–115.
(c) C.Tsiamis and L.C.Tzavellas, *Ibid.,* **207** (1993), no. 2, 179–190.
(d) B.D.Shrivastava, A.Mishra, and S.K.Joshi, *Jpn. J. Appl. Phys., Part 1,* **32** (1993), 839–841.
(e) G.Demunno, T.Poerio, M.Jule, *et al.*, *J. Chem. Soc., Dalton Trans.,* (1998), no. 10, 1679–1685.
218. M.A.S.Goher and A.E.H.Abdou, *Acta Chim. Hung.,* **128** (1991), no. 2, 219–231.
219. K.R.J.Thomas, P.Tharmaraj, V.Chandrasekhar, *et al.*, *Inorg. Chem.,* **33** (1994), no. 24, 5382–5390.
220. S.Debeukeleer, H.O.Desseyn, S.P.Perlepes, *et al.*, *Transition Met. Chem.,* **19** (1994), no. 4, 468–476.
221. R.Tomas, D.Rehorek, and H.Spindler, *Z. Chem.,* **12** (1972), 426–427.
222. P.Ray, *Inorg. Synth.,* **7** (1963), 6–8.
223. C.W.Carton, D.R.Lorenz, and J.R.Wasson, *Inorg. Nucl. Chem. Lett.,* **21** (1976), no. 5, 385–390.
224. M.Nöack and G.Gordon, *J. Chem. Phys.,* **48** (1968), no. 6, 2689–2699.
225. B.Morosin, *Acta Crystallogr., Sect. B: Struct. Sci.,* **31** (1975), no. 2, 632–634.
226. C.L.Raston and A.H.Wile, *J. Chem. Soc., Dalton Trans.,* (1974), no. 16, 1793–1796.
227. R.H.Hems and T.J.Cardwell, *Aust. J. Chem.,* **28** (1975), no. 2, 443.
228. S.J.Aschcroft and C.T.Mortomer, *Thermochemistry of Transition Metals*, New York: Academic Press, 1970.
229. A.Abragam and B.Bleaney, *Electron Paramagnetic Resonance of Transition Ions*, Oxford: Clarendon Press, 1970.
230. J.Gazo, K.Seratorova, and M.Seratorov, *Chem. Zvesti,* **13** (1959), no. 1, 3–13.
231. J.Gazo, *Pure Appl. Chem.,* **38** (1974), no. 3, 279–301.
232. J.Gazo, I.B.Bersuker, and J.Garaj, *Coord. Chem. Rev.,* **19** (1976), no. 3, 253–297.
233. M.Kabesova, J.Garaj, and J.Gazo, *Collect. Czech. Chem. Commun.,* **37** (1972), no. 3, 942–950.
234. D.Hall, A.J.McKinnon, and T.N.Waters, *J. Chem. Soc.,* (1965), no. 1, 425–430.
235. M.R.Kidd, R.S.Sager, and W.H.Watson, *Inorg. Chem.,* **6** (1967), no. 5, 946–951.
236. D.R.Johnson and W.H.Watson, *Ibid.,* **10** (1971), no. 6, 1281–1288.
237. B.Morosin, *Acta Crystallogr., Sect. B: Struct. Sci.,* **32** (1976), no. 4, 1237–1240.
238. M.E.Dyatkina and M.A.Porai-Koshits, *Dokl. Akad. Nauk SSSR,* **125** (1959), no. 5, 1030–1033.
239. I.B.Bersuker, *Zh. Fiz. Khim.,* **35** (1961), no. 2, 471–473.
240. (a) B.G.Vekhter, *Jahn–Teller Effect in Molecules and Crystals, Doctoral (Phys.-Math.) Dissertation*, Moscow, 1975.
(b) L.F.Chibotaru, *Mol. Phys.,* **81** (1994), no. 4, 891–920.
241. A.F.Wells, *Structural Inorganic Chemistry*, Oxford: Clarendon Press, 1986.

242. M.Gerloch, *Inorg. Chem.,* **20** (1981), no. 2, 638–640.

243. J.K.Burdett, *Ibid.,* **20** (1981), no. 7, 1959–1962.

244. J.Demuynk, A.Veillard, and A.Walhgren, *J. Am. Chem. Soc.,* **95** (1973), no. 17, 5563–5574.

245. M.Stebler and H.B.Burgi, *Ibid.,* **109** (1987), no. 5, 1395–1401.

246. R.Boca and P.Pelikan, *Inorg. Chem.,* **20** (1981), no. 5, 1618–1620.

247. C.J.Simmons, A.Clearfield, W.Fitzgerald, *et al., Ibid.,* **22** (1983), no. 17, 2463–2466.

248. M.A.Porai-Koshits, *Kristallokhimiya neorganicheskikh i organicheskikh soedinenii (Crystal Chemistry of Inorganic and Organic Compounds),* Kishinev: Shtiintsa, 1982.

249. M.A.Porai-Koshits and L.O.Atovmyan, *Kristallokhimiya i stereokhimiya koordinatsionnykh soedinenii molibdena (Crystal Chemistry and Stereochemistry of Molybdenum Coordination Compounds),* Moscow: Nauka, 1974.

250. Yu.V.Kokunov and Yu.A.Buslaev, *Coord. Chem. Rev.,* **47** (1981), no. 1, 15–40.

251. (a) M.A.Porai-Koshits and V.S.Sergienko, *Usp. Khim.,* **59** (1990), no. 1, 86–105.
 (b) J.Baldas, *Topics in Current Chem.,* **176** (1996), 37–74, 2995–3002.
 (c) F.A.Cotton and G.Schmid, *Inorg. Chem.,* **36** (1987), no. 11, 2267–2278.
 (d) G.Barea, A.Lledos, F.Maseras, *et al., Ibid.,* **37** (1998), no. 13, 3321–3325.

252. A.A.Levin and A.N.Klyagina, *Zh. Neorg. Khim.,* **25** (1980), no. 1, 97–109.

253. L.Karlsson, L.Mattson, and R.Jardny, *Phys. Scr.,* **14** (1976), no. 5, 230–231.

254. G.L.Gutsev and A.A.Levin, *Chem. Phys.,* **51** (1980), no. 3, 451–471.

255. G.L.Gutsev and A.A.Levin, *Chem. Phys. Lett.,* **57** (1978), no. 2, 235–238.

256. E.L.Rozenberg and M.E.Dyatkina, *Zh. Strukt. Khim.,* **11** (1970), no. 2, 323–330.

257. (a) R.Hubener, U.Abram, and J.Strahle, *Inorg. Chim. Acta,* **224** (1994), no. 1–2, 193–197.
 (b) H.J.Pietzsh, H.Spies, P.Leibnitz, and G.Reck, *Polyhedron,* **12** (1993), no. 24, 2995–3002.

258. N.Ueyama, H.Oku, M.Kondo, *et al., Inorg. Chem.,* **33** (1994), no. 3, 643–650.

259. K.Czarnecki, S.Nimri, Z.Gross, *et al., J. Am. Chem. Soc.,* **118** (1996), no. 12, 2929–2935.

260. L.Jean, A.Lledos, J.K.Burdett, and R.Hoffmann, *J. Chem. Soc., Chem. Commun.,* (1988), no. 2, 140–142.

261. E.Mayer and F.Sladky, *Inorg. Chem.,* **14** (1975), no. 3, 589–620.

262. A.S.Kanishcheva, Yu.N.Mikhailov, and A.A.Svetlov, *Zh. Neorg. Khim.,* **35** (1990), no. 7, 1760–1766.

263. A.S.Salomov, Yu.N.Mikhailov, A.S.Kanishcheva, A.A.Svetlov, M.N.Sinitsyn, and M.A.Porai-Koshits, *Zh. Neorg. Khim.,* **34** (1989), no. 2, 386–390.

264. A.S.Salomov, Yu.N.Mikhailov, A.S.Kanishcheva, A.A.Svetlov, M.N.Sinitsyn, and M.A.Porai-Koshits, *Zh. Neorg. Khim.,* **33** (1988), no. 10, 2608–2611.

265. M.N.Sinitsyn, A.A.Svetlov, A.S.Kanishcheva, Yu.N.Mikhailov, G.G.Sadikov, and Yu.V.Kokunov, *Zh. Neorg. Khim.,* **34** (1989), no. 11, 2795–2802.

266. A.S.Salomov, N.A.Parpiev, Kh.T.Sharipov, M.N.Sinitsyn, M.A.Porai-Koshits, and A.A.Svetlov, *Zh. Neorg. Khim.,* **29** (1984), no. 10, 2612–2618.

267. Yu.N.Mikhailov, A.S.Kanishcheva, and A.A.Svetlov, *Zh. Neorg. Khim.,* **34** (1989), no. 11, 2803–2806.

268. A.A.Levin, Yu.N.Mikhailov, A.S.Kanishcheva, A.A.Svetlov, *Zh. Neorg. Khim.*, **39** (1993), no. 6, 1503–1060.
269. A.A.Levin, *J. Mol. Struct.*, **272** (1992), 133–143.
270. A.A.Levin, *New J. Chem.*, **17** (1993), no. 1–2, 31–37.
271. E.A.Kravtsova and L.N.Mazalov, *Zh. Strukt. Khim.*, **28** (1987), no. 5, 68–72.
272. E.A.Kravchenko, M.Yu.Burtsev, M.N.Sinitsyn, A.A.Svetlov, Yu.V.Kokunov, and Yu.A.Buslaev, *Dokl. Akad. Nauk SSSR*, **294** (1987), no. 1, 130–136.
273. E.A.Kravchenko, M.Yu.Burtsev, V.G.Morgunov, A.A.Svetlov, M.N.Sinitsyn, Yu.V.Kokunov, *et al.*, *Koord. Khim.*, **14** (1988), no. 1, 49–56.
274. K.K.Pandey, R.B.Sharma, and P.K.Pandit, *Inorg. Chim. Acta*, **169** (1990), no. 2, 207–210.
275. A.A.Levin, *Koord. Khim.*, **19** (1993), no. 5, 368–390.
276. (a) W.D.Stohrer and R.Hoffmann, *J. Am. Chem. Soc.*, **94** (1972), no. 5, 1661–1668.
(b) W.D.Stohrer and R.Hoffmann, *J. Am. Chem. Soc.*, **94** (1972), no. 3, 779–786.
277. V.C.Gibson and M.McPartlin, *J. Chem. Soc., Dalton Trans.*, (1992), no. 67, 947–956.
278. (a) J.Chatt, Lj. Manojlovich-Muir, and K.W.Muir, *Chem. Commun.*, (1971), no. 13, 655–658.
(b) Lj. Manojlovich-Muir and K.W.Muir, *J. Chem. Soc., Dalton Trans.*, (1972), no. 5, 686–690.
279. (a) K.Yoon, G.Parkin, and A.L.Pheingold, *J. Am. Chem. Soc.*, **114** (1992), no. 6, 2210.
(b) P.J.Descrochers, K.W.Nebesny, and M.J.Labarre, *Inorg. Chem.*, **35** (1996), no. 1, 15–4.
280. Y.Jean, A.Lledos, J.K.Burdett, and R.Hoffmann, *J. Am. Chem. Soc.*, **110** (1988), no. 14, 4506.
281. P.N.D'yachkov and A.A.Levin, *Dokl. Akad. Nauk SSSR*, **280** (1985), no. 5, 1177–1180.
282. K.W.Bagnal, *Proc. 11e journees des actinides*, mai 1981, Jesolo; Lido, 1981.
283. A.Rosen, *Chem. Phys. Lett.*, **55** (1978), no. 2, 311–314.
284. D.D.Koelling, D.E.Ellis, and R.J.Bartlett, *J. Chem. Phys.*, **65** (1976), no. 8, 3331–3340.
285. D.A.Case and C.Y.Yang, *Ibid.*, **72** (1980), no. 6, 3443–3448.
286. G.L.Gutsev and A.A.Levin, *Dokl. Akad. Nauk SSSR*, **246** (1979), no. 6, 1366–1368.
287. Boring and H.G.Hecht, *J. Chem. Phys.*, **69** (1978), no. 1, 12.
288. (a) J.Thornton. N.Edelstein, and N.Rösch, *Ibid.*, **70** (1979), no. 11, 5218–5221.
(b) J.Thornton. N.Rösch, and N.Edelstein, *Inorg. Chem.*, **19** (1980), no. 5, 1304–1307.
289. A.R.Rossi and R.Hoffmann, *Inorg. Chem.*, **14** (1975), no. 2, 365–374.
290. P.N.D'yachkov, *Koord. Khim.*, **11** (1985), no. 4, 442–444.
291. V.Dyszmons, V.Staemmler, and W.Kutzelnigg, *Phys. Chem. Lett.*, **5** (1970), no. 6, 361–366.
292. A.Rauk, L.C.Allen, and K.Mislow, *J. Am. Chem. Soc.*, **94** (1972), no. 9, 3035–3040.
293. R.Hoffmann, J.M.Howell, and E.L.Muetterties, *Ibid.*, **94** (1972), no. 9, 3047–3058.

294. K.S.Krasnov, N.V.Filipenko, and V.A.Bobkova, *Molekulyarnye postoyannye neorganicheskikh soedinenii (Molecular Constants of Inorganic Compounds)*, Leningrad: Khimiya, 1979.
295. R.Hoffmann, B.Beier, E.L.Muetterties, and A.R.Rossi, *Inorg. Chem.,* **16** (1977), no. 3, 511–522.
296. M.D.Kaplan and B.G.Vekhter, *Cooperative Phenomena in Jahn-Teller Crystals*, New York: Plenum Press, 1995.
297. A.A.Levin and S.P.Dolin, *Koord. Khim.,* **24** (1998), no. 4, 287–292.
298. M.E.Lines and A.M.Glass, *Principles and Application of Ferroelectrics and Related Materials*, Oxford: Clarendon Press, 1977.
299. V.G.Vaks, V.I.Zinenko, and V.E.Shneider, *Usp. Fiz. Nauk,* **141** (1983), no. 3–4, 629–673.
300. A.A.Levin and S.P.Dolin, *Dokl. Ross. Akad. Nauk,* **341** (1995), no. 5, 638–640.
301. A.A.Levin and S.P.Dolin, *J. Phys. Chem.,* **100** (1996), no. 15, 6258–6261.
302. A.A.Levin and S.P.Dolin, *Dokl. Ross. Akad. Nauk,* **351** (1996), no. 4–6, 502–504.
303. A.A.Levin, S.P.Dolin, and V.L.Lebedev, *Zh. Neorg. Khim.,* **42** (1997), no. 8, 1321–1331.
304. A.A.Levin and S.P.Dolin, *Khim. Fiz.,* **17** (1998), no. 6, 97–107.
305. A.A.Levin, A.R.Zaitsev, and A.N.Isaev, *Zh. Neorg. Khim.,* **34** (1989), no. 9, 2418–2422.
306. A.A.Levin, S.P.Dolin, and A.R.Zaitsev, *Khim. Fiz.,* **15** (1996), no. 8, 84–93.
307. R.J.Nelmes, Z.Tun, and W.F.Kuhs, *Ferroelectrics,* **71** (1987), 125–131.
308. M.I.McMahon, R.J.Nelmes, W.F.Kuhs, and D.Z.Semmingsen, *Zs. Krystallogr.,* **195** (1991), 231–239.
309. A.A.Levin and S.P.Dolin, *Proc. Estonian Acad. Sci., Phys. Math.,* **44** (1995), no. 2–3, 144–152.
310. R.J.Baxter, *Exactly Solved Models in Statistical Mechanics*, London, New York: Academic Press, 1982.
311. A.A.Levin, S.P.Dolin, and V.L.Lebedev, *Khim. Fiz.,* **14** (1995), no. 9, 84–97.
312. V.G.Vaks, *Vvedenie v mikroskopicheskuyu teoriyu segnetoelektrikov (Introduction to Microscopic Theory of Ferroelectrics)*, Moscow: Nauka, 1973.
313. V.G.Vaks and V.I.Zinenko, *Zh. Eksp. Teor. Fiz.,* **64** (1973), no. 2, 650–658.
314. A.A.Levin, S.P.Dolin, V.L.Lebedev, and T.Yu.Mikhailova, *Khim. Fiz.,* **18** (1999) (in press).
315. A.A.Levin and P.N.D'yachkov, *Elektronnoe stroenie, struktura i prevrashcheniya geteroligandnykh molekul (The Electronic and Geometric Structure and Transformations of Heteroligand Molecules)*, Moscow: Nauka, 1990.
316. A.Wlodarczyk, S.J.Coles, M.B.Hursthouse, K.M.A.Malik, and H.F.Lieberman, *J. Chem. Soc., Dalton Trans.,* (1997), no. 17, 2921–2929.
317. A.E.Almaraz, L.A.Gentil, L.M.Baraldo, and J.A.Olabe, *Inorg. Chem.,* **35** (1996), no. 26, 7718–7727.
318. P.N.D'yachkov and N.N.Breslavskaya, *J. Mol. Struct. (Theochem),* **397** (1997), 199–211.
319. P.N.D'yachkov and N.N.Breslavskaya, *Mol. Mater.,* **7** (1996), 137–142.
320. A.N.Lazarev, B.F.Shchegolev, M.B.Smirnov, and S.P.Dolin, *Kvantovaya khimiya molekulyarnykh sistem i kristallokhimiya silikatov (Quantum Chemistry of Molecular Systems and Crystal Chemistry of Silicates)*, Leningrad: Nauka, 1988.

COMPOUND INDEX

SUBJECT INDEX